THEORY OF AUTOMATA AND COMPUTATION

By

Adesh Kumar Pandey

Deptt. of Computer Science
(Krishna Institute of Engg. and Tech.)
Ghaziabad

S.K. KATARIA & SONS®

Publishers of Engineering Books

4885/109, Prakash Mahal, Dr. Subhash Bhargav Lane,
Daryaganj, New Delhi-110002 (India)
Ph. : +91-11-43551243 Mobile : +91-9871775858, 7678425209
E-mail : katariabook@gmail.com, katariabooks@yahoo.com
Website: www.skkatariaandsons.com

(e-books, subscriptions and bulk buying also available on our website)

Published By :

S.K. Kataria & Sons®

4885/109, Prakash Mahal, Dr. Subhash Bhargav Lane,
Daryaganj, New Delhi-110002 (India)
Ph. +91-11-43551243, Mobile: +91-9871775858, 7678425209
E-mail: katariabook@gmail.com, katariabooks@yahoo.com

First Edition: 2011; Reprint: 2026

Laser Typeset by :
AISA Enterprises, Delhi-92.

Preface

The book addresses a full spectrum of automata theory concepts used in computer science. Each of these concepts is presented in detail, since these concepts are basis of computer science and provide the basics for developing software techniques like designing and checking the behaviour of digital circuits, lexical-analyser of the typical compiler, and software for Natural Language Processing. To introduce the more formal aspects of Computer Science, the author has presented automata as foundation. For students who are ambitious to achieve unparalleled goals in the field of computer science, and want to develop their programming skills towards system programming, it becomes necessary that their mathematical concept are very clear.

The book is written as a text, with problems and exercises, with particular emphasis on the examples. In every chapter of this book, due consideration has been given to the pattern and standard of various universities.

The focus of this book is on concepts, the "Why", "What" and "How" of the operations. The author recommend that these concept should be introduced to Computer Science students early in their course curriculum, so that they learn to use mathematical concepts to develop their skills.

To facilitate easy understanding, the text and problems have been put in simpler form. The emphasis is on simpler set of examples relating to every concept in each chapter. This will certainly help the readers to understand the subject in better way.

Suggestions are always welcome from the readers.

Author

Acknowledgement

The book itself is an acknowledgement to the intensity, drive and technical competence of many individuals, who have contributed to it.

I am thankful to Almighty God, without whose grace nothing is possible for any one. I am very thankful to my friends and students who have inspired me to write this book.

I extend my gratitude to my wife, Gunjan Pandey, my daughter, Shreya, my parents and my in-laws for their strength and support, failing which, this book wouldn't have been complete.

I convey my thanks to publisher of this book, for their constructive critism and suggestions for improvement in the manuscript.

Adesh K. Pandey

CONTENTS

SECTION-A

- Finite Automata
- Regular Expression
- Introduction to Machines

SYLLABUS

Finite Automata and Regular Expressions : Finite State Systems, Basic Definitions Non-Deterministic finite automata (NDFA), Deterministic Finite Automata (DFA), Equivalence of DFA and NDFA Finite Automata with E-moves, Regular Expressions, Equivalence of Finite Automata and Regular Expressions, Regular Expression Conversion and Vice Versa.

Introduction to Machines : Concept of Basic Machine, Properties and Limitations of FSM. Moore and Mealy Machines, Equivalence of Moore and Mealy Machines, Conversion of NFA to DFA by Arden's Method.

CHAPTER 1

Finite Automata

1.1. HISTORICAL ASPECT OF AUTOMATA

Automata theory is the study of abstract computing devices, or "machines". Before there were computers, in 1930's, Alen turing introduced an abstract machine that has all the capabilities of today's computers, at least as far as in what they could compute. In the 1940's and 1950's simpler kinds of machines, which we today call "finite-automata", were studied by a number of researchers. These automata, originally proposed to model brain function, turned out to be extremely useful for a variety of others purposes. Also in the late 1950's the linguist N. Chomsky began the study of formal "grammars". While not strictly machines, these grammars have close relationships to abstract automata and serve today on the basis of some important software components, including parts of compilers.

In 1969, S. Cook extended turing's study of what could and what could not be computed. Cook separated those problems that can be solved, efficiently by computer, from those problems that

INSIDE THIS CHAPTER

m problems can be solved, but in practice take so much time, that computers are useless for all but very small instances of the problem. The later class of problem's is called "intractable", or NP-hard. All of this theoritical development bears directly on what Computer Scientists do today.

In computer science we find many examples of finite state system and the theory of finite state systems, and the theory of finite automata as a useful design tool for these systems. A primary example is a switching circuit. A switching circuit is composed of a finite number of gates, each of which can be in one or two conditions, usually denoted by 0 and 1.

We can assume another situation, in case of electric switch only two situation are possible that is "on" and "off". Let us say 1 for on and 0 for off. We can interchange the situation by pushing the switch, it can be seen in Fig. 1.1.

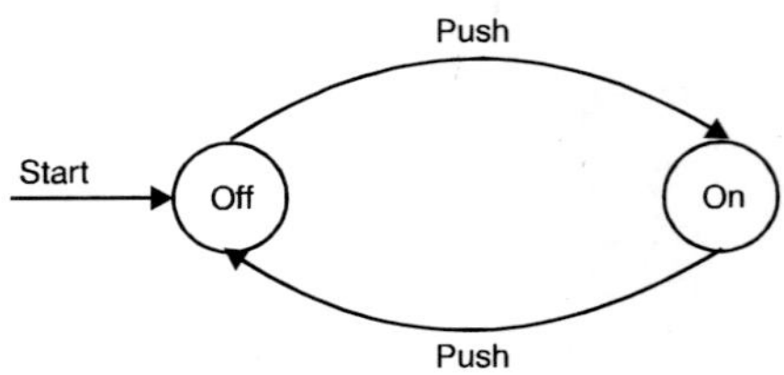

Fig. 1.1.

Automation is also tempting to view the human brain as a finite state system. The number of brain cell or neurons are limited, probably 2^{35} at most. Although there is evidence to the contrary, that the state of each neuron can be described by a small number of bits, if so then finite state theory applies to the brain also.

1.2. THE STUDY OF AUTOMATA THEORY IS FRUITFUL AND FUTURISTIC

There are several reasons why the study of automata and complexity is an important and core part of computer science. In this section we will introduce several reasons which will definitely motivate the computer science students to study automata theory. Let us list some important aspects of the automata theory.

(1) Automata theory plays an important role when we are making software for designing and checking the behaviour of digital circuit.

(2) The "Lexical analyser" of the typical compiler, that is the compiler component that breaks the input text into logical units, such as identifiers, keywords and punctuation.

(3) Software for scanning large bodies of text, such as collections of web pages, to find occurance of words, phrases or other patterns.

(4) Automata theory is key to software for verifying systems of all types that have a finite number of distinct states, such as communication protocols or protocol for secure exchange of information.

(5) Automata theory is most useful concept of software for natural language processing.

1.3. DEVELOP YOUR FEELINGS WITH THE AUTOMATA

Truly speaking it is very necessary to understand the subject like automata theory, that you should develop feelings about this subject.

And moreover you must try to implement it in your day to day life. In the whole book you will read about states so let us develop feelings for this term.

Sometime when you are sitting alone thinking about your previous life. Some events you certainly remember, which plays important role in your life. Some of them are not relevent and it is difficult for you to remember those events. Same theory applies in the case of finite state machines, when we design a system or machine, two kinds of points come into picture, some of them certainly affect the output of the system and others never affect the output of the system. So we can skip these irrelevent points.

TIPS

Those points which affect the output of any system are called states (with the assumption that system can design in finite number of steps). Clearly number of states will be finite. The advantage of having only a finite number of states is that we can implement the system with a fixed number of resources.

1.4. ALPHABETS

Everything in mathematics is based on symbols. This is also true for automata theory of computation. The symbols are generally letters and digits.

Definition "Alphabets are defined as a finite set of symbols"

An example of an alphabet is a set of decimal numbers.

$$\Sigma = \{0, 1, 2, 3, 4, 5, 6, 7, 8, 9\}$$

Here Σ is representation for the alphabet.

Another example is a binary alphabet that is :

$$\Sigma = \{0, 1\}$$

Similarly $\Sigma = \{A, B, ... Z\}$ is a alphabet.

$\Sigma = \{a, b, ... z\}$ is also a alphabet}.

1.5. STRINGS

Definition "A string (or sometimes word) is a finite sequence of symbols selected from some alphabet". For example, if Σ $\{a, b\}$ is an alphabet then *abab* is string over alphabet Σ. A string is generally denoted by w.

The "length" of the string is denoted by $|w|$ and it is, the number of positions for the symbol in the string. For instance string

$$w = 01101 \text{ has length}$$

$$|w| = 5.$$

TIPS

It is common to say that the length of the string is "the number of symbols" in the string; this statement is generally accepted but not strictly correct. Thus, there are only two symbols, 0 and 1, in the string 01101, but there are five positions for symbols, and its length is 5.

The "empty" string is the string with **zero occurance of symbols.** This string is represented by $\in$, or e or λ, or γ **but we will follow $\in$ in this particular book.**

The set of strings, including the empty, over an alphabet Σ is denoted by Σ^*, that is suppose $\Sigma = \{0, 1\}$ then Σ^* is a set containing empty and all combinations of 0's and 1's. Thus

$$\Sigma^* = \{\in, 0, 1, 01, 10, 000, 010, 0000, 1111000 ...\}$$

Similarly

$$\{0\}^* = \{\in, 0, 00, 000 ...\},$$

$$\{1\}^* = \{\in, 1, 11, 111 ...\}$$

Note that $\in$ is in Σ^*, regardless of what alphabet Σ is.

That is $\in$ is only string whose length is 0.

if

$$\Sigma = \{0, 1\}, \text{ then } \Sigma^1 = \{0, 1\}, \Sigma^2 = \{00, 01, 10, 11\},$$

$$\Sigma^3 = \{000, 001, 010, 011, 100, 101, 110, 111\}$$

and so on. Note that there is slight confusion between Σ and Σ^1. The formal is an alphabet; its members 0 and 1 are symbols. The latter is a set of strings; its members are the strings 0 and 1, each of which is of length 1. We shall not try to use seprate notations for the two sets, relying on context to make it clear whether {0, 1} or similar sets are alphabets or set of strings.

Sometimes, we wish to exclude the empty string from the set of strings. The set of non-empty strings from alphabet Σ is denoted by Σ^+. Thus, two appropriate equivalence are :

$$\Sigma^+ = \Sigma^1 \cup \Sigma^2 \cup \Sigma^3 \cup ...$$

$$\Sigma^* = \Sigma^+ \cup \{\in\}.$$

➤ **Concatenation of strings.** Let w_1 and w_2 be two strings. Then $w_1 w_2$ denotes the concatenation of w_1 and w_2, that is, the string formed by making a copy of w_1 and followed it by a copy of w_2.

if

$$w_1 = abc$$

$$w_2 = xyz$$

then

$$w_1 w_2 = abc\ xyz$$

➤ In the set form string w_2 is defined as :

$$w_2 = \begin{cases} \in & \text{for } i = 0 \\ w^{i-1}\ w & \text{for } i \geq 1 \end{cases}$$

that is if $w = aba$

then

$$w^3 = (aba)^3$$

$$w^3 = aba\ aba\ aba$$

$$= (aba)^2\ aba.$$

➤ **Reverse of the string.** Reverse of the string can be achieved by flipping over last symbols :

if

$$w = abc$$

then $w^R = cba$, where w^R is reverse of w, $|w| = |w^R|$

1.6. LANGUAGES

Definition "A set of strings all of which are chosen from some Σ^*, where Σ is a particular alphabet, is called a language". If Σ is an alphabet, and $L \subseteq \Sigma^*$, then L is said to be language over alphabet Σ.

Notice that a language over Σ need not include strings with all the symbols of Σ, so once we have established that L is a language over Σ, we also know it is a language over any alphabet which is a superset of Σ.

An example is english, where the collection of legal english words is a set of strings over the alphabet that consists of all the letters. Another example is C or any other programming language, where the legal programs are a subset of the possible strings that can be formed from the alphabet of the language. This alphabet is a subset of the ASCII characters. The exact alphabet may differ slightly among different programming languages, but generally includes the upper-and lower-case letters; the digits, punctuation, and mathmatical symbols.

However, there may be some other languages that appear when we study automata. Some examples are as follows :

(*a*) The language of all strings consisting of n 0's followed by n 1's, for some $n \geq 0$: $\{\in, 01, 0011, 000111, ...\}$

(*b*) The set of strings over 0's and 1's with equal number of each :

$\{\in, 01, 10, 0011, 0101, 1001 ...\}$.

(*c*) The set of binary numbers whose value is a prime :

$\{10, 11, 101, 111, 1011, ...\}$

(*d*) Σ^* is a language for any alphabet Σ.

(*e*) ϕ, the empty language, is a language over any alphabet.

(*f*) $\{\in\}$, the language consisting of only the empty string, is also a language over any alphabet. Notice that $\phi \neq \{\in\}$; the former has no strings and latter has one string.

The only important constraint on what can be a language is that all alphabet are finite. Thus languages, although they can have an infinite number of strings, are restricted to consist of strings drawn from one fixed, finite alphabet.

1.6.1. Concatenation of Languages

Concatenation of languages can be defined as :

if L_1 and L_2 are two languages, their concatenation

$L = L_1 . L_2$ where $L = \{w/w = xy, x \in L_1, y \in L_2\}$.

That is, the string in L_1 L_2 are formed by choosing a string in L_1 and followed by a string in L_2, in all possible combinations.

1.6.2. Languages in Set Forms

It is very common to describe a language using a "set-form" :

{*w*/some logical view about *w*}.

This expression is read "the set of words w such that (whatever is said about w to the right of the vertical bar)." For example :

(*a*) $\{w/w$ consists an equal number of 0's and 1's$\}$

(*b*) $L = \{w/w$ is a binary integer that is prime$\}$

It is common to replace w by some expression with parameters and describe the strings in the language by stating conditions on the parameters. Here are some examples :

$\{a^n b^n/n \geq 1\}$. Read "the set of a to the n, b to the n such that n is greater than or equal to 1," this language consists of the strings $\{ab, aabb, aaabbb, ...\}$. Notice that as with alphabet, we can raise a single symbol to a power n in order to represent n copies of that symbol.

1.7. KLEENE CLOSURE

Given an alphabet, we wish to define a language in which any string of letters form Σ is a word, even the null string. This language we shall call the closure of the alphabet. It is denoted by writing a star, after the name of the alphabet as a superscript :

$$\Sigma^*$$

This notation is some times known as the "kleene star" after the name of logician who was one of the founders of this subject.

Definition "If S is a set of words then by S^* we mean the set of all finite strings formed by concatenating words from S, where any word may be used as often we like, and where the null string is also included".

For example if $\Sigma = \{a\}$

then $\Sigma^* = \{\in, a, aa, aaa, aaaa, ...\}$

if $\Sigma = \{0, 1\}$

then $\Sigma^* = \{\in, 0, 1, 00, 01, 10, 11, 000, 001 ...\}$

We can think of the kleene star as an operation that makes an infinite language of strings of letters out of an alphabet. When we say "infinite language", we mean infinitely many words, each of which of finite length.

1.8. FINITE AUTOMATA

In this book, we are discussing the mathematical model of the computers and algorithms. In next two chapter we are going to define powerful models of computation, more and more sophisticated device for accepting the generating languages, which are restricted model of the actual computers, called finite automata or finite state machine. These machines are very similar to the Central Processing Unit of a computer. Absence of the memory makes these machines more restricted model.

Computer is also deterministic, by which we mean that, on reading one particular input instruction, the machine converts itself from the state it was, into some particular other state, where the resultant state is completely determined by the prior state and the input instruction. Some sequence of instruction may lead to success and some may not. Success is entirely determined by the sequence of inputs. Either program will work or not.

Before discussing the mathematical model let us discuss the pictorial representation of finite machine. Strings are fed into device by means of an input tape, which is divided into square, with one symbol in each square. The main part of the machine is a "black box", which is responsible for all proceedings. Let us say "black box" is the finite control, finite control can serve, that what symbol is written at any position on the input tape by means of a movable reading head. Initially the head is placed at the left most square of the tape and the finite control is set is designated initial state.

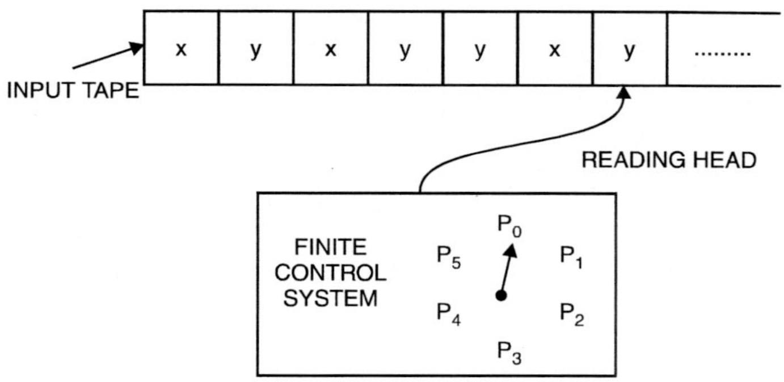

Fig. 1.2.

TIPS

Finite automation is called "finite" because number of possible states and number of letter in the alphabet are both finite, and "automation" because the change of the state is totally governed by the input. It is deterministic, since, what state is next is automatic not will-full, just as the motion of the hands of the clock is automatic, while the motion of hands of a human is presumably the result of desire and thought.

$P_0, P_1, P_2, P_3, P_4, P_5$ are states in finite control system x and y are input symbols.

At regular interval the automation reads one symbol from the input tape and then enters in a new state that depends only on the current state and the symbol just read.

After reading an input symbol, reading head moves one square to the right on the input tape, so that on the next move, it will read the symbol in next tape square. This process is repeated again and again.

The automation then indicates approval or disapproval.

If it winds up *in one of a set of final states* the input strings is considered to be accepted. The language accepted by the machine is the set of strings, it accepts.

1.9. DEFINITION OF DETERMINISTIC FINITE AUTOMATA

Definition When we analyse above discussion mathematically the following formal version obtained.

A deterministic finite automata is a quintuple

$$M = (Q, \Sigma, \delta, q_0, F)$$

where

Q : Is a non-empty finite set of states presents in the finite control. $(q_0, q_1, q_2, ...)$

Σ : Is a non-empty finite set of input symbols which can be passed to the finite state machine. $(a, b, c, d, e, ...)$

q_0 : Is a starting state, one of the state in Q.

F : Is a non-empty set of final states or accepting states, set of final states belongs to Q.

δ : Is a function called transition function that takes two arguments a state and a input symbol, it returns a *single* state.

Let 'q' is the state and 'a' be input symbol passed to the transition function as :

$$\delta(q, a) = q'$$

q' is the output of the function *i.e.,* a single state only, q' may be q. We can also view it graphically as :

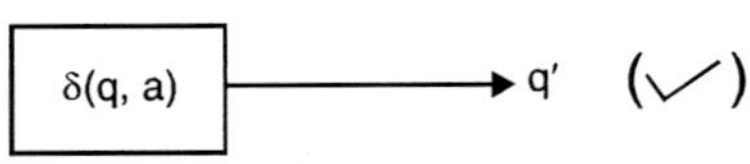

Fig. 1.3.

but q' may be same as q *i.e.* $q' = q$

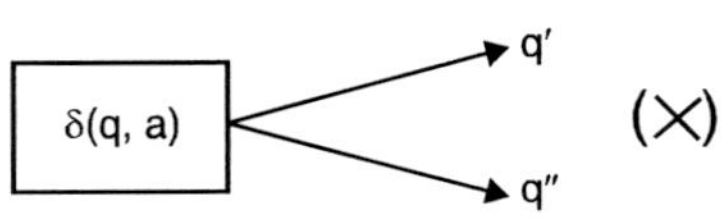

Fig. 1.4.

case discussed in Fig. 1.4 is not valid case in deterministic finite automata.

TIPS

Some people prefer to call this just a finite acceptor because its sole job is to accept certain input strings and rejects others. It does not do anything like printing machine. Some time it is also referred as language recognizer because FA merely recognizes whether the input strings is in the language as we can also recognize that some body is speaking *Russian* or *English* without necessarily understanding what it means.

1.10. PROCESSING OF STRINGS BY DFA

Suppose $a_1, a_2, a_3, ..., a_n$ is a sequence of input symbols. We start out with deterministic finite automata having $q_0, q_1, q_2, q_3, ..., q_n$ states where q_0 is the initial state and q_n is the final state and transition function and processed as

$$\begin{aligned}\delta(q_0, a_1) &= q_1\\ \delta(q_1, a_2) &= q_2\\ \delta(q_2, a_3) &= q_3\\ &\vdots\\ \delta(q_{n-1}, q_n) &= q_n\end{aligned}$$

Input $a_1, a_2, ..., a_n$ is said to be "**accepted**" since q_n is the member of the final states, and if not then it is "**rejected**". By adopting this approach we can define the language of a particular DFA, and can also check that particular string is accepted or rejected by given FA.

TIPS

"So a string is said to be accepted by the given FA only when it is processed by transition function δ in such a manner that it ends at the final state".

Before discussing the formal notations for the FA, let us know about the transition graph :

1.11. TRANSITION GRAPH

A transition graph is a collection of three things :

1. A finite set of states, at least one of which is designated as the start state 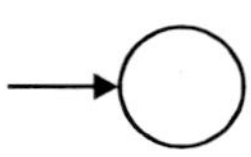 and some (may be none) of which are designated as final states 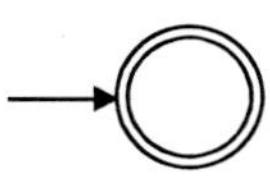.
2. An alphabet Σ of possible input letters from which input strings are formed.
3. A finite set of transitions (edge labels) that show how to go from some states to some others, based on reading specified substrings of input letters (possibly even the null string).

Transition graph were invented by John Myhill in 1957. A *successful path* through a transition graph is a series of edges forming a path beginning at some start state (there may be several start states) and ending at a final state. If we concatenate in order the strings of letters that label each edge in the path, we produce a word that is accepted by the this machine.

Let us consider some examples of TG's :

(*a*) 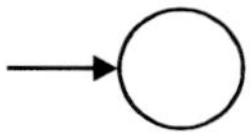

This picture above represents a TG that accepts nothing, not even the null string, to be accept any thing, it must be final state.

(*b*) The machine

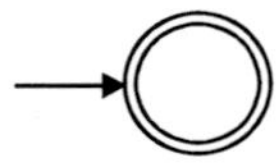

Fig. 1.5.

accept null string.

(*c*)

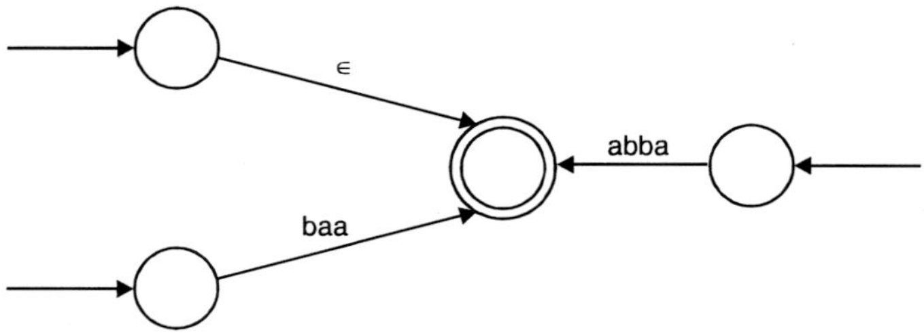

Fig. 1.6.

above TG accepts ∈, *baa* and *abba*.

(*d*)

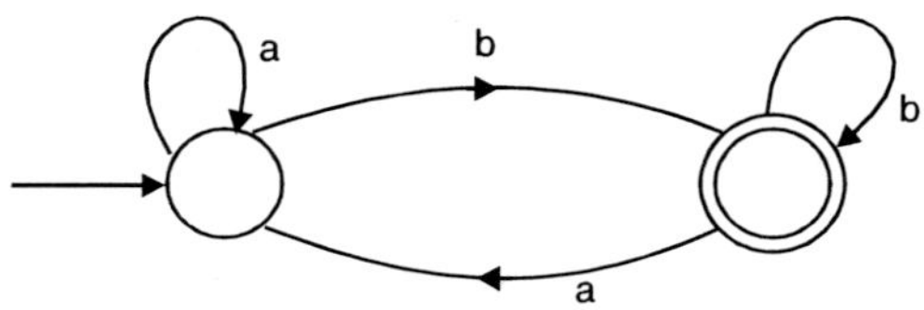

Fig. 1.7.

The language accepted by the TG is all words ending in *b* over alphabet $\Sigma = \{a, b\}$.

1.12. GENERALISED TRANSITION GRAPHS

The ultimate step liberating state-to-state transition is to allow the input to progress from one place to another by contributing a substring restricted to being a word in a predetermined language. For example :

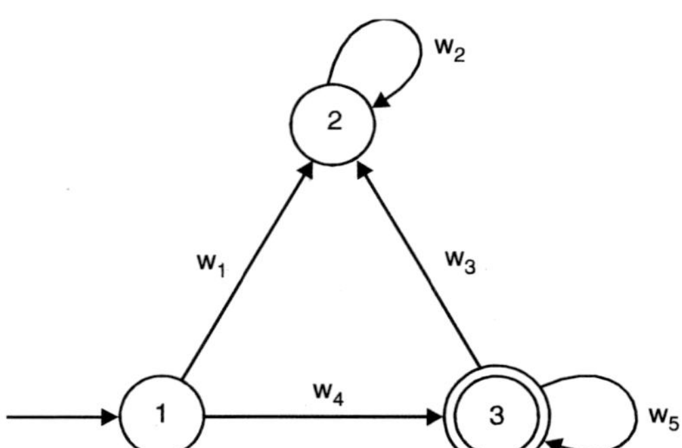

Fig. 1.8.

We can travel from start to state 2 by reading any (of course finite) word from the (possibly infinite) set of choices w_1 and, similarly, between all other states.

1.12.1. Definition of Generalised Transition Graph

A generalised transition graph (GTG) is a collection of three things :

1. A finite set of states, of which at least one is a start state and some (may be none) are final states.
2. An alphabet Σ of input letters.
3. Directed edges connecting some pair of states, each labelled with a regular expression.

Example :

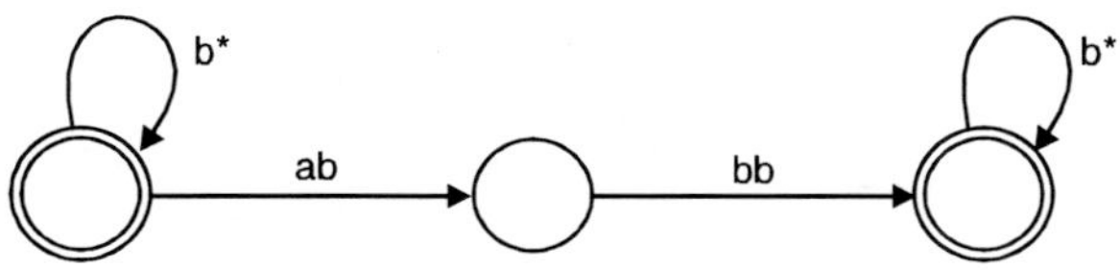

Fig. 1.9.

This machine accepts all strings without double *a*.

(Concept of Regular expression will be discussed in next chapter)

1.13. SIMPLER NOTATIONS FOR DFA's

There are two preferred notations for describing automata :

1. Transition Diagram,
2. Transition Table.

1. Transition Diagram Notations. A transition diagram for DFA, $M = (Q, \Sigma, \delta, q_0, F)$ is a graph defined as follows :

(*a*) For each state in Q there is a node represented by the circle.

(*b*) For each state q in Q and each input symbol a in Σ, let $\delta(q, a) = P$. Then transition diagram has an arc from node q to node P, labelled a. If there are several input symbols that cause transition from q to P, then the transition diagram can have one arc, labelled by the list of these symbols as :

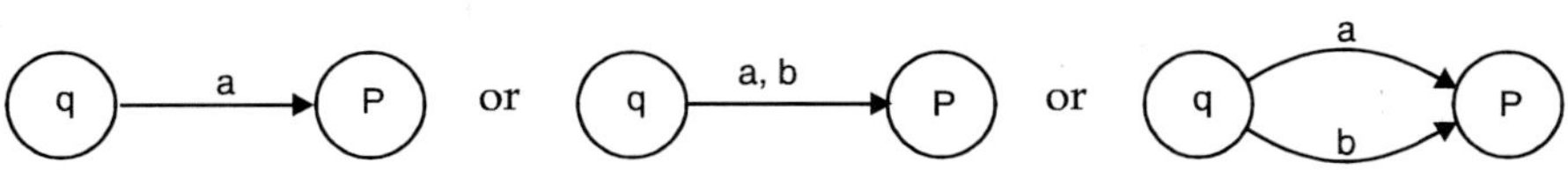

Fig. 1.10.

(*c*) If any state q in Q is the starting state then it represented by the circle with arrow as

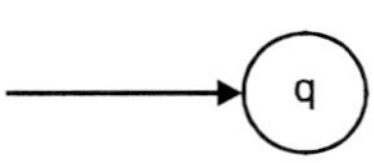

Fig. 1.11.

(*d*) Nodes corresponding to accepting states (those are in F) are marked by a double circle. States not in F have a single circle.

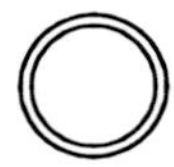

Fig. 1.12.

2. Transition Tables. A transition table is a Conventional, tabular representation of a function like δ that takes two arguments and returns a state. The rows of the table correspond to the states, and the columns correspond to the input. The entry for one row corresponding to state q and the column corresponding to input a is the state $\delta(q, a)$. For example :

δ/Σ	a	b
$\rightarrow q_0$	q_1	q_2
q_1	q_2	q_0
$*q_2$	q_2	q_2

Fig. 1.13.

q_0 is the starting state and q_2 is the final state,

$$\begin{aligned} Q &= \{q_0, q_1, q_2\} \\ F &= \{q_2\} \\ \delta(q_0, a) &= q_1, \delta(q_0, b) = q_2 \\ \delta(q_1, a) &= q_2, \delta(q_1, b) = q_0 \\ \delta(q_2, a) &= q_2, \delta(q_2, b) = q_2 \end{aligned}$$

Let us draw the transition diagram for the transition table shown in Fig. 1.13.

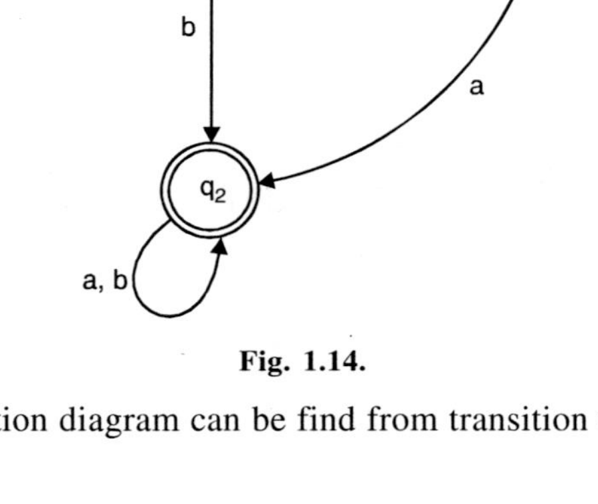

Fig. 1.14.

So clearly transition diagram can be find from transition table and viceversa is also true.

1.14. THE LANGUAGE OF A DFA

It have been explained informally that DFA defines a language : The set of all strings that result in a sequence of state transition from start state to an accepting state.

Now we can define the language of a DFA $M = (Q, \Sigma, \delta, q_0, F)$. This language is denoted by $L(M)$, and defined by

$$L(M) = \left\{w / \hat{\delta}(q_0, w) \text{ is in } F\right\}$$

* $\hat{\delta}$ (extended transition function) will be explained afterwords.

That is the language of M is the set of strings (w) that take the start q_0 to one of the accepting states.

TIPS

If L is $L(M)$ for some deterministic finite automata, then we say L is a regular language.

Now for practical understanding of FA, let us discuss the some examples.

Example 1.1. Design a FA that accepts strings containing exactly 1 over alphabet {0, 1}

Solution. String set according to the problem, which is accepted by the FA is like

1, 0001, 01000, 0001000, ...

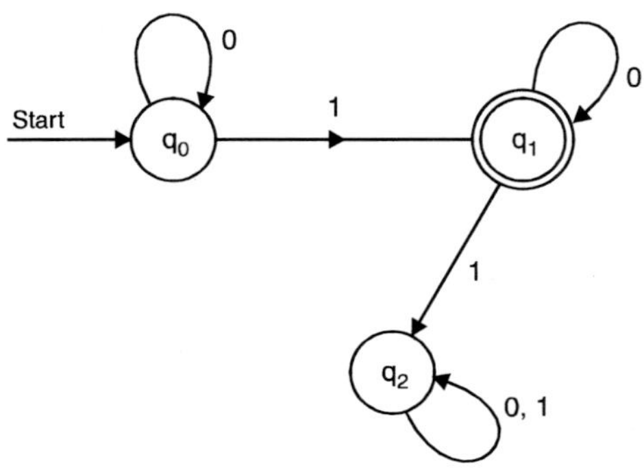

Fig. 1.15.

So clearly starting state to final state there is only one transition of 1.

Let suppose DFA be

$$M = (Q, \Sigma, \delta, q_0, F)$$
$$Q = \{q_0, q_1, q_2\}$$
$$F = \{q_1\}$$

q_2 is the dead state.

Transition table for above FA

δ/Σ	0	1
$\rightarrow q_0$	q_0	q_1
$* q_1$	q_1	q_2
q_2	q_2	q_2

$\rightarrow$ non final state which transit in self loop for every input symbol in Σ, so it is dead state.

Example 1.2. Design a FA which accepts the language

$L = \{w \in (0, 1)^*$/second symbol of w is '0' and fourth input is 1}.

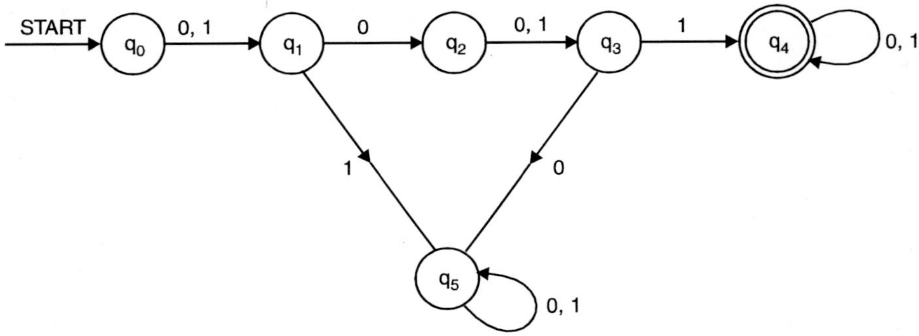

Fig. 1.16.

Solution. So after analysing the problem we find that every string of the language is like

0001, 1011, 10110000, ...

Let FA is

$$M = (Q, \Sigma, \delta, q_0, F)$$
$$Q = \{q_0, q_1, q_2, q_3, q_4, q_5\}$$
$$F = \{q_4\}$$

q_5 is dead state.

Example 1.3. For the given FA write the language and also give the transition table

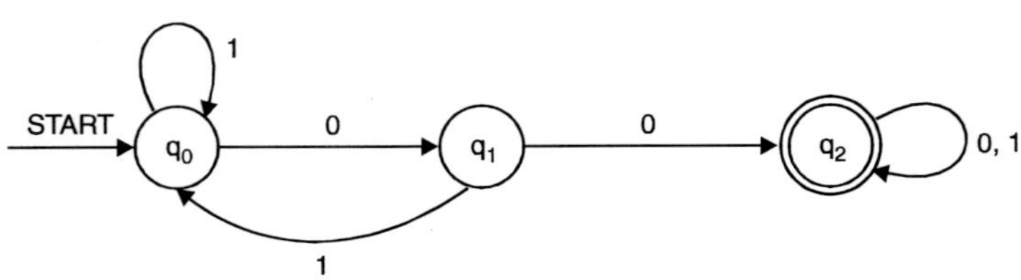

Fig. 1.17.

Solution. Let the language be L for FA shown in Fig. 1.17.

$L = \{w \in (0, 1)^*$/every string w of the language containing 00 as substring}.

Transition table :

δ/Σ	0	1
$\rightarrow q_0$	q_1	q_0
q_1	q_2	q_0
$* q_2$	q_2	q_2

Example 1.4. Design FA for the language

$$L = \{(01)^i \, 1^{2j} \mid i \geq 1, J \geq 1\}.$$

Solution. By analysing language L, it is clear that FA will accepts strings start with any number of 01 (not empty) and end with even number of 1's.

Let FA be $M = (Q, \Sigma, \delta, q_0, F)$

q : initial state.

δ : is transition function.

$\Sigma = \{0, 1\}$ given.

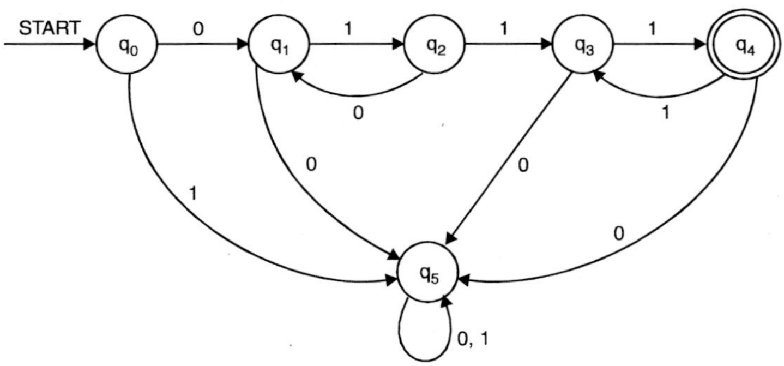

Fig. 1.18.

$$Q = \{q_0, q_1, q_2, q_3, q_4, q_5\}$$
$$F = \{q_4\}$$

q_5 is dead state.

Example 1.5. Design DFA for the language

$$L = \{w \in (a, b)^*/n_b(w) \bmod 3 > 1\}.$$

Solution. Let us first analyse the language L.

$n_b(w)$: number of b's in every string of language.

$L_b(w)$ mod 3 : number of b's in string is divided by three and remainder is result of $n_b(w)$ mod 3.

$n_b(w)$ mod 3 > 1 means that remainder is only 2 since when any number is divided by 3 then remainder may be only 0, 1 and 2.

Let $M = (Q, \Sigma, \delta, q_0, F)$ be FA

$\Sigma = \{a, b\}$

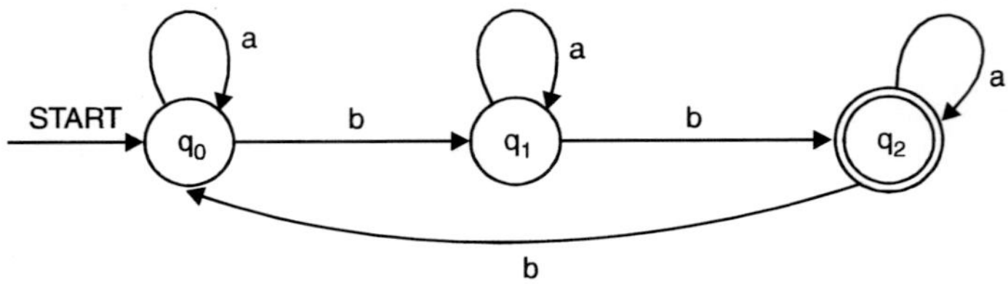

Fig. 1.19.

It will accept string, having $2b$'s, $5b$'s and $8b$'s so remainder is always two.

$$Q = \{q_0, q_1, q_2\}$$
$$F = \{q_2\}$$

Example 1.6. Design a FA over alphabet $\Sigma = \{0, 1\}$, which accepts the set of strings either start with 01 or end with 01.

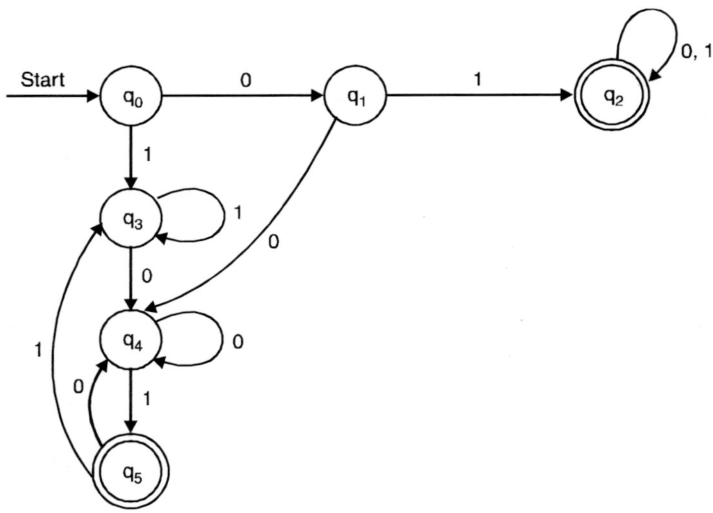

Fig. 1.20.

Solution. By the analysis of problem, it is clear that FA will accepts the strings such as 01, 011111, 01000, 000101, 00011101, 111101..... .

Let FA $M = \{Q, \Sigma, \delta, q_0, F)$

$$\Sigma = \{0, 1\} \text{ given}$$

q_0 is initial state.

$$Q = \{q_0, q_1, q_2, q_3, q_4, q_5\}$$
$$F = \{q_2, q_5\}$$

Example 1.7. Give the DFA accepting the set of strings over alphabet $\Sigma = \{0, 1\}$, such that in each string number of 0's is divisible by five *and* number of 1's is divisible by 3.

Solution. Here two conditions *i.e.* number of 0's is divisible by five *and* number of 1's is divisible by 3 must satisfied simultaneously. So clearly only those state of FA will be final which satisfied both the condition simultaneously.

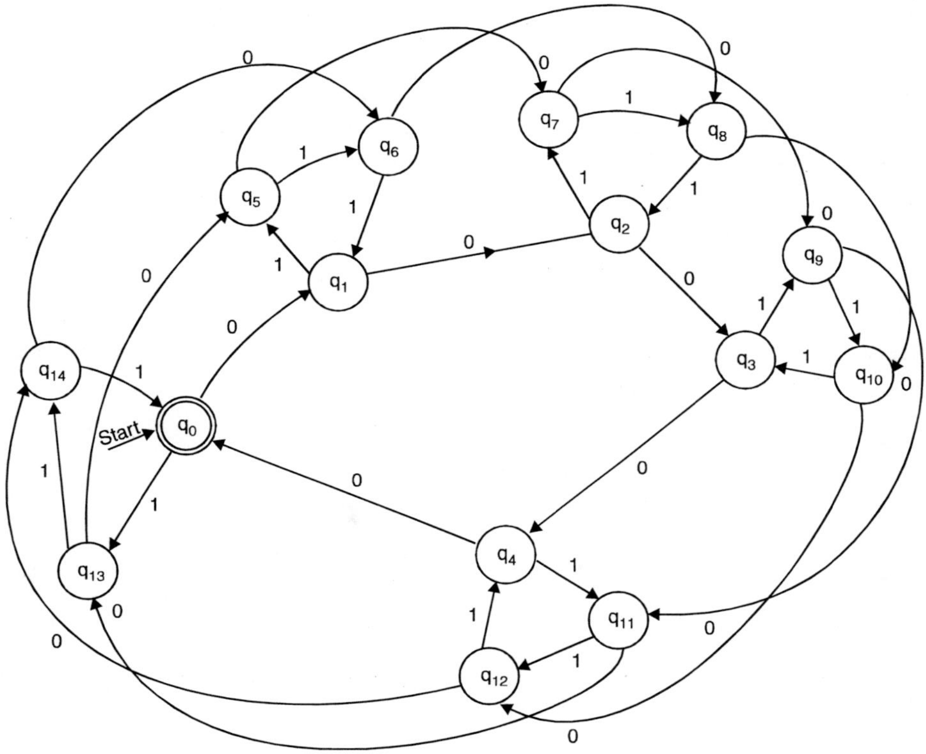

Fig. 1.21.

Let initial state is q_0, number of 0's passed at q_0 are zero which is divisible by five similarly number of 1's passed at q_0 are zero which is also divisible by three so clearly q_0 is final state.

Let FA be

$$M = (Q, \Sigma, \delta, q_0, F)$$
$$Q = \{q_0, q_1, q_2, q_3, q_4, q_5, q_6, q_7, q_8, q_9, q_{10}, q_{11}, q_{12}, q_{13}, q_{14}\}$$
$$F = \{q_0\}$$

1.15. EXTENDING TRANSITION FUNCTION TO STRINGS

In terms of the transition diagram, the language of a DFA is the set of labels along the paths that lead from the start state to any accepting state.

Now, we need to make the notation of the language of FA precise. To do so we define extended transition function that describes what happens when we start in any state and follow any sequence of inputs. If δ is our transition function, the extended function constructed from δ will be called $\hat{\delta}$.

The extended transition function is the function that takes a state q and a string w and returns a state P, the state that automation reaches when starting in state q and processing the sequence of inputs w. We define $\hat{\delta}$ by induction on the length of the input string as follows :

Basis : $\hat{\delta}(q, \in) = q$. That is, if we are in state q and read no input, then we are still in state q.

Induction. Suppose w is a string of the form xa, that is a is the last symbol of w and x is the substring of w, consisting of all except the last symbol 'a'. For example

$w = 1101$ is broken into $x = 110$ and $a = 1$ then

$$\hat{\delta}(q, w) = \delta(\hat{\delta}(q, x), a)$$

i.e., to compute $\hat{\delta}(q, w)$, first compute $\hat{\delta}(q, x)$, the state that the automation is in after processing all but the last symbol of w. Suppose this state is P'; that is $\hat{\delta}(q, x) = P'$. Then $\hat{\delta}(q, w)$ is what we get by making a transition from state P' on input a, the last symbol of w.

$$\hat{\delta}(q, w) = \delta(P', a) = P$$

Let see this concept through an example.

Example 1.8. Design a FA which accepts the language $L = \{w/w$ has both an even number of 0's and an even number of 1's over alphabet $\Sigma = \{0, 1\}\}$.

Solution. Let q_0 be the initial state, since '0' is even so at initial state number of 0's and number of 1's both are even so clearly q_0 will be final state

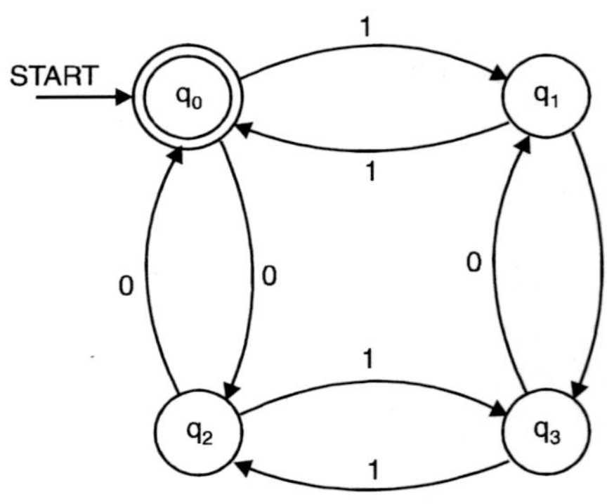

Fig. 1.22.

Let FA be $M = (Q, \Sigma, \delta, q_0, F)$

$\Sigma = \{0, 1\}$

q_0 : both 1 and 0's are even

q_1 : 0's are even and 1's are odd

q_2 : 1's are even and 0's are odd

q_3 : both are odd.

Check 110101 string on above FA *i.e.* it is accepted by FA or rejected.

Suppose extended transition function is $\hat{\delta}$,

$$\hat{\delta}(q_0, \in) = q_0$$

$$\hat{\delta}(q_0, 1) = \delta(\hat{\delta}(q_0, \in), 1) = \delta(q_0, 1) = q_1$$

$$\hat{\delta}(q_0, 11) = \delta(\hat{\delta}(q_0, 1), 1) = \delta(q_1, 1) = q_0$$

$$\hat{\delta}(q_0, 110) = \delta(\hat{\delta}(q_0, 11), 0) = \delta(q_0, 0) = q_2$$

$$\hat{\delta}(q_0, 1101) = \delta(\hat{\delta}(q_0, 110), 1) = \delta(q_2, 1) = q_3$$

$$\hat{\delta}(q_0, 11010) = \delta(\hat{\delta}(q_0, 1101), 0) = \delta(q_3, 0) = q_1$$

$$\hat{\delta}(q_0, 110101) = \delta(\hat{\delta}(q_0, 11010), 1) = \delta(q_1, 0) = q_0$$

and q_0 is final state so string 110101 is accepted by FA in Fig. 1.22.

1.16. NONDETERMINISM

Nondeterminism means a choice of moves for an automaton. Rather than prescribing a unique move in each situation, we allow a set of possible moves. Formally, we achieve this by defining the transition function so that its range is a set of possible states.

1.16.1. Why Nondeterminism

A "nondeterministic" finite automation (NFA) has the power to be in several states at once. This ability is often expressed as an ability to "guess", something about its output. Each NFA accepts a language that is also accepted by some DFA. That is the NFA's accept exactly the regular language, just as DFA's do. We can always convert a NFA to DFA. It is clear now, we can permit several possible "next states" for a given combination of current state and input symbol. The automation, as it reads the input string, may choose at each step to go into any one of these legal next states, the choice is not deterministic by anything in our model, and is therefore said to be nondeterminstic.

TIPS

Though nondeterministic is a feature we do not normally associate with real computers, it arises naturally as an extension of the behaviour of deterministic finite automata and in some cases greatly simplifies the description of these automata. Moreover, nondeterminism is an essential feature of FA, every NFA is equivalent to FA. Thus we profit from the greater flexibility of nondeterminism, when we wish to design language recognizers without actually extending the class of language we are able to describe.

Definition Now let me introduce the formal notions associated with nondeterministic finite automata. The difference between DFA's and NFA's will be pointed out as we do.

Let M be NFA then M is defined as

$$M = (Q, \Sigma, \delta, q_0, F)$$

where Q : is a non empty finite set of states in NFA.

Σ : is a non-empty finite set of input symbols.

q_0 : initial state of NFA and it is the member of Q.

F : a non-empty finite set of final states in Q.

δ : is the transition function that takes a states and Q and an input symbol in Σ as arguments and returns a subset of Q.

that is $Q \times (\Sigma \cup \{\in\}) \to 2^Q$

So in nondeterministic accepter, the range of δ is in the power set 2^Q, so its value is not a single element of Q, but a subset of it. This subset defines the set of possible states that can be reached by transition.

Let q be an state of Q and $a \in \Sigma$ then

We can transit from a state q on same input a to different states $q_0, q_1, ..., q_n$ in Q. This not possible in the case of FA.

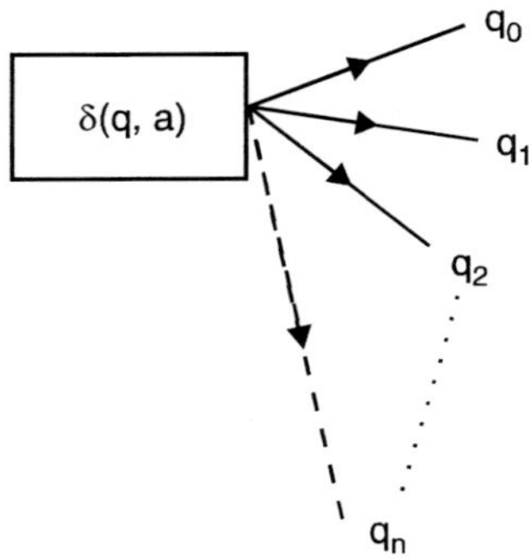

Fig. 1.23.

Example 1.9. Design a NFA for the language L = all strings over {0, 1} that have at least two consecutive 0's or 1's.

Solution. By the analysis it is clear that NFA will accepts the strings of the pattern 00, 11, 101000, 101100, ...

Let NFA be

$$M = (Q, \Sigma, \delta, q_0, F)$$
$$\Sigma = \{0, 1\}$$

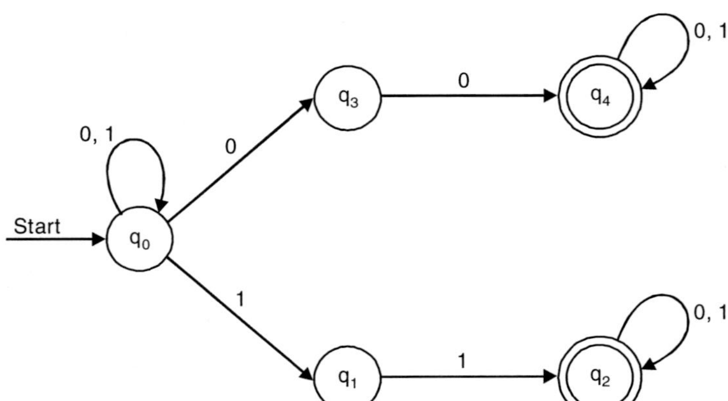

Fig. 1.24.

Note that state q_3 does not go anywhere for an input of 1. **We use the terminology that the path "dies" if in q_3 getting an input 1.**

The tree of states this NFA is in for the input 0100011.

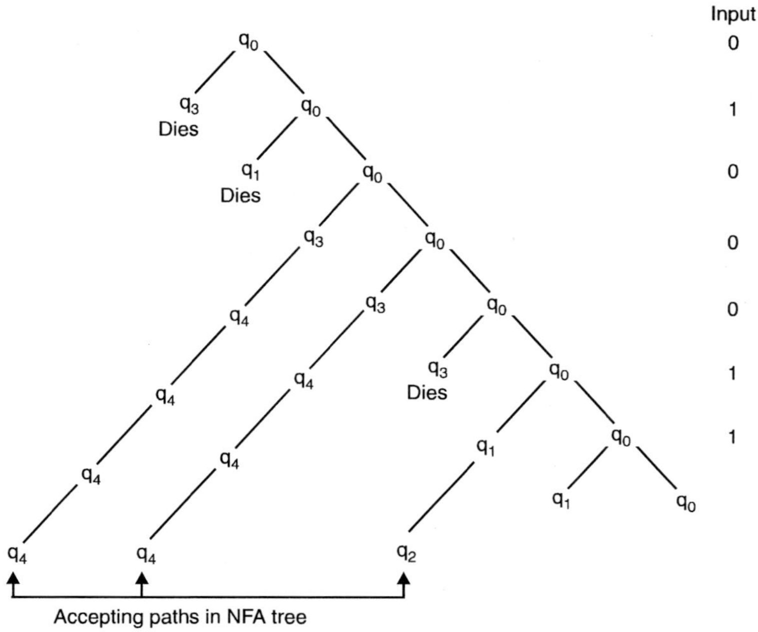

Fig. 1.25.

So it is clear from Fig. 1.25 that for a particular string there may be various paths, some may lead it to accepted states, and some may lead it to non accepted states and some may lead it to the blocked paths as $q_0 \rightarrow q_3$ as q_3 dies.

Example 1.10. Design a NFA for the language $L = (ab \cup aba)^*$.

Solution. Let us first analyse the given language *i.e.*

$L = (ab \cup aba)^*$ means either *ab*, *aba* or any combination of *ab* and *aba* should be accepted by the NFA. null string ($\in$) is also accepted.

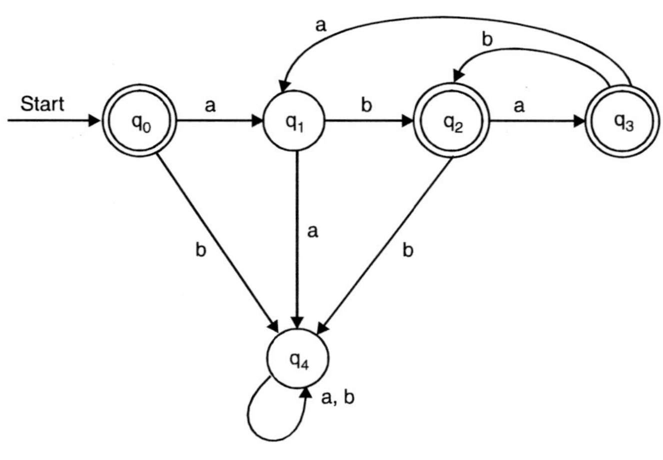

Fig. 1.26. DFA for the *L*.

If DFA for language L.

Let us now design NFA for L.

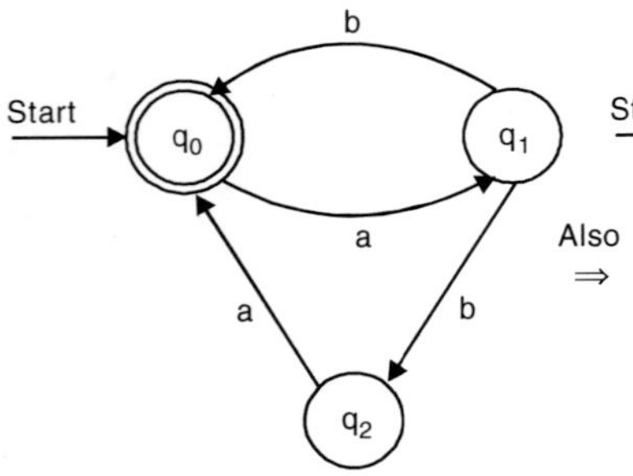

Fig. 1.27.

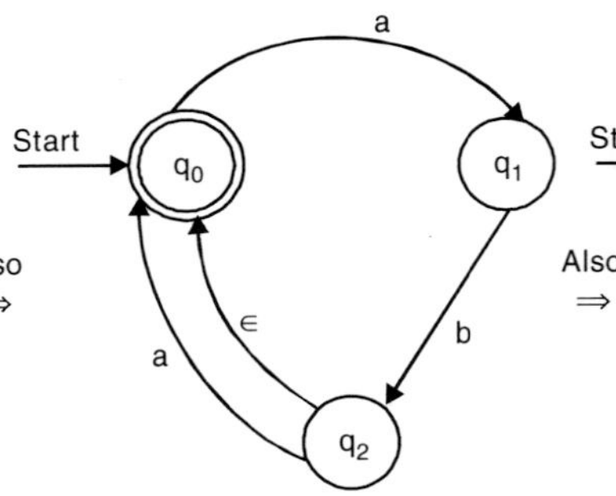

Fig. 1.28.

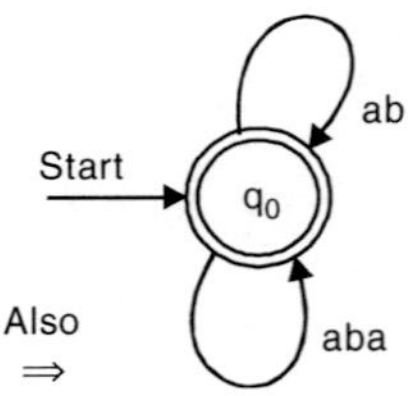

Fig. 1.29.

Figure 1.27, Fig. 1.28 and Fig. 1.29 are the NFA for the language L. Thus we can have different NDFA for same language.

Example 1.11. Draw the state diagram for NFA accepting language

$$L = (ab)^* (ba)^* \cup aa^*.$$

Solution. We can construct NFA for the language L in two paths *i.e.*

$$L = L_1 \cup L_2$$
$$L_1 = (ab)^* (ba)^*$$
$$L_2 = aa^*$$

First we can construct NFA for L_1

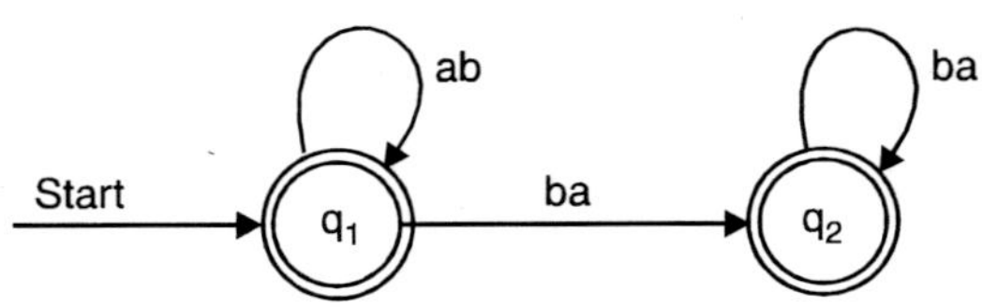

Fig. 1.30. NFA for L_1.

Secondly we construct NFA for L_2.

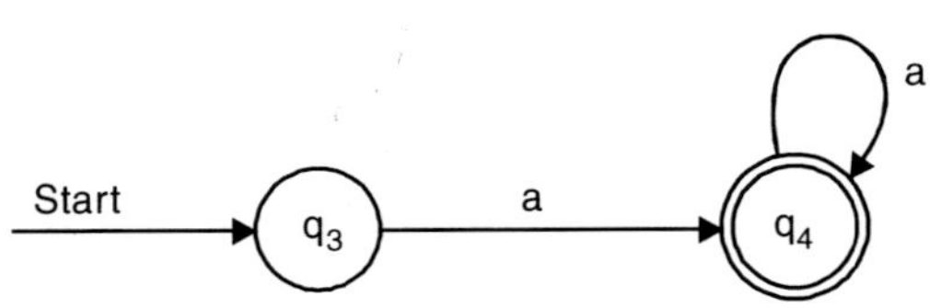

Fig. 1.31.

Now we can combine NFA in Fig. 1.30 and 1.31. For combining we will introduce another state q_0 and will connect both NFA with Null transitions as follows :

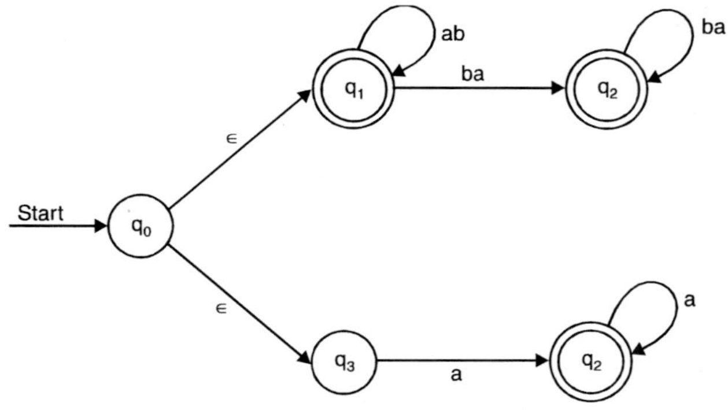

Fig. 1.32.

NFA in Fig. 1.32 is NFA for the language L.

Example 1.12. Find the NFA with four state for the language

$$L = \{(a^n : n \geq 0) \cup (b^n a : n \geq 1)\}.$$

Solution. Again we will adopt the same approach as in example 1.14.

$$L = L_1 \cup L_2$$
$$L_1 = a^n : n \geq 0$$
$$L_2 = b^n a : n \geq 1$$

NFA for the language L_1 is 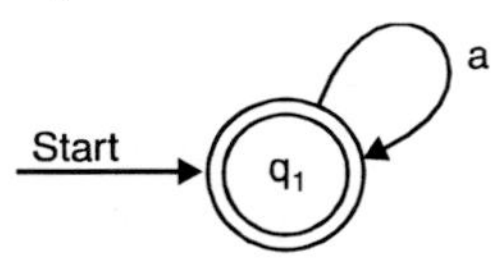

Fig. 1.33.

NFA for the language L_2 is

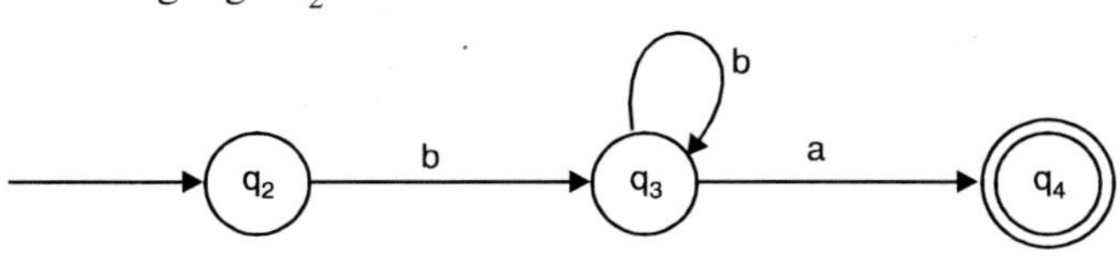

Fig. 1.34.

Now combine NFA in Fig. 1.33 and Fig. 1.34 as

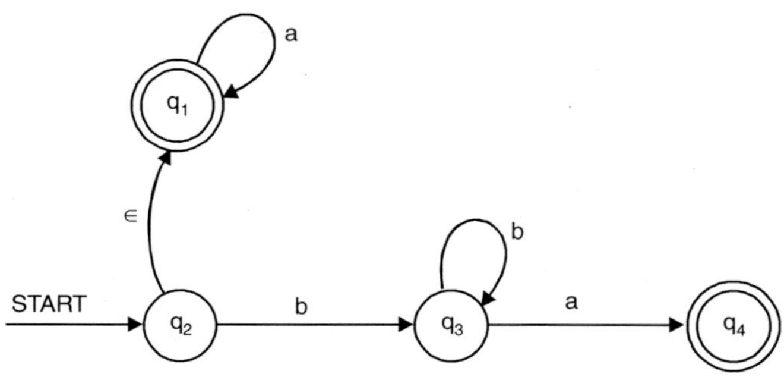

Fig. 1.35. NFA for L.

Finally Fig. 1.35 is the NFA for the language L.

Extended Transition Function for NFA

Similar to DFA's, we need to extend the transition function δ of an NFA to a function $\hat{\delta}$ that takes a state and string and return the set of states.

Basis. $\hat{\delta}(q,\in) = \{q\}$. That is without reading any input symbols, we are only in the same state.

Induction. Suppose w is of the form $w = xa$, where a is the last symbol and x is the substring containing rest of w. Let us suppose that

$$\hat{\delta}(q,x) = \{p_1, p_2, \ldots p_k\},$$

Let
$$\bigcup_{i=1}^{k} \delta(p_i, a) = \{r_1, r_2, r_3, \ldots, r_m\}$$

Then $\hat{\delta}(q, w) = \{r_1, r_2, \ldots r_m\}$. Less formally, we compute $\hat{\delta}(q, w)$ by first computing $\hat{\delta}(q, x)$, and by then following any transition from any of these states that is labeled a. Let us see an example for better under standing of the concept.

Example 1.13. Consider NFA in Fig. 1.36, and describe the processing of string 00101 by it.

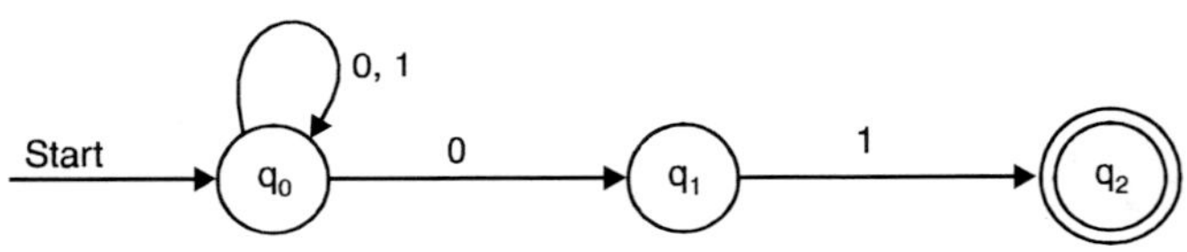

Fig. 1.36.

Solution. It is NFA accepting all strings end in 01. Let us use $\hat{\delta}$ to describe the processing of input 00101 by NFA. A summary of the steps is.

Step 1. $\hat{\delta}(q_0,\in) = \{q_0\}$

Step 2. $\hat{\delta}(q_0, 0) = \delta(q_0, 0) = \{q_0, q_1\}$

Step 3. $\hat{\delta}(q_0, 00) = \delta(q_0, 0) \cup \delta(q_1, 0) = \{q_0, q_1\} \cup \phi = \{q_0, q_1\}$

Step 4. $\hat{\delta}(q_0, 001) = \delta(q_0, 1) \cup \delta(q_1, 1) = \{q_0\} \cup \{q_2\} = \{q_0, q_2\}$

Step 5. $\hat{\delta}(q_0, 0010) = \delta(q_0, 0) \cup \delta(q_2, 0) = \{q_0, q_1\} \cup \phi = \{q_0, q_1\}$

Step 6. $\hat{\delta}(q_0, 00101) = \delta(q_0, 1) \cup \delta(q_1, 1) = \{q_0\} \cup \{q_2\} = \{q_0, q_2\}$.

Step 1 is the basis rule. We obtain step 2 by applying δ to the done state q_0, that is in the previous set, and get $\{q_0, q_1\}$ as a result. Step 3 is obtained by taking the union over two states in previous set of what we get when we apply δ to the input 0. That is, $\delta\ (q_0, 0) = \{q_0, q_1\}$, while $\delta\{q_1, 0\} = \phi$.

For step 4, we take union of $\delta(q_0, 1) = q_0$ and $\delta(q_1, 1) = \{q_2\}$. So step 5 and step 6 are obtained in similar way.

1.17. THE LANGUAGE OF NFA

Definition: As we discussed, an NFA accepts a string w if it is possible to make any sequence of choice of next state while reading the characters of w, and go from the start state to any accepting state.

If M is a NFA

$$M = (Q, \Sigma, \delta, q_0, F) \text{ then}$$

$$L(M) = \left\{w / \hat{\delta}(q_0, w) \cap F \neq \phi\right\}$$ is the Language of M *i.e.* NFA. $L(M)$ is the set of strings w in Σ^* such that $\hat{\delta}(q_0, w)$ contains at least one accepting state.

1.18. TRANSFORMATION OF NFA TO DFA

For every non deterministic finite automata there exist an equivalent deterministic finite automata. The equivalence is defined in terms of language acceptance. Since a NFA is a nothing but a finite automata in which zero, one or more transition on an input symbol is permitted, we can always construct a finite automata which will simulate all the moves of NFA on a particular input symbol in parallel, then get a finite automata in which there will be exactly one transition on every input symbol, hence it will be DFA equivalent to NFA.

Since the DFA equivalent of NFA is to simulate the moves in parallel state of a DFA will be combination of one or more states of NFA, hence every state of a DFA will be represented by some subset of set of states of NFA, **and therefor the transition of NFA to DFA is normally called a subset construction.**

So transformation of NFA to DFA in value, finding subset of set of states of NFA. Let us implement this concept in the form of set of steps as algorithm through an example.

1.18.1. Procedure For Converting NFA to Equivalent DFA

Let $M_2 = \{Q_2, \Sigma, q'_0, \delta_2, F_2\}$ be NFA that recoginzes the languages L. Then the DFA $M = \{Q, \Sigma, q_0, \delta, F\}$ that satisfies the following conditions recoginzes L :

$Q = 2^{Q_2}$, that is the set of all subsets of Q_2.

$q_0 = \{q'_0\}$

$\delta(q, a) = \bigcup_{p \in q} \delta_2(p, a)$ for each state q in Q and each symbol a in Σ and

$F = \{q \in Q \mid q \cap F_2 \neq \phi\}$.

To obtain a DFA $M = \{Q, \Sigma, \delta, q_0, F\}$ which accepts the same language as given NFA $M_2 = \{Q_2, \Sigma, q'_0, \delta_2, F_2\}$ does, we may proceed as follows :

Step 1. Intially $Q = \phi$.

Step 2. Put $\{q'_0\}$ into Q. $\{q'_0\}$ is the initial state of the DFA M.

Step 3. Then for each state q in Q do the following : add this new state, add $\delta(q, a) = \bigcup_{p \in q} \delta_2(p, a)$ to δ, where the δ_2 on the right hand side is that of NFA M_2.

Step 4. Repeat step-3 till new states are there to add in Q, if there is new state found to add in Q, the process terminates. All the states of Q that contain accepting states of M_2 are accepting states of M.

Note : The states that are not reached from the initial state are not included in Q obtained by this procedure. Thus the set of states Q thus obtained is not necesarily equal to 2^{Q_2}.

Example 1.14. Convert the following NFA in to DFA.

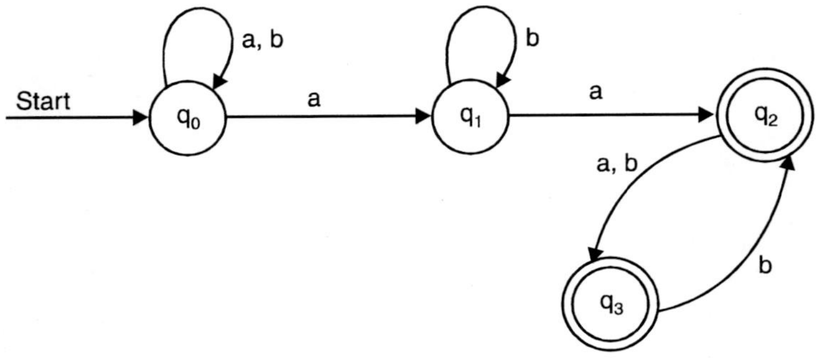

Fig. 1.37.

Solution.

Step 1. Seek all the transition from starting state q_0 for every symbol in Σ *i.e.* (a, b). If you get a set of states for same input then consider that set as new single state as :

$$\delta(q_0, a) = \{q_0, q_1\}$$

in Fig. 1.37 q_0 transit to both q_0 and q_1 for input a.

$$\delta(q_0, b) = \{q_0\}$$

Step 2. In step 1 we are getting a new state $\{q_0, q_1\}$. Now repeat step 1 for this new state only *i.e.* check all transitions of a and b (that is Σ) from $\{q_0, q_1\}$ as :

$$\delta(\{q_0, q_1\}, a) = \delta(q_0, a) \cup \delta(q_1, a) = \{q_0, q_1\} \cup \{q_2\}$$
$$= \{q_0, q_1, q_2\}$$
$$\delta(\{q_0, q_1\}, b) = \delta(q_0, b) \cup \delta(q_1, b) = \{q_0, q_1\}$$

Step 3. Repeat step 2 till you are getting any new state. All those states which consists at least one accepting state of given NFA as member state will be considered as final states.

$[q_0, q_1, q_2]$ is new state.

$$\delta(\{q_0, q_1, q_2\}, a) = \delta(q_0, a) \cup \delta(q_1, a) \cup \delta(q_2, a)$$
$$= \{q_0, q_1\} \cup \{q_2\} \cup \{q_3\}$$
$$= \{q_0, q_1, q_2, q_3\} \quad \text{(New state)}$$
$$\delta(\{q_0, q_1, q_2\}, b) = \delta(q_0, b) \cup (q_1, b)\ \delta(q_2, b)$$
$$= q_0 \cup q_1 \cup q_3 = \{q_0, q_1, q_3\} \quad \text{(New state)}$$
$$\delta(\{q_0, q_1, q_2, q_3\}, a) = \{q_0, q_1, q_2, q_3\} \quad \text{(Old state)}$$
$$\delta(\{q_0, q_1, q_2, q_3\}, b) = \{q_0, q_1, q_3, q_2\} \quad \text{(Old state)}$$
$$\delta(\{q_0, q_1, q_3\}, a) = \{q_0, q_1, q_2, q_3\} \quad \text{(Old state)}$$
$$\delta(\{q_0, q_1, q_3\}, b) = \{q_0, q_1, q_3, q_2\} \quad \text{(Old state)}$$

Let draw transition table

δ/Σ	a	b
$\rightarrow q_0$	$\{q_0, q_1\}$	$\{q_0\}$
$\{q_0, q_1\}$	$\{q_0, q_1, q_2\}$	$\{q_0, q_1\}$
$*\{q_0, q_1, q_2\}$	$\{q_0, q_1, q_2, q_3\}$	$\{q_0, q_1, q_3\}$
$*\{q_0, q_1, q_2, q_3\}$	$\{q_0, q_1, q_2, q_3\}$	$\{q_0, q_1, q_2, q_3\}$
$*\{q_0, q_1, q_3\}$	$\{q_0, q_1, q_2, q_3\}$	$\{q_0, q_1, q_2, q_3\}$

Let us say

$$q_0 \rightarrow A$$
$$\{q_0, q_1\} \rightarrow B$$
$$\{q_0, q_1, q_2\} \rightarrow C$$
$$\{q_0, q_1, q_2, q_3\} \rightarrow D$$
$$\{q_0, q_1, q_3\} \rightarrow E$$

A is initial state and C, D and E are final states since they contain q_2 and q_3 as member which are final states of NFA.

So

δ/Σ	a	b
$\rightarrow A$	B	A
B	C	B
$*C$	D	E
$*D$	D	D
$*E$	D	D

Let us draw transition diagram

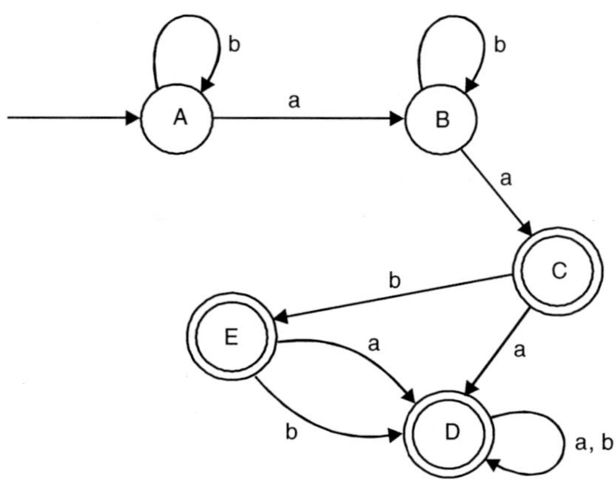

Fig. 1.38. DFA equivalent of given NFA.

It is a DFA.

Example 1.15. Given NFA is

δ/Σ	a	b
$\rightarrow q_0$	$\{q_0, q_1\}$	$\{q_2\}$
q_1	$\{q_0\}$	$\{q_1\}$
$*q_2$	ϕ	$\{q_0, q_1\}$

Convert it in to DFA.

Solution. Do the same process as in example 1.16.

$$\delta(q_0, a) = \{q_0, q_1\} \quad \text{(New)}$$
$$\delta(q_0, b) = \{q_2\} \quad \text{(New)}$$
$$\delta(\{q_0, q_1\}, a) = \delta(q_0, a) \cup \delta(q_1, a) = \{q_0, q_1\} \cup \{q_0\}$$
$$= \{q_0, q_1\} \quad \text{(Old)}$$
$$\delta(\{q_0, q_1\}, b) = \delta(q_0, b) \cup \delta(q_1, b) = \{q_1\} \cup \{q_1\}$$
$$= \{q_1, q_2\} \quad \text{(New)}$$
$$\delta\{q_2, a) = \{\phi\} \quad \text{(Empty set)}$$
$$\delta(q_2, b) = \{q_0, q_1\} \quad \text{(Old)}$$
$$\delta(\{q_1, q_2\}, a) = \delta(q_1, a) \cup \delta(q_2, a) = \{q_0\} \cup \{\phi\}$$
$$= \{q_0\} \quad \text{(Old)}$$
$$\delta(\{q_1, q_2\}, b) = \delta(q_1, b) \cup \delta(q_2, b) = \{q_1\} \cup \{q_0, q_1\}$$
$$= \{q_0, q_1\} \quad \text{(Old)}$$

Let say

$$q_0 \rightarrow A \quad \text{(Initial state)}$$
$$q_2 \rightarrow B \quad \text{(Final state)}$$
$$\{q_0, q_1\} \rightarrow C$$
$$\{q_1, q_2\} \rightarrow D \quad \text{(Final state)}$$

Transform table for DFA

δ/Σ	a	b
$\rightarrow A$	C	B
$*B$	$\{\phi\}$	C
C	C	D
$*D$	A	C

Transition diagram for DFA

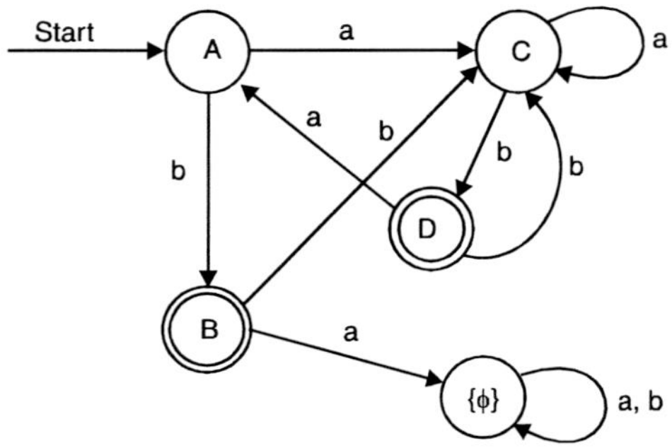

Fig. 1.39.

Here {ϕ} state is nothing but interpreted as if there is a input '*a*' on state *B* then there is no path for it and machine will be hang.

1.19. NFA WITH ∈–TRANSITIONS

TIPS

If a FA is modified to permit transition without input symbols, along with zero, one or more transition on input symbols, then we get a NFA with ∈-transitions, because the transition made without symbols are called as ∈-transitions.

Consider the NFA given by following diagram.

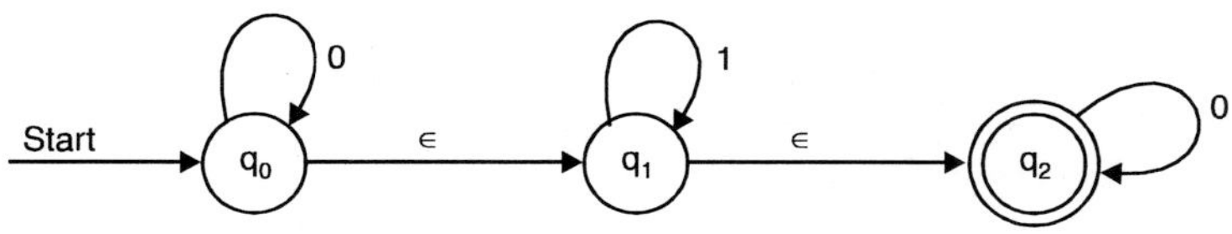

Fig. 1.40.

This is a NFA with ∈-transitions because it is possible to make transition from state q_0 to q_1 without consuming any of the input symbols. Being a finite Automata, NFA with ∈-transitions will also be denoted as five tuple.

$$M = (Q, \Sigma, \delta, q_0, F)$$

where Q, Σ, q_0 and F are having usual meaning, and transition function δ defines a mapping from $Q \times (\Sigma \cup \in)$ to 2^Q.

1.20. ACCEPTANCE OF A STRING BY NFA WITH ∈-TRANSITIONS

A string *w* in Σ^* will be accepted by NFA with ∈-transition, if there exist at least one path corresponding *w*, which starts in an initial state, and ends in one of the final states. Since this path may formed by ∈-transition as well non ∈-transitions to find out wheather *w* is accepted or not by the NFA with ∈ moves, we require to defined a function ∈-closure (*q*), where *q* is a state of the automata.

The function ∈-closure (*q*) is defined as follows :

∈-closure (*q*) = set of all those states of the automata (NFA with ∈-transitions) which can be reached from *q* on a path labeled by ∈ *i.e.* without consuming any input symbol.

Consider the NFA discussed in Fig. 1.40 then

$$\in\text{-closure } (q_0) = \{q_0, q_1, q_2\}$$
$$\in\text{-closure } (q_1) = \{q_1, q_2\}$$
$$\in\text{-closure } (q_2) = \{q_2\}$$

1.21. CONVERSION OF NFA WITH ∈-TRANSITIONS TO NFA WITHOUT ∈-TRANSITIONS

There is a conversion algorithm from NFA with ∈-transition *s* to *a* NFA without ∈-transitions !

We will learn this algorithm by solving an example.

Example 1.16. Consider the NFA with $\in$-transition $M = (Q, \Sigma, \delta, q_0, F)$

$$Q = \{q_0, q_1, q_2\}$$
$$\Sigma = \{a, b, c\} \text{ and } \in \text{ moves.}$$
$$\text{Initial state} = \{q_0\}$$
$$F = \{q_2\}$$

Transition Table

δ/Σ	a	b	c	$\in$
$\rightarrow q_0$	$\{q_0\}$	$\{\phi\}$	$\{\phi\}$	$\{q_1\}$
q_1	$\{\phi\}$	$\{q_2\}$	$\{\phi\}$	$\{q_2\}$
$* q_2$	$\{\phi\}$	$\{\phi\}$	$\{q_2\}$	$\{\phi\}$

and transition diagram is following

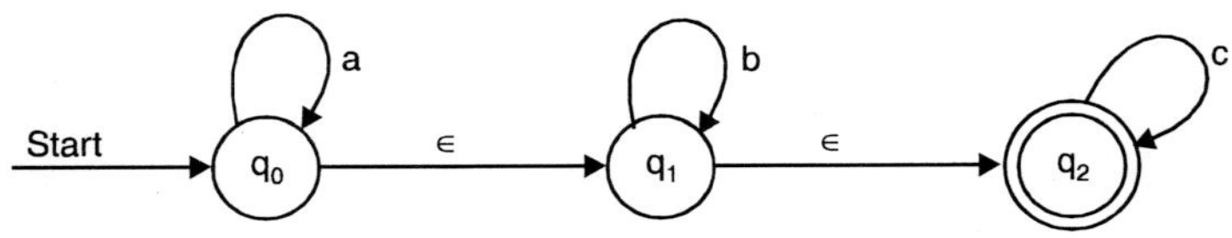

Fig. 1.41. NFA with $\in$-transition.

Solution.

Step 1. Find the states of NFA without $\in$-transition including initial state and final states as follows :

(*i*) The language accepted by the above NFA with $\in$-transform is the set of strings over $\{a, b, c\}$ including the null string and all strings with any number of a's followed by any number of b's and follows by any number of c's.

(*ii*) There will be the same number of states but the names can be constructed by writing the state name as the set of states in the $\in$-closure.

(*iii*) Initial state will be $\in$-closure of initial state of NFA with $\in$-transition in Fig. 1.41 as

$$\in\text{-closure } (q_0) = \{q_0, q_1, q_2\} \quad \text{(new initial state for NFA, without } \in \text{ transition)}$$

(*iv*) Rest of the states are

$$\in\text{-closure } (q_1) = \{q_1, q_2\} \quad \text{(New state)}$$
$$\in\text{-closure } (q_2) = \{q_2\} \quad \text{(New state)}$$

(*v*) The final states of NFA without $\in$-transition are all those new states which contains final state of NFA with $\in$-transition as member.

So $\{q_0, q_1, q_2\}$, $\{q_1, q_2\}$ and $\{q_2\}$ all are final states.

So if NFA without $\in$-transition is

$$M' = (Q', \Sigma, \delta', q_0', F')$$
$$Q' = (\{q_0, q_1, q_2\}, \{q_1, q_2\}, \{q_2\})$$

$$q_0' = \{q_0, q_1, q_2\}$$
$$F' = (\{q_0, q_1, q_2\}, \{q_1, q_2\}, \{q_2\})$$

Step 2. Now we have to decide δ' to find out the transitions as follows :

Very carefully consider each old machine transition in each row, you can ignore ϕ enteries and $\in$-transitions column. In the old machine $\delta(q_0, a) = q_0$ thus in new machine *i.e.* in M', $\delta'(\{q_0, q_1, q_2\}, a) = \{q_0, q_1, q_2\}$ this is just because the new machine accepts the same language as the old machine and must at least have the same transitions for the new state names. We can allow interpret it as follows :

$$\begin{aligned}\delta'(\{q_0, q_1, q_2\}, a) &= \in\text{-closure } (\delta(q_0, q_1, q_2), a) \\ &= \in\text{-closure } (\delta(q_0, a) \cup \delta(q_1, a) \cup \delta(q_1, a)) \\ &= \in\text{-closure } (q_0 \cup \phi \cup \phi) \\ &= \in\text{-closure } (q_0) \\ &= \{q_0, q_1, q_2\}\end{aligned}$$

Similarly
$$\begin{aligned}\delta'(\{q_0, q_1, q_2\}, b) &= \in\text{-closure } (\delta(q_0, q_1, q_2), b) \\ &= \in\text{-closure } (\delta(q_0, b) \cup \delta(q_1, b) \cup \delta(q_2, b)) \\ &= \in\text{-closure } (\phi \cup q_1 \cup \phi) \\ &= \in\text{-closure } (q_1) \\ &= \{q_1, q_2\}.\end{aligned}$$

$$\begin{aligned}\delta'(\{q_0, q_1, q_2\}, c) &= \in\text{-closure } (\delta(q_0, q_1, q_2), c) \\ &= \in\text{-closure } (\delta(q_0, c) \cup \delta(q_1, c) \cup \delta(q_1, c) \\ &= \in\text{-closure } (\phi \cup \phi \cup q_2) \\ &= \in\text{-closure } (q_2) \\ &= \{q_2\}.\end{aligned}$$

In the same fashion we can write

$$\begin{aligned}\delta'(\{q_1, q_2\}, a) &= \phi \\ \delta'(\{q_1, q_2\}, b) &= \{q_1, q_2\} \\ \delta'(\{q_1, q_2\}, c) &= \{q_2\} \\ \delta'(q_2, a) &= \phi \\ \delta'(q_2, b) &= \phi \\ \delta'(q_2, c) &= \{q_2\}\end{aligned}$$

So transition table for NFA without $\in$-transition will be as follows :

δ'/Σ	a	b	c
$\rightarrow * \{q_0, q_1, q_2\}$	$\{q_0, q_1, q_2\}$	$\{q_1, q_2\}$	$\{q_2\}$
$* \{q_1, q_2\}$	ϕ	$\{q_1, q_2\}$	$\{q_2\}$
$* \{q_2\}$	ϕ	ϕ	$\{q_2\}$

Let us say $\{q_0, q_1, q_2\}$ as q_x

$\{q_1, q_2\}$ as q_y

and $\{q_2\}$ as q_z

So transition diagram is as follows :

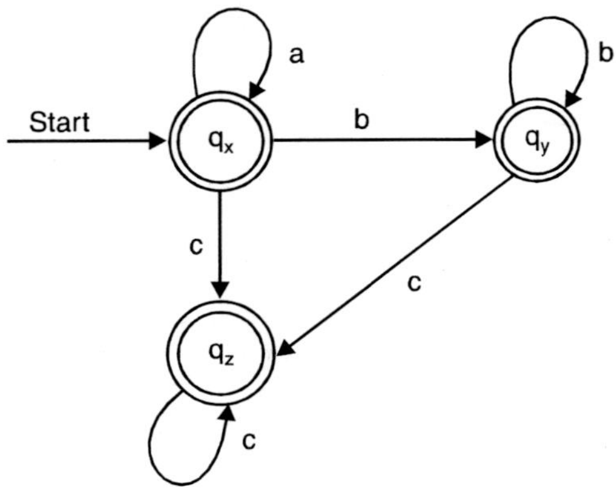

Fig. 1.42. NFA without ∈-transitions.

Let us see one more example.

Example 1.17. Consider the NFA given by following diagram

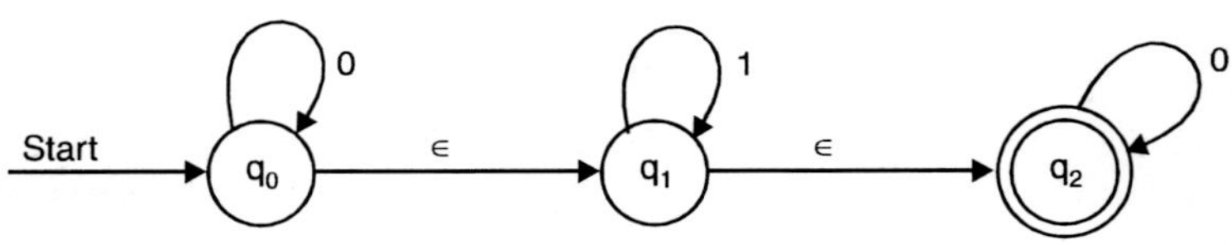

Fig. 1.43.

Find the equivalent NFA without ∈-transitions.

Solution. Let NFA without ∈-transition be

$$M' = \{Q', \Sigma, q_0', \delta', F')$$

So as we see in example 1.16.

$$Q' = (\in\text{-closure } (q_0), \in\text{-closure } (q_1), \in\text{-closure } (q_2))$$

$$= (\{q_0, q_1, q_2\}, \{q_1, q_2\}, \{q_2\}) \cong (q_a, q_b, q_c)$$

$$\Sigma = \{0,1\}$$

$$F' = (q_a, q_b, q_c) \text{ as } q_a \to \{q_0, q_1, q_2\}$$

$$q_b \to \{q_1, q_2\}$$

$$q_c \to \{q_2\}$$

and

$$\delta'(q_a, 0) = q_a$$

$$\delta'(q_a, 1) = q_b$$

$$\delta'(q_b, 0) = q_c$$

$$\delta'(q_b, 1) = q_b$$

$$\delta'(q_c, 0) = q_c$$

$$\delta'(q_c, 1) = \phi$$

So transition table is as follows :

δ'/Σ	0	1
$\rightarrow * q_a$	q_a	q_b
$* q_b$	q_c	q_b
$* q_c$	q_c	ϕ

and transition diagram will be as follows :

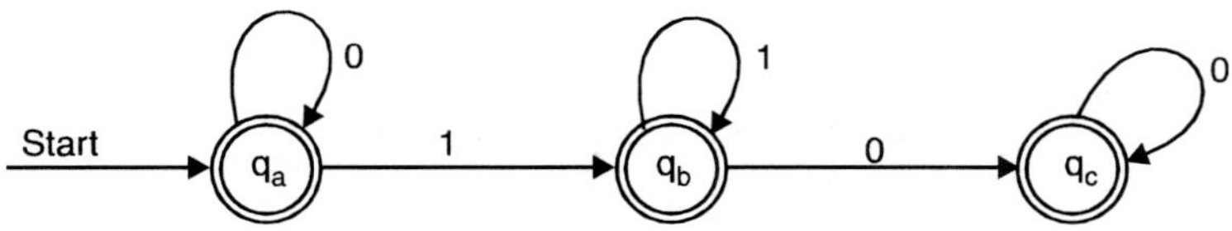

Fig. 1.44. NFA without ∈-transitions.

1.22. PROPERTIES OF FINITE STATE MACHINES

(*i*) Recognition problem can be solved in linear time (independent of the size of the automaton).

(*ii*) There is an algorithm to transform each automaton into a unique equivalent automaton with the least number of states.

(*iii*) Finite acceptors provide a way to describe potentially infinite languages in finite ways.

(*iv*) Determining whether a string is accepted or rejected by acceptor is linear in the length of the string.

(*v*) Acceptors can be both determinised and minimised.

1.23. APPLICATION OF FINITE AUTOMATA

1. A DFA-Based Text Filter and Lexical Analyzers

DFA is used to design text filtering software. DFA also used to design lexical analyzers which is very important component of compiler to produce tokens.

2. Finite Machines as Acceptors and Recognizers

Acceptors and recognizers (also sequence detectors) produce a binary output, saying either yes or no to answer whether the input is accepted by the machine or not. All states of the FSM are said to be either accepting or not accepting.

The machine can also be described as defining a language, which would contain every word accepted by the machine but none of the rejected ones; we say then that the language is accepted by the machine. By definition, the languages accepted by FSMs are the regular languages - that is, a language is regular if there is some FSM that accepts it.

3. Finite Machines as Transducers

Transducers generate output based on a given input and/or state using actions. They are used for control applications and in the field of computational linguistics. The two types of transducers are Moore machine and Mealy machine.

4. Hardware Applications of Finite Machines

In a digital circuit, an FSM may be built using a programmable logic device, a programmable logic controller, logic gates and flip flops or relays. More specifically, a hardware implementation requires a register to store state variables, a block of combinational logic which determines the state transition, and a second block of combinational logic that determines the output of an FSM. One of the classic hardware implementations is the Richards controller.

Mealy and Moore machines produce logic with asynchronous output, because there is a propagation delay between the flip-flop and output. This causes slower operating frequencies in FSM. A Mealy or Moore machine can be changed to a FSM which output is directly from a flip-flop, which makes the FSM run at higher frequencies. A counter is the simplest form of this kind of FSM.

1.24. LIMITATIONS OF FINITE AUTOMATA

(*a*) The defining characteristic of FA is that they have only a finite number of states. Hence, finite automata can only "count" a finite number of input scenarios.

(*b*) Finite automata don't have any auxiliary memory so it works in "process and forget" mode. It can store anything at particular state. There is no finite automaton that recognizes these strings:

(*i*) The set of binary strings consisting of an equal number of 1's and 0's

(*ii*) The set of strings over '(' and ')' that have "balanced" parentheses

(*c*) Finite automata can recognize and process regular languages.

Example 1.18. Design a DFA for the language

$$L = \{w : n_a = 1, w \in (a, b)^*\}$$

Solution. Let DFA be $M = (Q, \Sigma, \delta, q_0, F)$

When we analyse the language, then it contains a set of string, every string contains exactly one '*a*' only, that is *a*, *bbb* ... *a*, *bbb* ... *abbb* ..., and *abbb*... . Then transition diagram will be as follows :

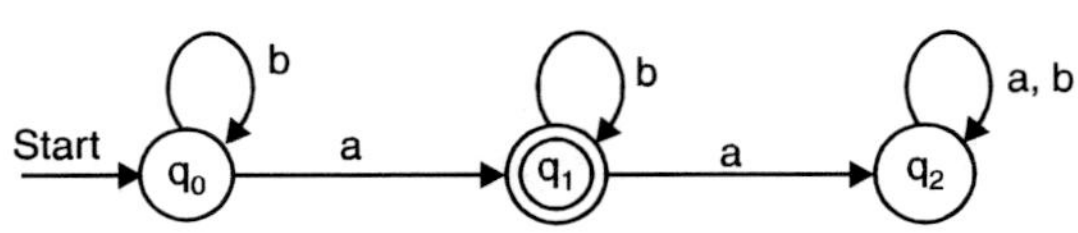

Fig. 1.45.

Here q_2 is dead state

$$Q = \{q_0, q_1, q_2\}$$
$$\text{intial state} = \{q_0\}$$
$$\Sigma = \{a, b\}$$
$$F = \{q_1\}.$$

Example 1.19. Design a DFA for the language

$$L = \{w : n_a(w) \geq 1, w \in (a, b)^*\}.$$

Solution. Given language is $L = \{w : n_a(w) \geq 1, w \in (a, b)^*\}$

Clearly language is the set of strings, in which every string containing at least one '*a*', any where in the string such as *a, bbbb ... a, aaaa, bababa, aaabbb ...* .

Let DFA be $M = (Q, \Sigma, \delta, q_0, F)$

$\Sigma = (a, b)$

$Q = (q_0, q_1\}$

$F = \{q_1\}$

Transition Diagram is

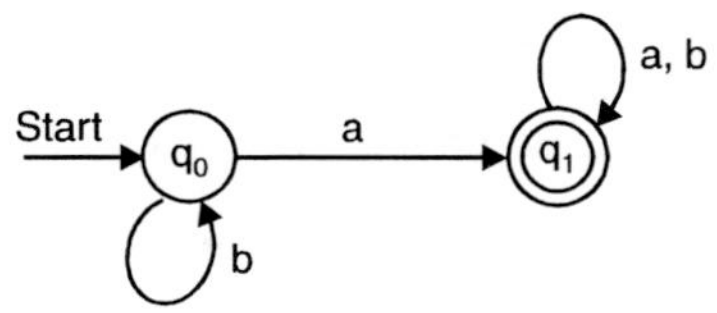

Fig. 1.46.

Example 1.20. Design the determinstic finite automata for the given language

$$L = \{w : n_a(w) \leq 3, w \in \{a, b\}^*\}.$$

Solution. Let us analyse the language first, clearly language *L* contains set of strings in which maximum number of *a*'s can be three only. It means strings included no *a*'s , one *a*, two *a*'s and three *a*'s are the part of the language, empty string is also included.

Strings may be $\in$, *a, aa, aaa, bbbb ... , ba, baba, baaabbb ...* .

Let DFA be $M = (Q, \Sigma, \delta, q_0, F)$

$\Sigma = (a, b)$ given

$Q = (q_0, q_1, q_2, q_3, q_4)$

$F = (q_0, q_1, q_2, q_3)$

Transition diagram is shown in Fig. 1.47.

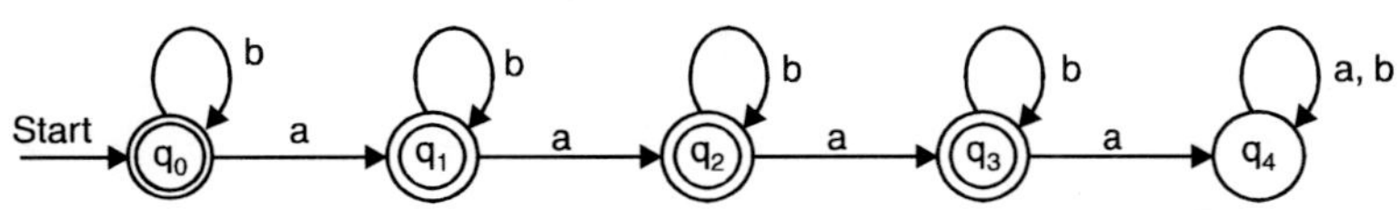

Fig. 1.47.

Here q_4 is dead state.

Example 1.21. Design a DFA for the given language

$$L = \{w : n_a(w) \geq 1, n_b(w) = 2, w \in \{a, b\}^*\}.$$

Solution. Let us analyse the language first, since here conditions are defined as $n_a(w) \geq 1$ and $n_b(w) = 2$ means every string of the language must contain at least one '*a*' exactly two *b*'s such as *abb, bab, baaaaa ... b, aaaaa ... baaab, bba* and many more which are satisfying both the conditions.

Let us assume that DFA be $m = (Q, \Sigma, \delta, q_0\ F)$

where $Q = (q_0, q_1, q_2, q_3, q_4, q_5, q_6, q_7, q_8)$

$F = (q_3, q_8)$

$\Sigma = (a, b)$

Transition diagram is shown in Fig. 1.48.

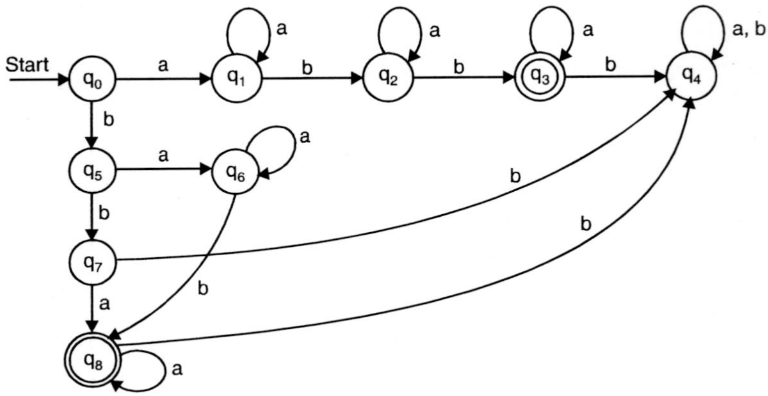

Fig. 1.48.

Example 1.22. Design a DFA for the language

$$L = \{w : n_a(w) = 2, n_b(w) > 2, w \in \{a, b\}^*\}.$$

Solution. Given language is $L = \{w : n_a(w) = 2, n_b(w) > 2, w \in (a, b)^*\}$

Now it is clear from $n_a(w) = 2$ and $n_b(w) > 2$ that strings, which are belongs to language must satisfied both the condition simultaneously that is every string must contain exactly two a's in it (in any position) and number of b's must be more then two (in any position), such as *abbbbabbb*, *aabbbbb*, *aabbbb*, *bbbaa*.

Let DFA be $M = (Q, \Sigma, \delta, q_0, F)$

$Q = (q_0, q_1, q_2, q_3, q_4, q_5, q_6, q_7, q_8, q_9, q_{10}, q_{11}, q_{12})$

$\Sigma = (a, b)$

$F = (q_5)$

Transition diagram is shown in Fig. 1.49.

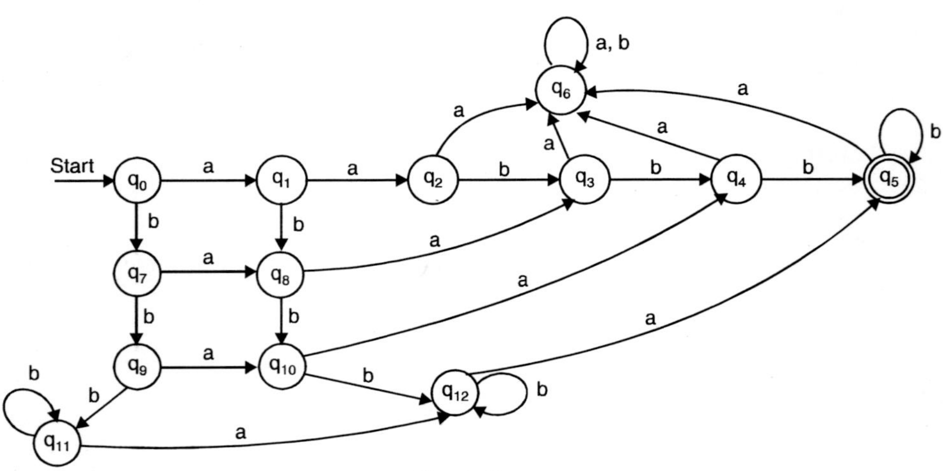

Fig. 1.49.

Example 1.23. Given DFA is

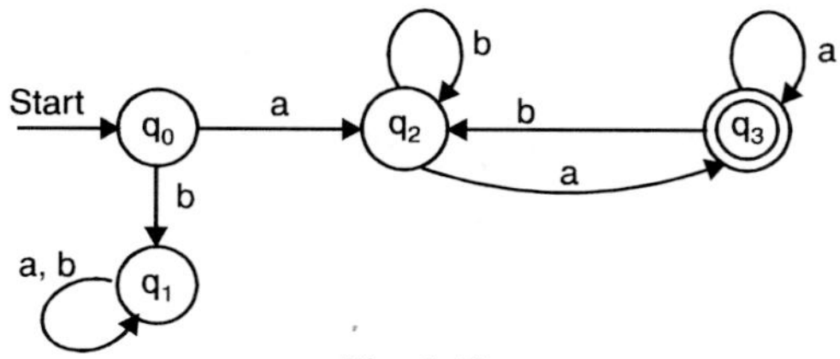

Fig. 1.50.

(*i*) Give the language of DFA shown in Fig. 1.50.

(*ii*) Prove that there exist a DFA for $\overline{L}$ if *L* is the language of given DFA.

Solution. (*i*) Let

$$M = (Q, \Sigma, \delta, q_0, F) \text{ be given DFA}$$
$$Q = (q_0, q_1, q_2, q_3)$$
$$F = (q_3)$$
$$\Sigma = (a, b)$$

Here q_1 is the dead state, clearly no string of the language will start from *b* character. So every string starts with '*a*' and end with '*a*'.

(*ii*) Now make q_3 as non-final state and q_0, q_1, q_2 as final states and prove the new automation obtained accepts $\overline{L}$ *i.e.* it does not accepts *awa*.

Now it can be show through the following transition diagram.

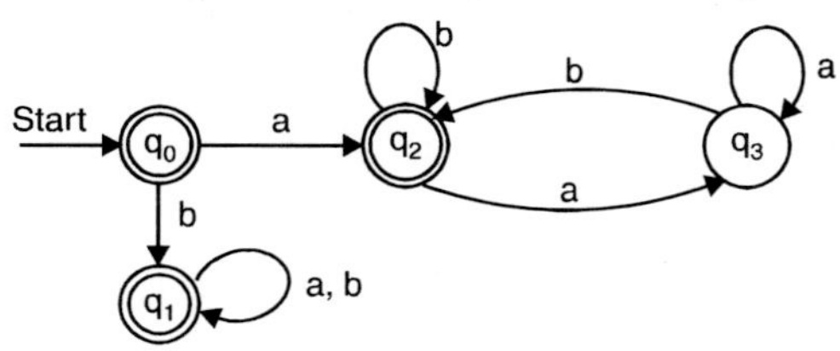

Fig. 1.51.

This automation accepts $b(a, b)^*$, $\in$, and $ab^*\ aa^*\ bb^*$ or ab^*. Clearly it accepts all strings which starts with either *b* or if it starts with an *a*, it ends with ab^* and if it encounters any *a* after b^*, it must end with *b*.

So, this automation accepts $\overline{L}$.

Example 1.24. Show that if $M = (Q, \Sigma, \delta, q_0, F)$ and $\overline{M} = (Q, \Sigma, \delta, q_0, Q\text{-}F)$ are two DFA's, then $\overline{L(M)} = L(\overline{M})$.

Solution. Let there be an automation

$$M = (Q, \Sigma, \delta, q_0, F)$$

where Q = non-empty set of state in automation
Σ = non-empty set of input symbols
δ = transition rule (function) from $Q \times \Sigma$ to Q
q_0 = start state (where $q_0 \in Q$)
F = non-empty set of final states.

If we consider another automation

$$\overline{M} = (Q, \Sigma, \delta, q_0, Q\text{-}F)$$

where all Q, Σ, δ and q_0 are same as previous, but the final states are the other state than *F*, then the new automation can not reach finally at the state (*F*) in this automation as it used to be in the automation M. So, this automation will accept other language, but it will not accept the language which is accepted by M.

So $\overline{L(M)} = L(\overline{M})$

We can also understand this by an example.

Let us consider a DFA for the language

$$L = \{a^n b^0 : n \geq 0\}.$$

Transition diagram as follows :

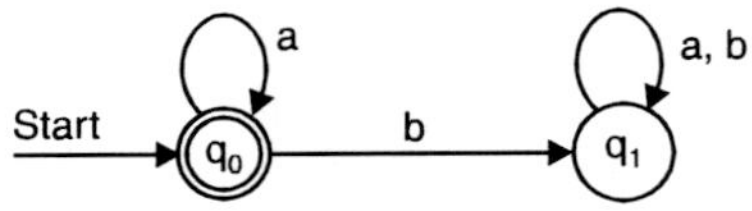

Fig. 1.52.

Let say $M = \{Q, \Sigma, \delta, q_0, F)$

$Q = (q_0, q_1)$

$q_0 = q_0$

Now making $F = \{q_0, q_1\} - \{q_0\} = \{q_1\}$

So we will get

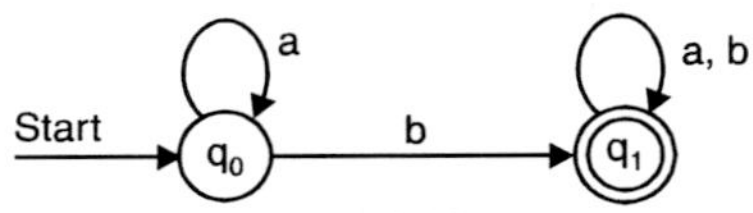

Fig. 1.53.

This automation (shown in Fig. 1.53) accepts all strings, except $a^n b^0$, $n \geq 0$.

Example 1.25. Design a DFA for the language

$$L = \{ab^5 w b^4 : w \in \{a, b\}^*\}.$$

Solution. Let DFA be $M = (Q, \Sigma, \delta, q_0, F)$

where $Q = (q_0, q_1, q_2, q_3, q_4, q_5, q_6, q_7, q_8, q_9, q_{10}, q_{11})$

$\Sigma = \{a, b\}$

$q_0 = \{q_0\}$

$F = \{q_{10}\}$

Transition diagram is following :

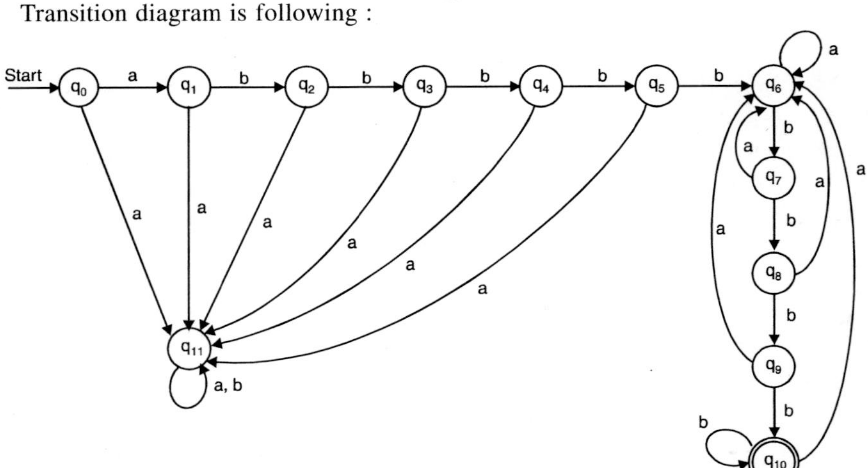

Fig. 1.54.

Example 1.26. Design a DFA for the language

$$L = \{w_1\, ab\, w_2 : w_1, w_2 \in (a, b)^*\}.$$

Solution. It is clearly known after analysis of language L that every string at least one ab in string at any position.

Let DFA be $M = (Q, \Sigma, \delta, q_0, F)$

where
$$Q = \{q_0, q_1, q_2\}$$
$$\Sigma = \{a, b\}$$
$$q_0 = \{q_0\}$$
$$F = \{q_2\}$$

Transition diagram for the given language is shown in Fig. 1.55, as follows :

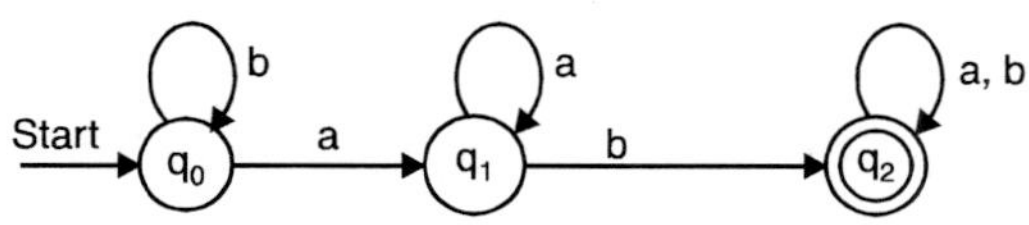

Fig. 1.55.

Example 1.27. Give the description of the language for the DFA, shown in Fig. 1.56.

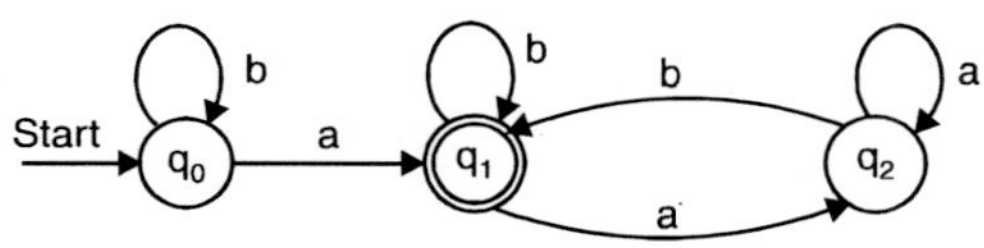

Fig. 1.56.

Solution. Given DFA $M = (Q, \Sigma, \delta, q_0, F)$
$$Q = \{q_0, q_1, q_2\}$$
$$\Sigma = \{a, b\}$$
$$q_0 = \{q_0\}$$
$$F = \{q_1\}$$

Language of DFA can be defined as follows :

$$L(M) = \{b^*a\, \{wb\}^* : w \in \{a, b\}^*\}$$

or
$$L(M) = \{b^*a\, \{a^*\, b\}^*\}.$$

Example 1.28. Design a determinstic finite automation over alphabet $\Sigma = \{a, b\}$, such that every string, accepted by automation contains no runs of length less than four.

Solution. Let us first undetstand the problem properly, every string contains no runs of length less than four means that string may start either from a or from b but, automation will accept only those strings which contains a's or b's in the runs of four such as *aaaa* or *bbbb*.

Let DFA be $M = (Q, \Sigma, \delta, q_0, F)$

where $Q = (q_0, q_1, q_2, q_3, q_4, q_5, q_6, q_7, q_8, q_9\}$

$$\Sigma = \{a, b\}$$
$$q_0 = \{q_0\}$$
$$F = \{q_4, q_8\}$$

Transition diagram is shown in Fig. 1.57, as follows :

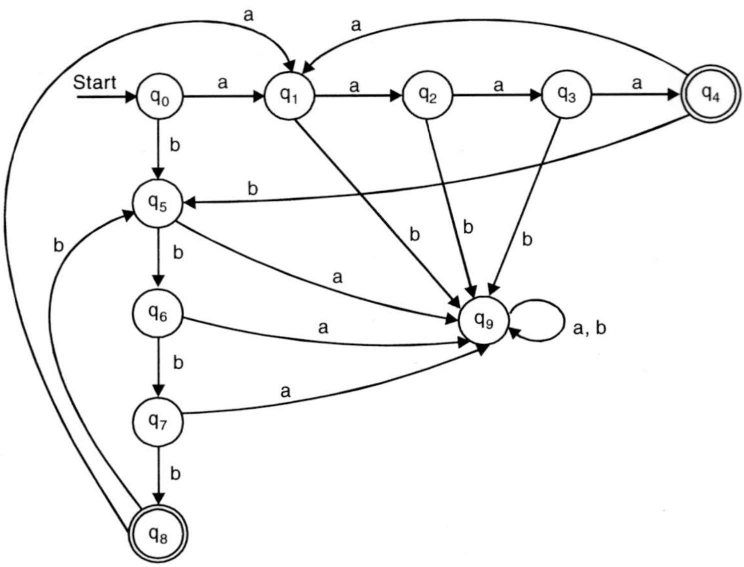

Fig. 1.57.

Here q_0 is dead state.

Example 1.29. Design a DFA for the language

$L = \{w$: every run of a's has length either two or three}.

Solution. We can solve this problem with same logic as we have solve Example 1.28.

Automation will accepts only those strings, if an 'a' encountered then it must follow an 'a' or two a's, that is the meaning of "every run of a's has length either two or three.

Let DFA be $$M = (Q, \Sigma, \delta, q_0, F)$$

Here this automation can have three different transition diagrams, as follows :

(*a*)

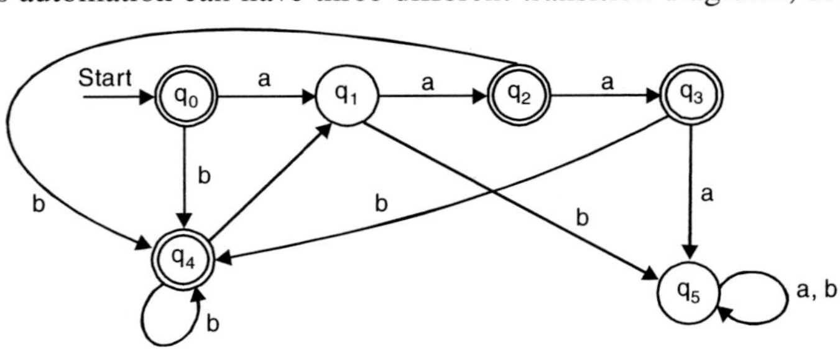

Fig. 1.58.

$$Q = \{q_0, q_1, q_2, q_3, q_4, q_5\}$$
$$\Sigma = \{a, b\}$$
$$q_0 = \{q_0\}$$
$$F = \{q_2, q_3, q_4, q_0\}$$

(*b*)

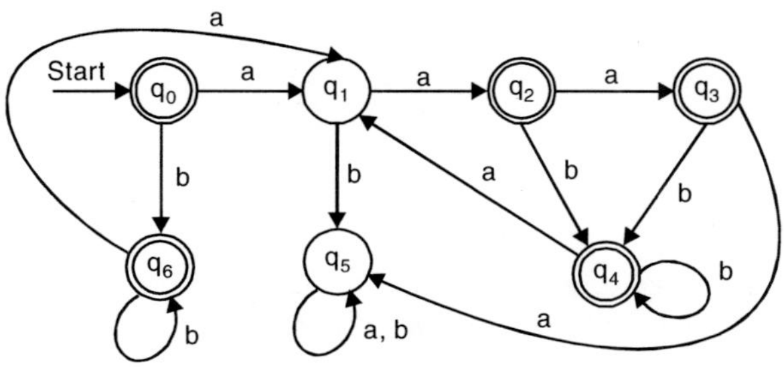

Fig. 1.59.

Here
$$Q = \{q_0, q_1, q_2, q_3, q_4, q_5, q_6\}$$
$$\Sigma = \{a, b\}$$
$$q_0 = \{q_0\}$$
$$F = \{q_2, q_3, q_6, q_0\}.$$

(*c*)

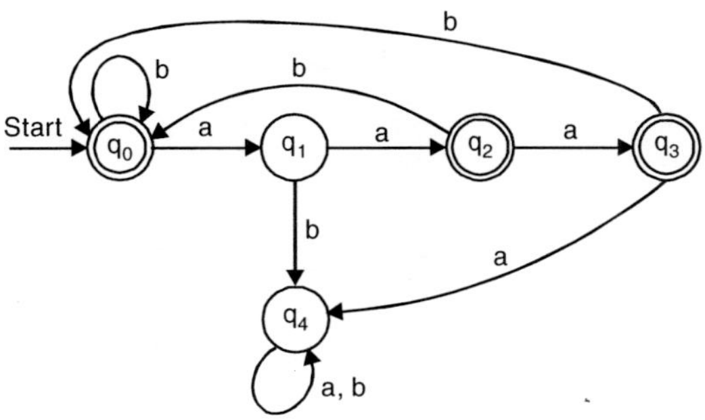

Fig. 1.60.

Here
$$Q = \{q_0, q_1, q_2, q_3, q_4\}$$
$$\Sigma = \{a, b\}$$
$$q_0 = \{q_0\}$$
$$F = \{q_2, q_3, q_0\}.$$

Example 1.30. Design a DFA, which accepts strings, in which every 00 is followed immediately by a 1. For example, the strings 001, 0010, 00100111001 are in the language, but 0001 and 00100 are not.

Solution. Let DFA be
$$M = (Q, \Sigma, \delta, q_0, F)$$
where
$$Q = \{q_0, q_1, q_2, q_3, q_4\}$$
$$\Sigma = \{0, 1\}$$
$$q_0 = \{q_0\}$$
$$F = \{q_0, q_1\}$$

Transition diagram is show in Fig. 1.61.

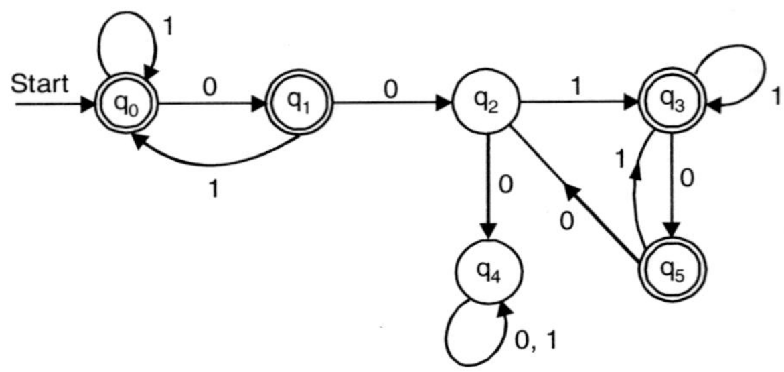

Fig. 1.61.

Here q_4 is dead state.

Example 1.31. Design a DFA for the language, contains strings in which left most symbol differ from right most symbol. Σ is given {0, 1}.

Solution. Here problem is straight forward. Language is the set of strings such as 01110001101 , 01, 10, or in general we can say that strings are either 0 w 1 or 1 w 0 where $w \in (0, 1)^*$.

Let DFA be

$$M = (Q, \Sigma, \delta, q_0, F)$$
$$Q = \{q_0, q_1, q_2, q_3, q_4\}$$
$$\Sigma = \{0, 1\}$$
$$q_0 = \{q_0\}$$
$$F = \{q_2, q_4\}.$$

Transition diagram is described in Fig. 1.62.

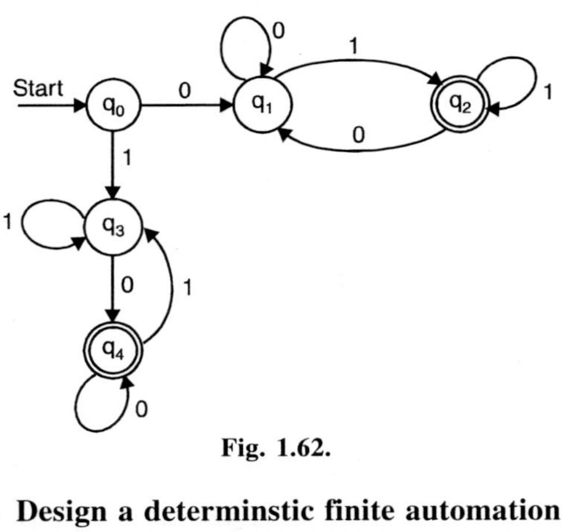

Fig. 1.62.

Example 1.32. Design a determinstic finite automation which accepts set of strings such that every string containing 00 as a substring but not 000 as sub-string.

Solution. Let DFA be

$$M = (Q, \Sigma, \delta, q_0, F)$$

where

$$Q = \{q_0, q_1, q_2, q_3, q_4, q_5\}$$
$$\Sigma = \{0, 1\}$$
$$q_0 = \{q_0\}$$
$$F = \{q_2, q_3, q_5\}$$

Transition diagram is defined in Fig. 1.63 as follows :

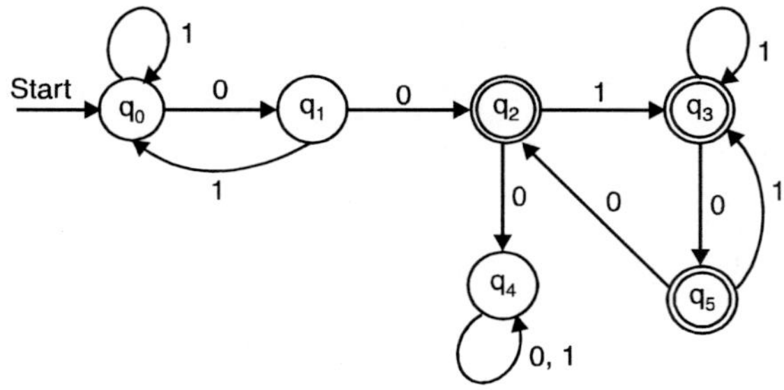

Fig. 1.63.

Here q_4 is dead state.

Example 1.33. Construct a DFA that accepts strings on {0, 1}, if and only if the value of the string, interpreted as a binary representation of an integer, is zero module five. For example, 0101 and 1111, representing the integers 5 and 15, respectively, are to be accepted.

Solution. Let us undetstand problem first. We have to design a automation which accepts those binary strings, when we interpreted them in integer then those integers are divisible by 5 that means after dividing 5 reminder is 0, it is the only meaning of the "is zero module five". So clearly integer interpretation should be 0, 5, 15, 20, 25, ... So binary strings which are accepted by automation are 0, 101, 1111 ...

Let DFA be $M = (Q, \Sigma, \delta, q_0, F)$

where

$$Q = \{q_0, q_1, q_2, q_3, q_5\}$$
$$\Sigma = \{0, 1\}$$
$$q_0 = \{q_0\}$$
$$F = \{q_0\}$$

and finally transition diagram is defined as follows :

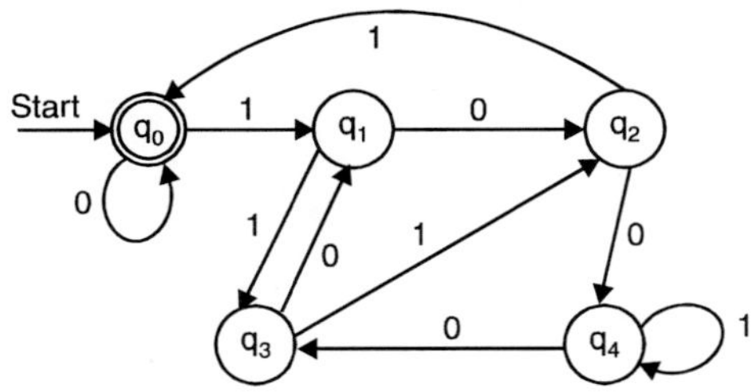

Fig. 1.64.

Example 1.34. Show that if L is regular, so is L–{∈}.

Solution. Let us assume that L is regular and w is the string such that $w \in L$.

So automation will be

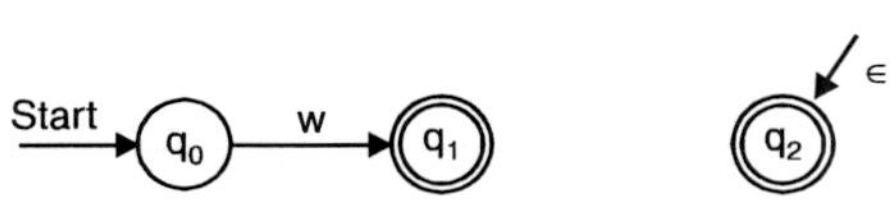

Fig. 1.65.

If we substracted $\in$ from L then we have $L - \in = L^+$

Transition diagram can be defined as :

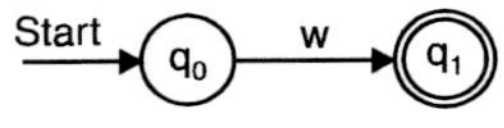

Fig. 1.66.

So, $L - \in$ is also regular.

Example 1.35. Here DFA given in Fig. 1.67 for the language L find the DFA for L^2.

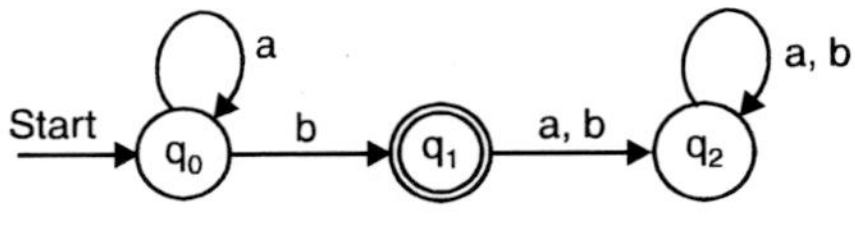

Fig. 1.67. DFA for language L.

Solution. We have given DFA for language L in Fig. 1.67(a).

The language for DFA given is

$$L = \{a^n b : n \geq 0\}$$

Let as assume that DFA for L^2 is

$$M = (Q, \Sigma, \delta, q_0, F)$$
$$Q = \{q_0, q_1, q_2, q_3\}$$
$$\Sigma = \{a, b\}$$
$$q_0 = \{q_0\}$$
$$F = \{q_2\}$$

Then transition diagram for L^2 is defined as follows :

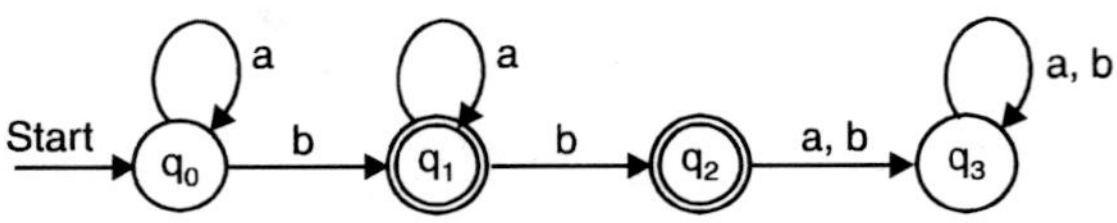

Fig. 1.67(a). DFA for L^2.

Here q_3 is dead state.

Example 1.36. Show that the automata M_1 and M_2 is given figure are equivalent or not.

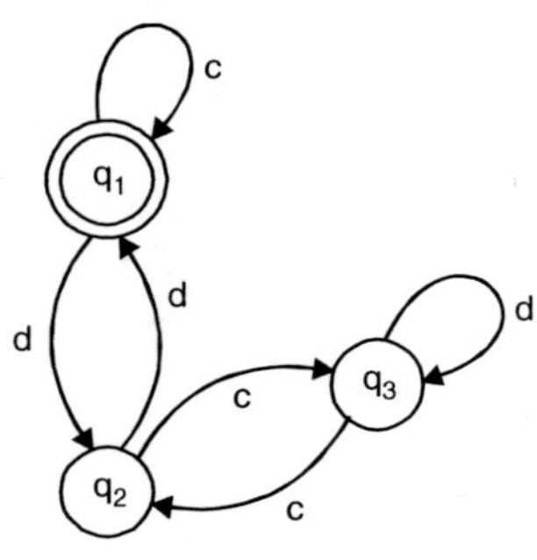

Fig. 1.68. M_1

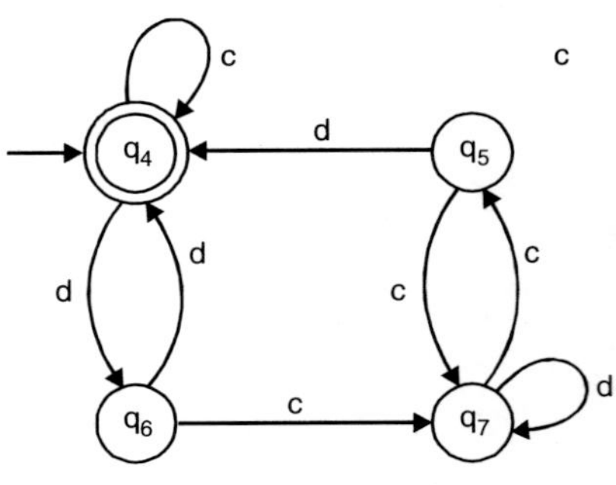

Fig. 1.69. M_2

Solution. The initial states in M_1 and M_2 are q_1 and q_4. Let us see that combined transition of M_1 and M_2 on inputs c and d as follows :

δ/Σ	c	d	
(q_1, q_4)	(q_1, q_4)	(q_2, q_6)	$\rightarrow (q_2, q_6)$ is new state
(q_2, q_6)	(q_3, q_7)	(q_1, q_4)	$\rightarrow (q_3, q_7)$ is new state
(q_3, q_7)	(q_2, q_5)	(q_3, q_7)	$\rightarrow (q_2, q_5)$ is new state
(q_2, q_5)	(q_3, q_7)	(q_1, q_4)	

Let us analyse the above transition table carefully, then we find that, the behaviour of M_1 and M_2 is similar.

We do not get a pair (q, q') where q is a final state and q' is a non final state (or vice versa) at every row. Therefore, M and M' are equivalent.

Example 1.37. Let $M = (\{q_1, q_2, q_3\}, \{0, 1\}, \delta, \{q_1\}, \{q_3\})$ is NFA where δ is given as :

$$\begin{aligned}
\delta(q_1, 0) &= \{q_1, q_3\}\\
\delta(q_1, 1) &= \{q_1\}\\
\delta(q_2, 0) &= \{q_1, q_2\}\\
\delta(q_2, 1) &= \{q_1, q_2\}\\
\delta(q_3, 0) &= \{q_2\}\\
\delta(q_3, 1) &= \{q_1, q_2\}
\end{aligned}$$

Construct the transition diagram corresponding to NFA and find its equivalent DFA, show all the intermediate step also.

Solution. Transition diagram is as follows :

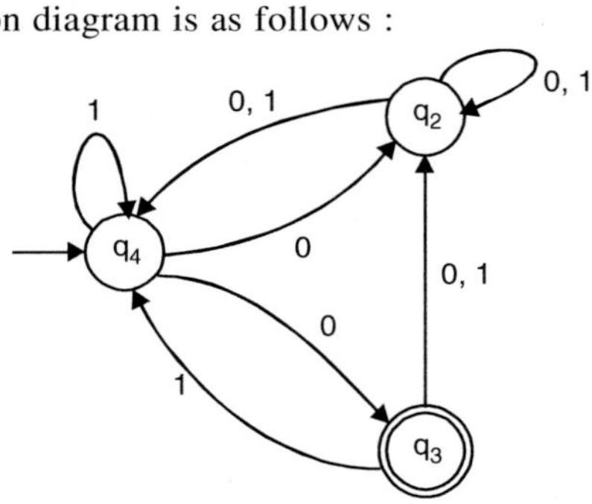

Fig. 1.70. NFA.

The DFA equivalent of above NFA is follows :

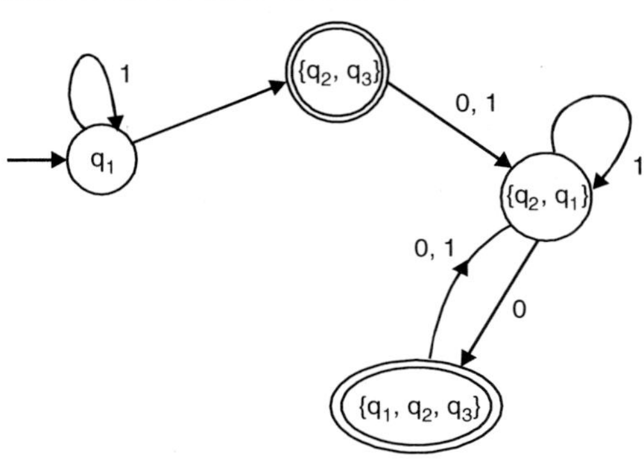

Fig. 1.71. DFA.

δ/Σ	*0*	*1*
$\rightarrow q_1$	$\{q_2, q_3\}$	q_1
$\{q_2, q_3\}*$	$\{q_1, q_2\}$	$\{q_1, q_2\}$
$\{q_1, q_2\}$	$\{q_1, q_2, q_3\}$	$\{q_1, q_2\}$
$\{q_1, q_2, q_3\}*$	$\{q_1, q_2\}$	$\{q_1, q_2\}$

Example 1.38. Prove that for every non-deterministic finite automata (NDFA) there is an equivalent DFA that has only one final state.

Solution. A DFA can simulate the behaviour of NDFA by increasing the number of states. Any NDFA is more general machine without being more powerful. We now give a theorem on equivalence of DFA and NDFA.

Theorem : For every NDFA, there exists a DFA which simulates the behaviour of NDFA. Alternatively, if *L* is the set accepted by NDFA then exists a DFA which also accepts *L*.

Proof. Let $M = \{Q, \Sigma, \delta, q_0, F\}$ be an NDFA accepting language L. We construct a DFA M' as follows :

$$M' = \{Q', \Sigma, \delta', q_0', F'\}$$

where

(*i*) $Q_1 = 2^Q$ (any state in Q is denoted by $\{q_1, q_2, ..., q_i\}$ where $q_1, q_2, ..., q_j \in Q$);

(*ii*) $q'_0 = [q_0]$

(*iii*) F' is the set of all subsets of Q containing an element of F.

(*iv*) $\delta'(q_1, q_2, ..., q_i], a) = \delta(q_1, a) \cup \delta(q_2, a) \cup ... \cup \delta(q_i, a)$ equivalently $\delta'([q_1, q_2, ..., q_i]\ a) = [P_1 ... P_j]$ if and only if $\delta[\{q_1, q_2, ..., q_i\}, a) = \{P_1, P_2, ..., P_j\}$

if and only if $\delta(q_0, x) = \{q_1 ... q_i\}$ for all x in Σ

First we are going to prove by induction that $\delta^*(q_0, w) = \delta'^*(q_0', w)$ for any string w, when it is proven, it obviously implies that NDFA M and DFA M' accept the same strings.

Now we have a task to prove $\delta^*(q_0, w) = \delta'^*(q_0', w)$. This is going to be proven by induction on w.

Note : δ^* represents the transition function for processing strings.

Basic step : For $w = t$

$$\begin{aligned} \delta'^*(q'_0, t) &= q'_0 \text{ by the definition of } \delta'^* \\ &= q_0 \text{ by construction of DFA } M' \\ &= \delta^*(q_0, t) \text{ by the definition of } \delta^*. \end{aligned}$$

Inductive Step :

Assume that $\delta^*(q_0, w) = \delta'^*(q_0', w)$ for an arbitrary string w. ... Induction hypothesis.

For the string w and an arbitrary symbol a in Σ,

$$\begin{aligned} \delta^*(q_0, wa) &= \text{U}_{P \in \delta^*(q_0, w)} \delta(P, a) \\ &= \delta'(\delta^*(q_0, w), a) \end{aligned}$$

$$= \delta'(\delta'^*(q_0', w), a)$$
$$= \delta^*(q_0', wa)$$

Thus for any string w $\delta^*(q_0, w) = \delta'^*(q_0', w)$ holds.

So we can say that $L(M) = L(M')$.

Example 1.39. An NFA with states 1–5 and input alphabet [a, b] has the following transition table :

q	$\delta(q, a)$	$\delta(q, b)$
1	{1, 2}	{1}
2	{3}	{3}
3	{4}	{4}
4	{5}	{ϕ}
5	{ϕ}	{5}

(*i*) Calculate $\delta^*(1, ab)$.

(*ii*) Calculate $\delta^*(1, abaab)$.

Solution. (*i*) $\delta^*(1, ab)$

$\Rightarrow$ $\delta^*(1, ab)$

$\Rightarrow$ $\delta^*(1, b)$ or $\delta^*(2, b) = \{1\}$ or $\{3\}$

(*ii*) $\delta^*(1, abaab)$
↓

$\Rightarrow$ $\delta^*(1, abaab)$

$\Rightarrow$ $\delta^*(1, baab)$ or $\delta^*(2, baab)$
↓ ↓ ↓

$\Rightarrow$ $\delta^*(1, aab)$ or $\delta^*(3, aab)$
↓ ↓ ↓

$\Rightarrow$ $\delta^*(1, ab)$ or $\delta^*(2, ab)$ or $\delta^*(4, ab)$
↓ ↓ ↓ ↓

$\Rightarrow$ $\delta^*(1, b)$ or $\delta^*(2, b)$ or $\delta^*(3, b)$ or $\delta^*(5, b)$

$\Rightarrow$ {1} or {3} or {4} or {5}

Example 1.40. Design a FA for $L = \{a^m b^n\}$ m and n are positive integer.

Solution. Let FA is $M = (Q, \Sigma, \delta, q0, F)$

$$Q = \{q_0, q_1, q_2, q_3\}$$
$$\Sigma = \{a, b\}$$

q_0 is initial state

$$F = \{q_1\}$$

δ is defined as follows :

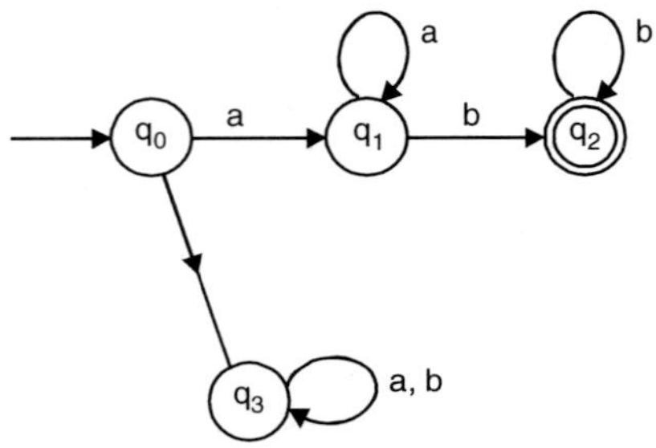

Fig. 1.72.

Note : Here q_3 is dead state.

Example 1.41. Find the finite automation that accepts $L = \{(ab)^n : n > 0\}$.

Solution. Let DFA is $M = (Q, \Sigma, \delta, q_0, F)$.

δ is defined as follows :

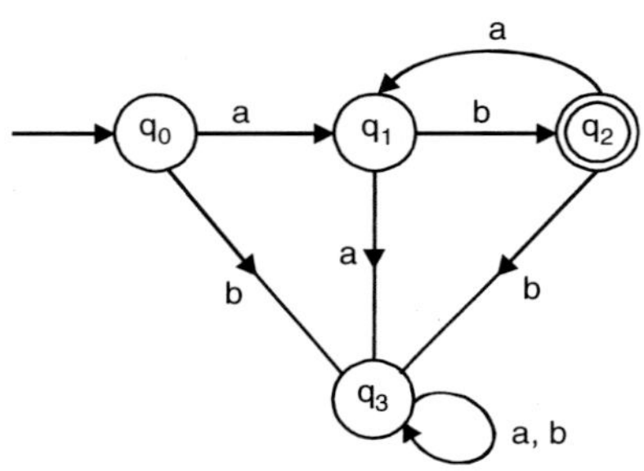

Fig. 1.73.

Example 1.42. Design a NFA accepting all strings over {0, 1} which ends in 1 but does not contain the substring 00.

Solution. Let NFA be $M = (Q, \Sigma, \delta, q_0, F)$.

δ is defined as follows :

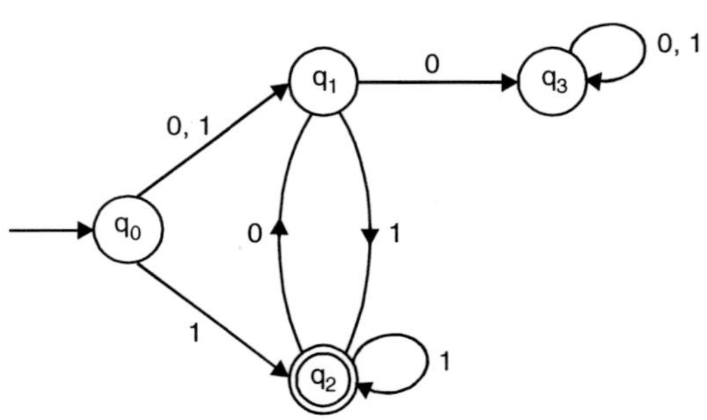

Fig. 1.74.

Example 1.43. Obtain an NFA with accepts $L = \{w \in (a, b)^* \mid |w| \geq 3$ and third symbol of w from right is 'a'}.

Solution. Let NFA be $M = (Q, \Sigma, \delta, q_0, F)$.

The δ is defined as follows :

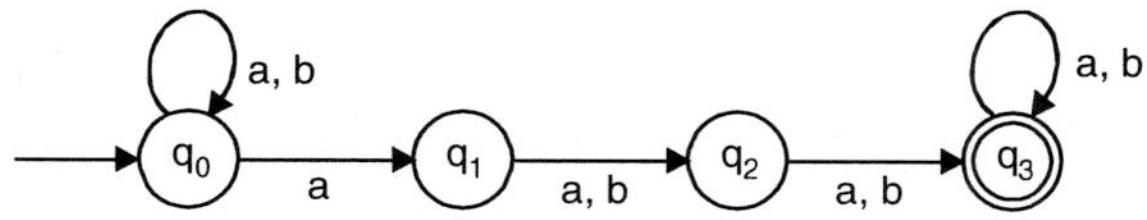

Fig. 1.75.

Example 1.44. Design NFA for $L = \{(aa^*(a + b))\}$.

Solution. Let NFA is $M = (Q, \Sigma, \delta, q_0, F)$.

δ is defind as follows :

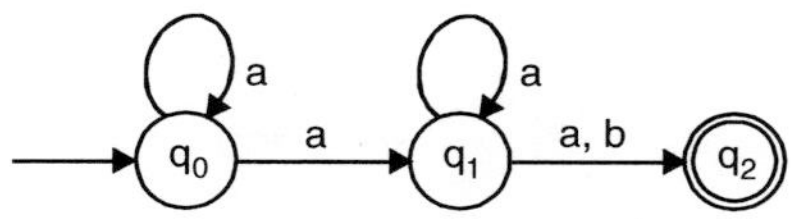

Fig. 1.76.

Example 1.45. Design NFA for $L = \{abab^n : n \geq 0\} \cup \{aba^n : n \geq 0\}$.

Solution. Let NFA is $M = (Q, \Sigma, \delta, q_0, F)$.

Given language is $L = \{abab^n/n \geq 0\} \cup \{aba^n/n \geq 0\}$, *ab* is common in all strings either they belongs to $abab^n : n \geq 0$ or $aba^n : n \geq 0$.

So the δ can be defined as follows :

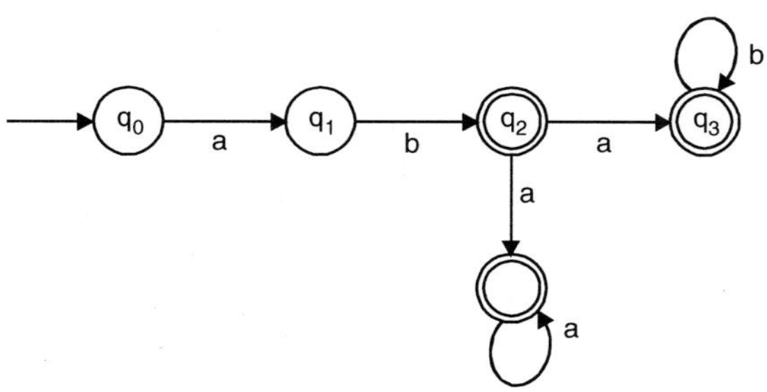

Fig. 1.77.

Example 1.46. Design a FA for $L = \{a, b\}^* baaa$.

Solution. Let FA is $M = (Q, \Sigma, \delta, q_0, F)$.

δ is defined as follows :

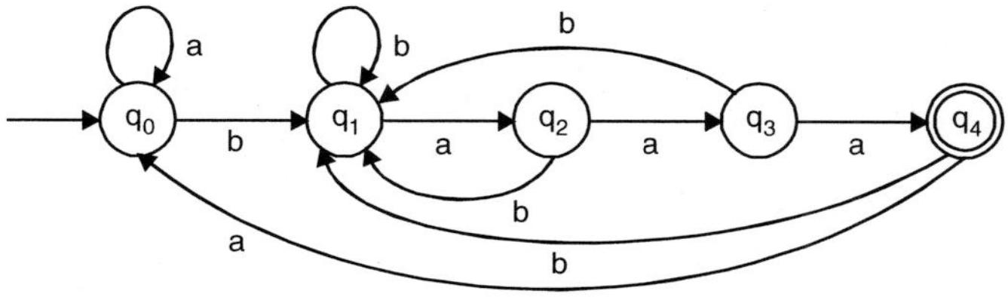

Fig. 1.78.

Example 1.47. Design a FA for $r = (a + b)^* (ab + bba) (a + b)^*$.

Solution. The FA for given language is

$$M = (Q, \Sigma, \delta, q_0, F).$$

The transition relation δ is defined as follows :

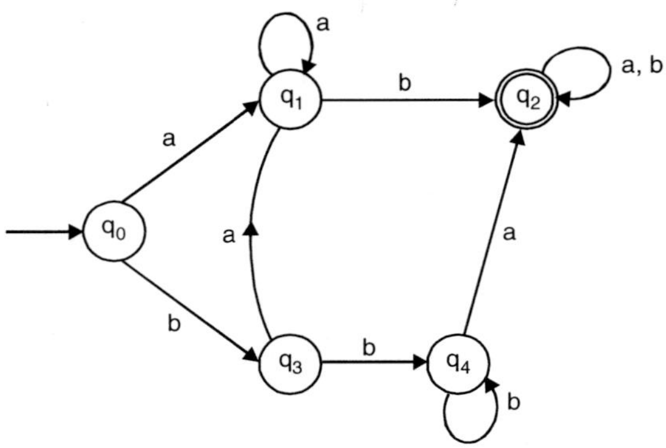

Fig. 1.79.

Example 1.48. If L_1 and L_2 are languages over alphabet Σ, then design an FA that accept $L_1 - L_2$.

Solution. Let FA corresponding L_1 is $M_1 = (Q_1, \Sigma, \delta_1, q_1, F_1)$ and FA corresponding L_2 is $M_2 = (Q_2, \Sigma, \delta_2, q_2, F_2)$.

Now let us assume FA for $L_1 - L_2$ is M. M should be rember at each step both the information that 'M_1' rembers and information 'M_2' rembers. That is, it amount to rembering the ordered pair (p, q), where P is the current of M_1 and q is not in F_2. Accordingly, we construct :

$$M = (Q, \Sigma, \delta, q_0, F)$$

where $Q = Q_1 \times Q_2$

$q_0 = (q_1, q_2)$

and the transition function is defined as follows. $\delta((p, q), a) = (\delta_1 (p, a), \delta(q, a)$ for any $p \to Q_1$, $q \to Q_2$, $a \in \Sigma$. If

$$F = \{(p, q) : p \in F_1 \text{ and } q \notin F_2\},$$

then M accepts the language $L_1 + L_2$.

EXERCISES

1. Let $L = \{ab, aa, baa\}$. Which of the following strings are in L^* : *abaab aaabaa, aaaabaaaa, baaaaabaaaab, baaaaabaa.*
2. Give some examples of strings in, and not in these sets, where $\Sigma = \{a, b\}$.
   (a) $\{w : \text{for sum } x \in \Sigma, w = x\, x^R x\}$
   (b) $\{w : ww = www\}$
   (c) $\{w : \text{for some } \mu, v \in \Sigma^*, \mu v w = w v \mu\}$
   (d) $\{w : \text{for some } \mu \in \Sigma^*, www = \mu\mu\}$.
3. Which of the following are true? Explain.
   (a) $baa \in a^* b^* a^* b^*$ (b) $b^* a^* \cap a^* b^* = a^* \cup b^*$
   (c) $a^* b^* \cap b^* c^* = \phi$ (d) $abcd \in (a(cd)^* b)^*$.

4. Build an FA that accepts only the works *baa*, *ab* and *abb* and no other strings longer or shorter.
5. For the $\Sigma = \{a, b\}$
   (*i*) Build an FA that accepts only those words that have more than four letters.
   (*ii*) Build an FA that accepts only those words that have fewer than four letters.
   (*iii*) Build an FA that accepts only those words with exactly four letters.
   (*iv*) Build an FA that accepts only those words that do not end with *ba*.
   (*v*) Build an FA that accepts only those words that begin or end with a double letter.
6. Describe the language accepted by the following FA's :

(*i*)

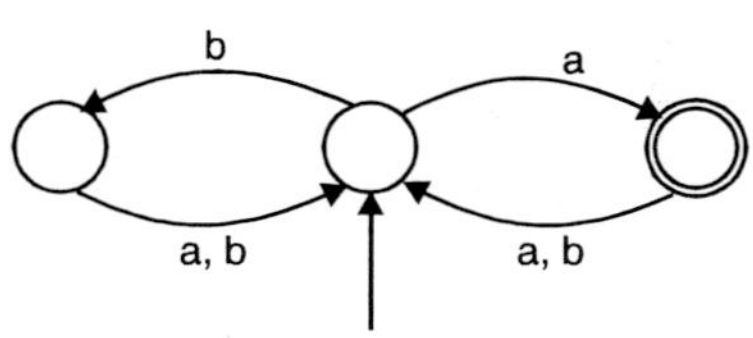

Fig. 1.80.

(*ii*)

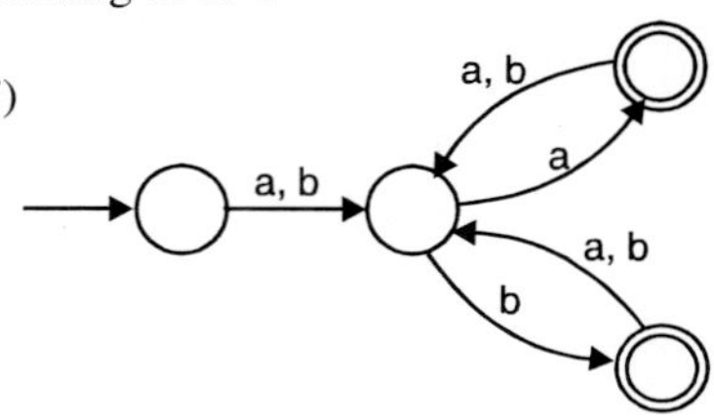

Fig. 1.81.

7. Convert each of the following NFA's in to FA's

(*i*)

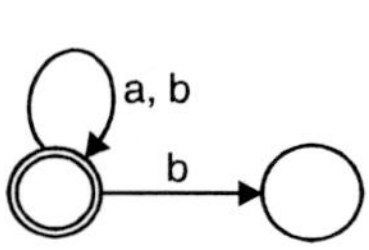

Fig. 1.82.

(*ii*)

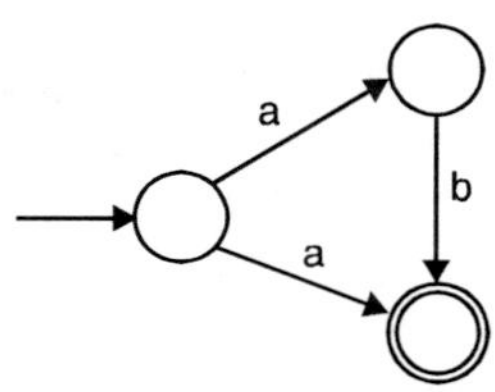

Fig. 1.83.

(*iii*)

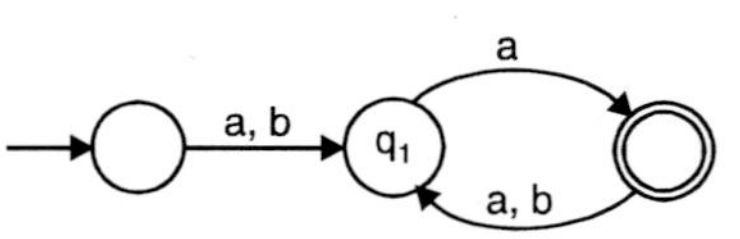

Fig. 1.84.

(*iv*)

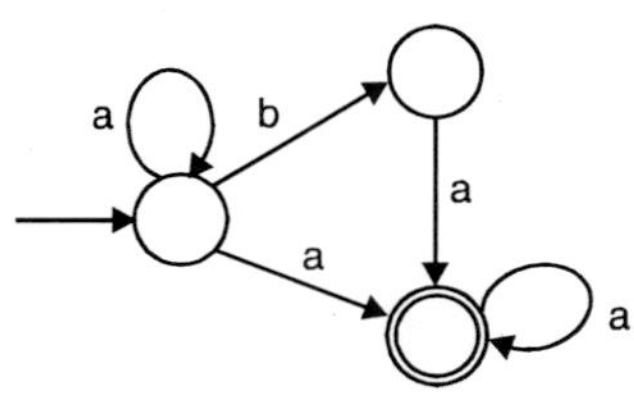

Fig. 1.85.

(*v*)

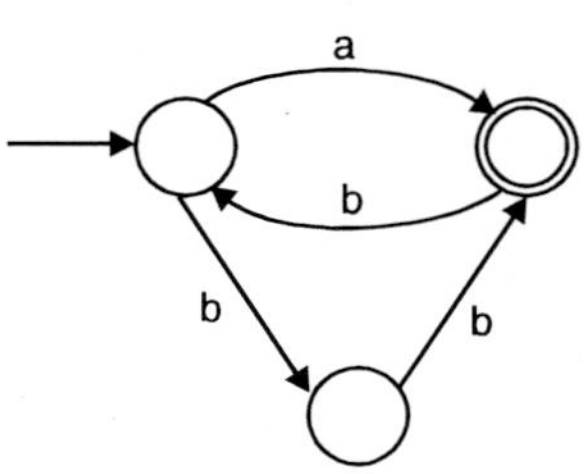

Fig. 1.86.

(*vi*)

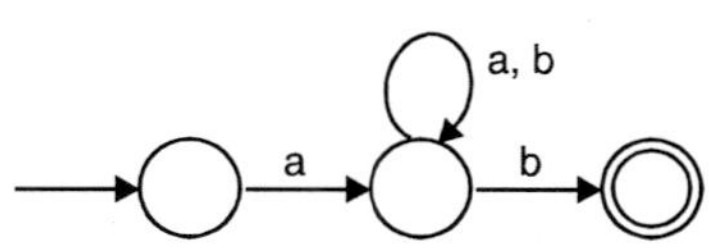

Fig. 1.87.

(*vii*)

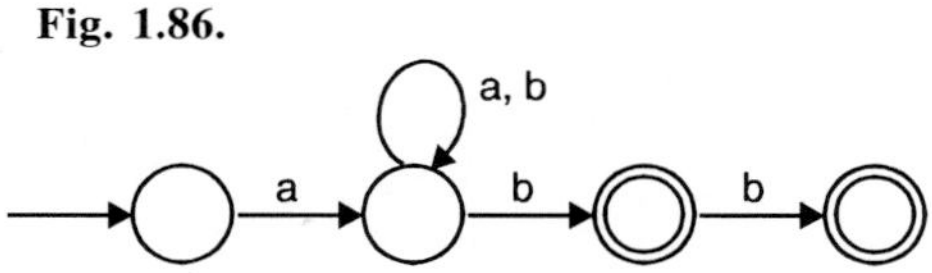

Fig. 1.88.

(viii)

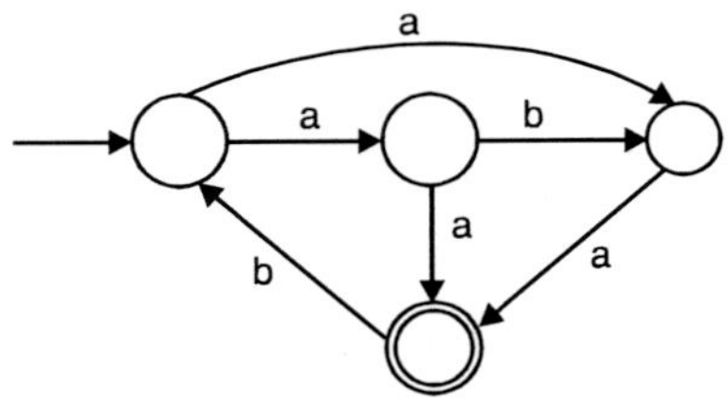

Fig. 1.89.

8. Convert the following NFA–∈'s into NFA without ∈-transitions.

(i)

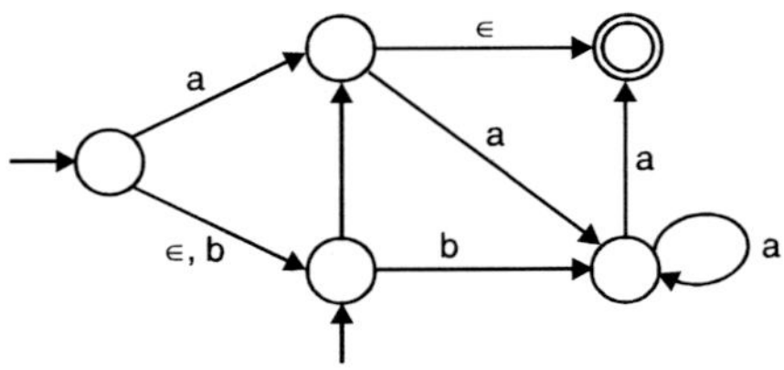

Fig. 1.90.

(ii)

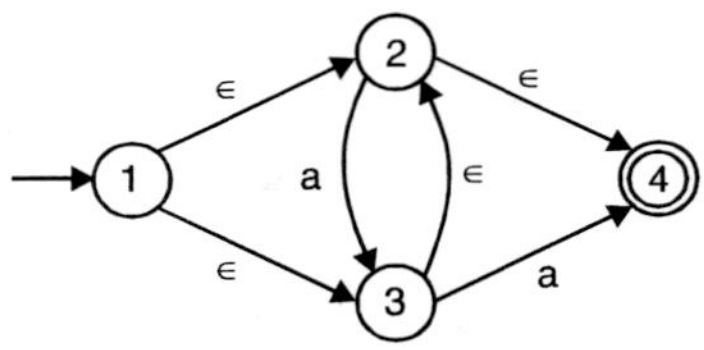

Fig. 1.91.

9. For the language accepted by the following machine, find a different FA with four states. Find an NFA that accepts the same languages has only seven edges (where edges with two labels are counted twice).

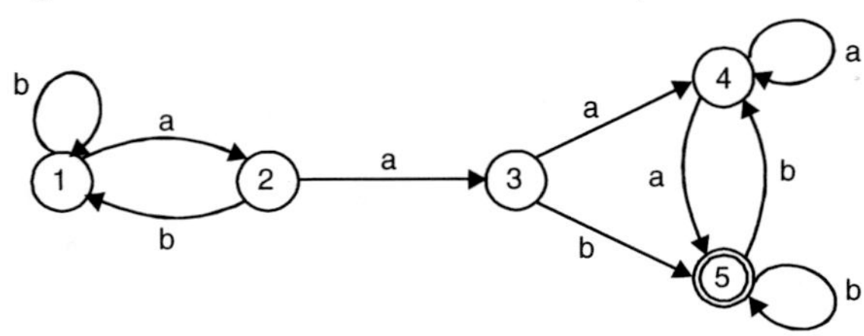

Fig. 1.92.

10. Design a NFA with no more than five states for the set $\{abab^n : n \geq 0\} \cup \{aba^n : n \geq 0\}$.

11. Find an NFA that accepts $\{a\}^*$ and is such that if in its transition graph a single edge is removed, the resulting automation accepts $\{a\}$.

12. Find an NFA without ∈-trantion and with final state that accepts the set $\{a\} \cup \{b^n : n \geq 1\}$.

13. Write a computer program that produces a minimal DFA for any given DFA.

14. Find the minimal DFA's for the following languages below :

(i) $L = \{a^n b^m : n \geq 2, m \geq 1\}$

(ii) $L = \{a^n : n \geq 0, n \neq 3\}$

(iii) $L = \{a^n : n \neq 2 \text{ and } n \neq 4\}$

*****

CHAPTER 2

Regular Expression

2.1. REGULAR EXPRESSION

INSIDE THIS CHAPTER

We have already studied Finite Automata. Now we switch our attention from machine like description of language to an algebraic description : "the regular expression." We shall find that, regular expressions can define exactly the same languages that the various forms of automata describe : the regular languages.

However, regular expression offer something that automata do not : a declarative way to express the strings, we want to accept. Thus regular expression as the input language for many system. For example, lexical-analyser generators, such as Lex or Flex.

2.1.1. The Operators of Regular Expression

Before describing the regular notation, we need to learn the three operations on languages that operators of regular expression represent. These operations are :

1. The union of two languages L_1 and L_2, denoted $L_1 \cup L_2$, is the set of string that are in either L_1 or L_2, or both.

   For example, if $L_1 = \{001, 10, 111\}$ and $L_2 = \{\in, 001\}$ then $L_1 \cup L_2 = \{\in, 10, 001, 111\}$.

2. The concatenation of languages L_1 and L_2 is a set of strings that can be formed by taking any string in L_1 and concatenating it with any strings in L_2. We denote concatenation of languages with a dot.

   For example, if $L_1 = \{001, 10, 111\}$ and $L_2 = \{\in, 001\}$, then $L_1 . L_2 = \{001, 10, 111, 001001, 10001, 111001\}$.

3. The closure (or star, or kleene closure) of a language L is denoted by L^* and represents the set of those strings that can be formed by taking any number of strings from L, possibly with repetitions (that is same string can be repeated more then once) and concatenating all of them. For example, $L = \{0, 1\}$, then L^* is all strings of 0's and 1's including null string.

   More formally, L^* is the infinite union $U_i \geq L^i$, where $i \geq 0$.

4. The positive closure of a language L is denoted by L^+ and represent the set of those strings that can be formed by taking any number of strings from L, possibly with repetitions and concatenating all of them, excluding null string *i.e.*

$$L^+ = L^* - \in$$

More formally, L^+ is the infinite union $U_i \geq 1\ L^i$, where $i > 0$.

2.1.2. Definition of Regular Expression

Definition

The set of regular expression is defined by the following rules :

1. Every letter of Σ can be made into a regular expression, null string, $\in$ itself is a regular expression.
2. If r_1 and r_2 are regular expression, then

   (*i*) (r_1) (*ii*) $r_1 r_2$

   (*iii*) $r_1 + r_2$ (*iv*) $r_1{}^*$

   (*v*) r_1^+ are also regular expression.
3. Nothing else is regular expression.

2.1.3. Building Regular Expression

The algebra of regular expressions follows this path using constants and variables that denote languages, and operators for three operations of union, dot and star. We can describe the regular expression recursively, as follows :

In this definition, we not only describe what are the legal regular expression, but for each regular expression r, we describe the language, it represents, which we denote as $L(r)$. We can consider some facts :

1. The constants $\in$ (null string) and ϕ (empty set) are regular expression, denoting the languages $\{\in\}$ and ϕ, respectively.

That is, $L(\in) = \{\in\}$, and $L(\phi) = \phi$.

2. If a is any symbol, then a is regular expression. This expression denotes the language $\{a\}$. That is $L(a) = \{a\}$.

3. A variable, usually capitalized and italic such as L is a variable, representing any language.

2.1.4. Inductive Step

There are four parts to the inductive step, one for each of three operators and one for the introduction of parantheses.

1. If r_1 and r_2 are regular expressions, then $r_1 + r_2$ is a regular expression denoting the union of $L(r_1)$ and $L(r_2)$.

That is, $$L(r_1 + r_2) = L(r_1) \cup L(r_2)$$

2. If r_1 and r_2 are regular expression, then r_1r_2 is a regular expression denoting the concatenating of $L(r_1)$ and $L(r_2)$

That is, $$L(r_1r_2) = L(r_1) . L(r_2)$$

3. If r is a regular expression then r^* is a regular expression denoting the closure of $L(r)$. That is,

$$L(r^*) = (L(r))^*$$

4. If r is a regular expression, then (r), a parenthesized r, is also a regular expression, denoting the same language as r.

Formally, $$L((r)) = L(r).$$

Now let us discuss some examples of building regular expression for better understanding.

Example 2.1. Write the regular expression over alphabet (a, b, c) containing at least one a and at least one b.

Solution. Let us first analyse the language of regular expression.

Problem is very simple and the language will have set of strings, like *ab*, *abbbb*, *abab, aaaab, baaaa* .... So strings are $\underline{1}\ a\ \underline{2}\ b\ \underline{3}$, any combination of *a*'s, *b*'s and *c*'s is possible at the places 1, 2, and 3 any combination of *a*, *b*, *c* means $(a + b + c)^*$.

So regular expression is

$$r = (a + b + c)^* a(a + b + c)^* b(a + b + c)^* + (a + b + c)^* b(a + b + c)^* a(a + b + c)^*$$

If $r = r_1 + r_2$ (say)

then minimum string of r_1 is *ab* and minimum string of r_2 is *ba*.

Example 2.2. Write the regular expression for the set of strings of 0's and 1's whose tenth symbol from the right end is 1.

Solution. $\Sigma = \{0, 1\}$ given the set of strings, according to the required regular expression are 11<u>1</u>000000000, 1001<u>1</u>010101010 ..... So clearly we concern about only tenth position and it should be 1, rest 9 places from the right are either 0 or 1. Similarly positions 11th onward from right end may be either 0 or 1.

So regular expression is

$$r = (0 + 1)^* 1(0 + 1) (0 + 1) (0 + 1) (0 + 1) (0 + 1) (0 + 1) (0 + 1) (0 + 1) (0 + 1)$$

$$r = (0 + 1)^* 1(0 + 1)^9.$$

Example 2.3. Write the regular expression for the set of strings of an equal number of 0's and 1's such that in every prefix, the number of 0's differs from the number of 1's by at most 1.

Solution. Clearly the set of strings is having strings like 1010101010, 10101010, 0110, 1001 ...

So regular expresion will be

$$r = (01 + 10)^*.$$

Example 2.4. Write a regular expression for the set of strings of 0's and 1's not containing 101 as a substring.

Solution. When we analyse the set of strings then it becomes clear that string may start with '0' and 0's may be repeated, wherever one is encountered then no single 0 must follow it so strings are like 0000010, 00000111111, 1001001 ....

Regular expression is $r = (0^*1^*00)^*\ 0^*\ 1^*$.

Example 2.5. Write the regular expression over alphabet {*a*, *b*} for the set of strings with even number of *a*'s followed by odd number of *b*'s that is for the language

$$L = \{a^{2n}b^{2m+1} : n \geq 0, m \geq 0\}.$$

Solution. Clearly every string of language L will start with even number of a's. String should end with odd number of b's.

Regular expression is $r = (aa)^*\ (bb)^*\ b$.

Example 2.6. Write the regular expression for the language

$$L = \{w \in \{0, 1\}^* : w \text{ has no pair of consecutive zeros}\}.$$

Solution. Clearly whenever a 0 occurs, it must be followed by *a* 1. Such a substring may be proceeded and followed by any number of 1's, that is $(1^*011^*)^*$. But strings ending in 0 or consisting of all 1's are unaccounted here, that is $1^*\ (0 + \in)$.

So regular expression is $r_1 = (1^*011^*)^*\ (0 + \in) + 1^*(0 + \in)$

Another possible solution is $r_2 = (1 + 01)^*\ (0 + \in)$

"Although the two expression look different, both answers are correct, as they denote the same language. Generally, there are an unlimited number of regular expressions for any given language."

Example 2.7. Write the regular expression for the language

$$L = \{a^n b^m / n \geq 4, m \leq 3\}$$

Solution. Language contain the set of strings such that every string start with atleast 4 a's and end at the most 3 b's.

So regular expression will be $r = a^4\ a^*\ (\in + b + b^2 + b^3)$.

Example 2.8. Write the regular expression for the language

$$L = \{a^n\ b^m : (n + m) \text{ is even}\}$$

Solution. We know $(n + m)$ can be even in two cases

Case 1. n and m both are even.

Case 2. n and m both are odd.

Let regular expression for case 1 is r_1 then

$$r_1 = (aa)^*\ (bb)^*$$

Let regular expression for case 2 is r_2 then

$$r_2 = (aa)^*\ a\ (bb)^*\ b$$

Let regular expression for language L is r

then

$$r = r_1 + r_2$$

$$r = (aa)^*\ (bb)^* + (aa)^*\ a\ (bb)^*\ b.$$

Example 2.9. Write the regular expression for the language

$$L = \{ab^n\ w : n \geq 3, w \in (a, b)^+\}$$

Solution. Clearly every string of language L starts with a followed by atleast three b's, followed by atleast one a or one b, that is strings are like ab^3a, *abbbbba*, *abbbbbbbb*, *abbbb....a*

So regular expression is : $r = ab^3b^*(a + b)^+$

Here + is a positive closure *i.e.* $(a + b)^+ = (a + b)^* - \in$.

Example 2.10. Write the regular expression for the language

$$L = \{w : |w| \bmod 3 = 0\},\ w \in (a, b)^*\}$$

Solution. Let us first understand the meaning of $|w| \bmod 3 = 0$, when length of string belongs to the language L, is divided by 3 then reminder should be 0, that is length of w must be 0, 3, 6, 9 ... . Clearly multiple of three.

Regular expression is : $r = ((a + b)^3)^*$

Example 2.11. Write the regular expression for the language

$$L = \{w \in (a, b)^* : n_a(w) \bmod 3 = 0\}$$

Solution. $n_a(w) \bmod 3 = 0$ means, number of a's in string should be 0, 3, 6, 9 ... .

So regular expression is : $r = (b^*ab^*ab^*ab^*)^* + b^*$

2.2. COMPARATIVE STUDY OF REGULAR EXPRESSION, REGULAR SETS AND FINITE AUTOMATA

Regular expression	*Regular set*	*Finite automata*
ϕ	{ }	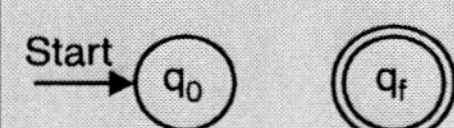 **Fig 2.1.** q_f **: is the final state.**
$\in$	$\{\in\}$	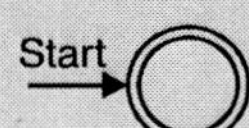 **Fig. 2.2.**
Every a in Σ is a regular expression.	$\{a\}$	**Fig. 2.3.**
If r_1 and r_2 are regular expression then $(r_1 + r_2)$ or r_1/r_2 is regular expression.	$R_1 \cup R_2$ (where R_1 and R_2 are regular sets corresponding to r_1 and r_2 respectively).	Where N_1 is FA accepting R_1 and N_2 is a finite automata accepting R_2. **Fig. 2.4.**
r_1r_2 is a regular expression.	R_1R_2 (where R_1 and R_2 are regular sets corresponding to r_1 and r_2 respectively).	Where N_1 is finite automata for R_1 and N_2 is FA for R_2. **Fig. 2.5.**
r^* is a regular expression if r is regular expression.	R^* (where R is a regular set corres-ponding to r).	where N_1 is finite automata accepting R. **Fig. 2.6.**

2.3. CONSTRUCTION OF FA FOR REGULAR EXPRESSION

We can construct NFA with $\in$ transmission by considering following four facts.

1. The expression is $r + s$ for some smaller expression r and s. The following automation serves, where R is automation for r and S is automation for s. That is, starting at the new start state, we can go to the start state of either the automation for r or the automation for S. We can reach the accepting state of one of these automata, following a path labelled by some string in $L(r)$ and $L(s)$, respectively. Once we reach the accepting state of the automation for r or s, we can follow one of the $\in$-arcs to the accepting state of the new automation. The language for the automation in Fig. 2.7 (a) is $L(r) \cup L(s)$.

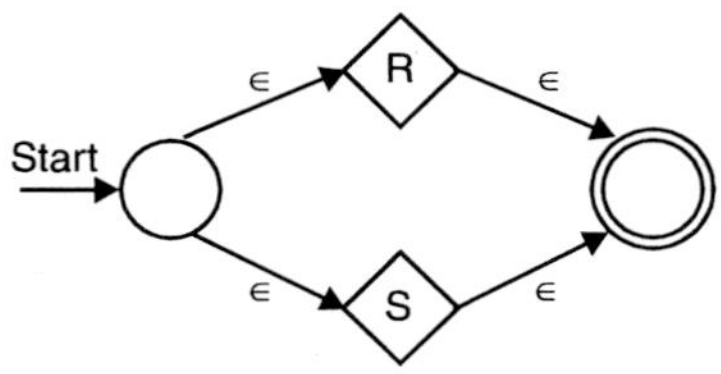

Fig. 2.7 (*a*).

2. The expression rs for some smaller expression r and s. The automation for the concatenation is show in Fig. 2.7 (b).

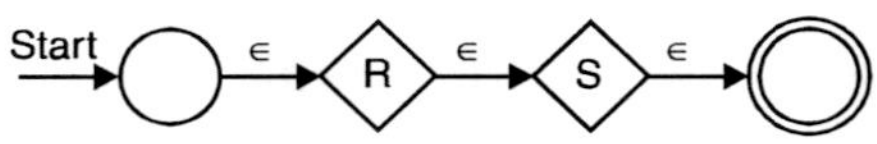

Fig. 2.7 (*b*).

TIPS

Note that the start state of the first automation becomes the start state of the whole, and the accepting state of the second automation becomes the accepting of the whole. The idea is that the only paths from start to accepting state to go first through the automation of r, where it must follow a path labelled by a string in $L(r)$, and then through the automation for s, where it follows a path labelled by a string in $L(s)$. Thus, the path in the automation of Fig. 2.7 (b) are all and only those labelled by strings in $L(R)$ $L(s)$.

3. The expression is r^* for some smaller expression r. Then we use automation of Fig. 2.7 (c). That automation allow us to go either :

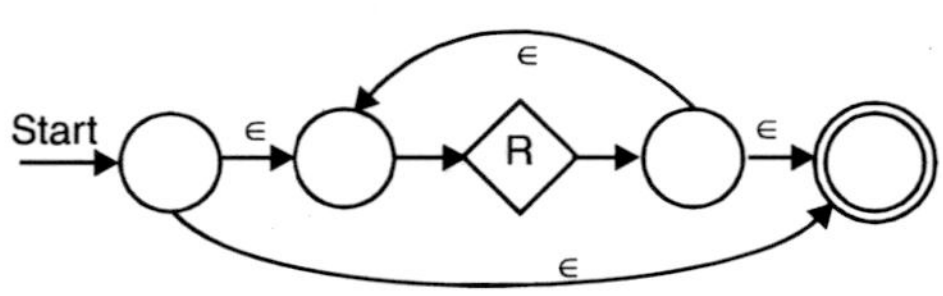

Fig. 2.7 (*c*).

(i) Directly from the start state to the accepting state along a path labelled $\in$.

(ii) To the start state of automation for r, through that automation one or more times, and then to accepting state. This set of paths allow us to accept strings in $L(r)$, $L(r)$, $L(r)$, $L(r)$, $L(r)$, $L(r)$, and so on, thus covering all strings in $L(r^*)$ except perhaps $\in$ which was covered by the direct arc to the accepting state.

where R is automation for r.

4. The expression is (r) for the smaller expression r. The automation for r also serves as the automation for (r), since the parentheses do not change the language defined by the expression.

Example 2.12. One the basis of above discussion find the automation for regular expression $a \cdot (a + b)^* \cdot b \cdot b$.

Solution. Let us solve this problem step by step :

Step 1. The basic regular expression involved are a and b, we start with automation for a and automation for b. Since brackets are evaluated first. We first constructed automation for $(a + b)$, using automation for a and automation b shown in Fig. 2.8(a)

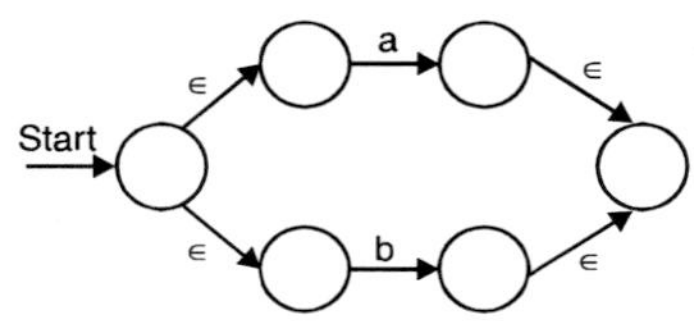

Fig. 2.8(a). Automation for $(a + b)$.

Step 2. Since closure is required to take next, we construct automation for $(a + b)^*$ using automation for $(a + b)$ as shown in Fig. 2.8(b).

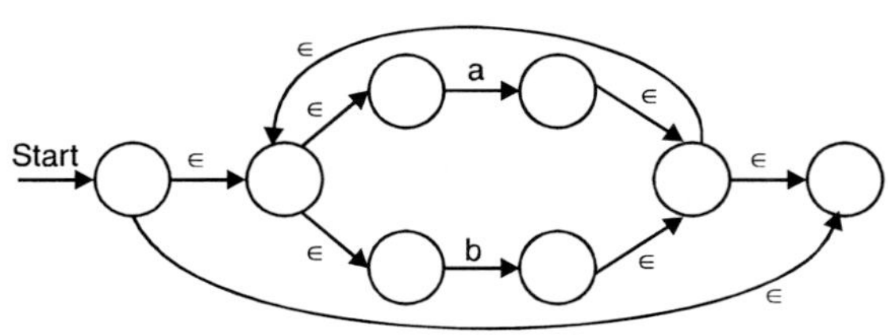

Fig. 2.8(b). Automation for $(a + b)^*$

Step 3. Next we construct the automation for $a \cdot (a + b)^*$ as shown in Fig. 2.8 (c).

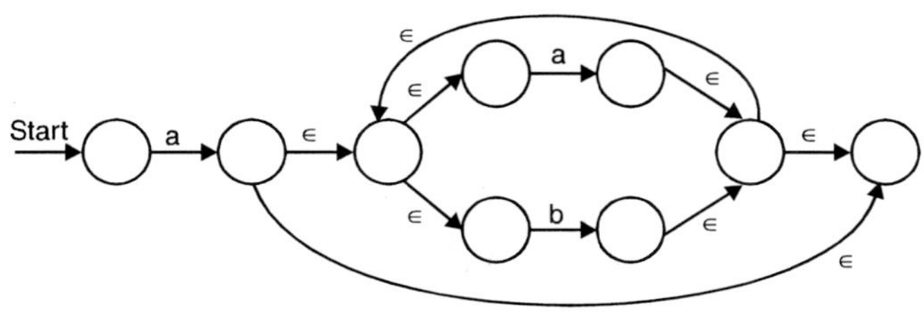

Fig. 2.8(c). Automation for $a \cdot (a + b)^*$.

Step 4. Next we construct the automation for $a \cdot (a + b)^* \cdot b$ by using automation for $a \cdot (a + b)^*$.

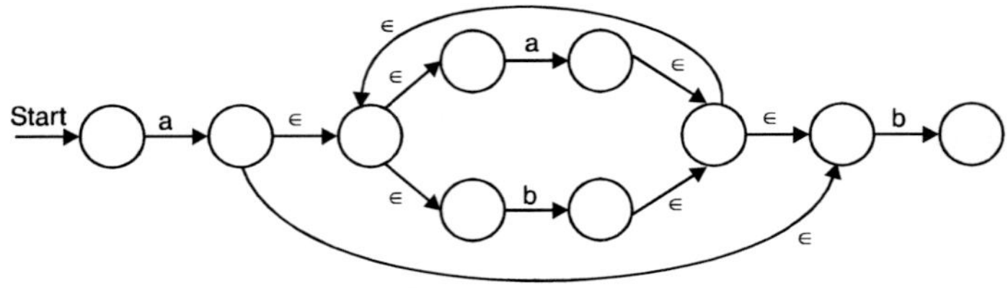

Fig. 2.8(*d*). Automation for *a* . (*a* + *b*)* . *b*.

Step 5. Now finally we can construct automation for *a* . (*a* + *b*)* . *b* . *b* by using automation for *a* . (*a* + *b*)* . *b*.

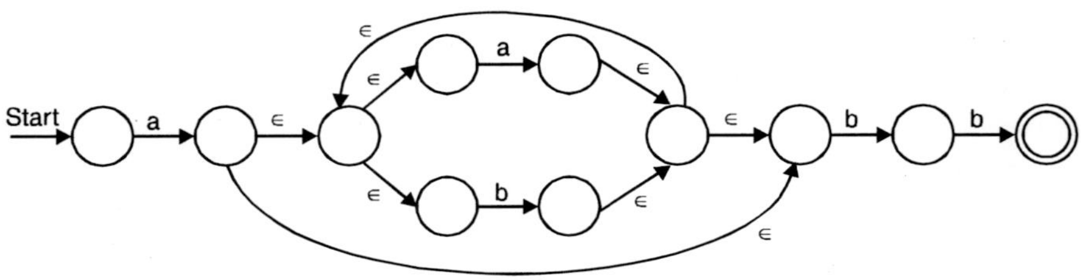

Fig. 2.8(*e*). Automation for *a* . (*a* + *b*)* . *b* . *b*.

It is required automation for given regular expression.

Example 2.13. Construct the ∈-NFA for the regular expression (0 + 1)* 1 (0 + 1).

Solution. Here automation for 0 and 1 is basic automation. First we will make automation for (0 + 1) as in Fig. 2.9 (*a*).

Now we will convert it into automation for (0 + 1)* as shown in Fig. 2.9(*b*).

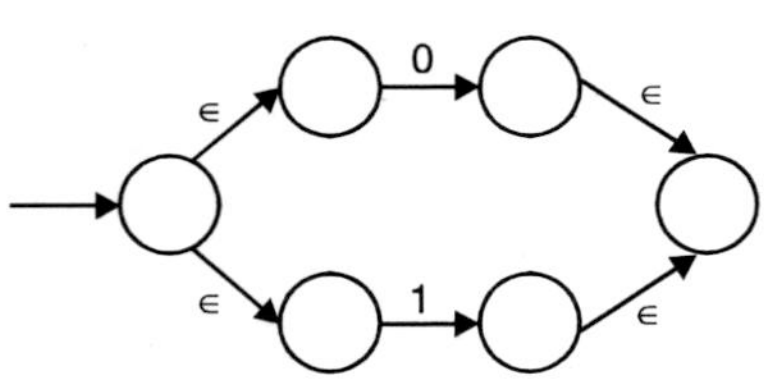

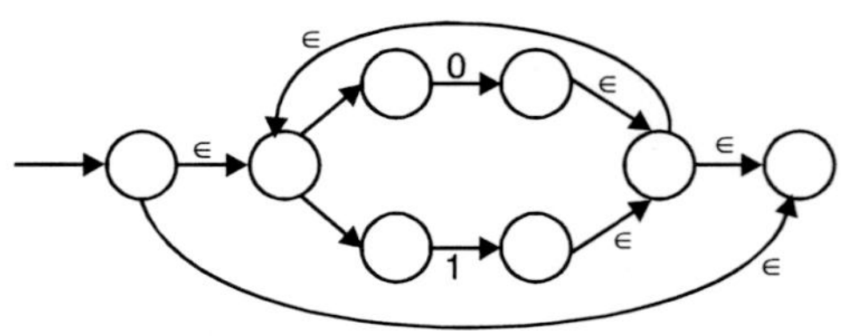

Fig. 2.9(*a*). Automation for (0 + 1). **Fig. 2.9(*b*). Automation for (0 + 1)*.**

Now we will construct automation for 1 . (0 + 1) as shown in Fig. 2.9(*c*).

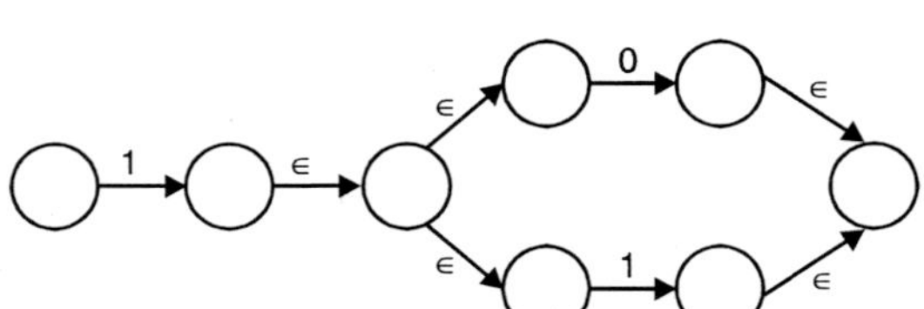

Fig. 2.9 (*c*). Automation for 1 . (0 + 1).

Now we have to connect ∈-NFA of Fig. 2.9(*b*) and ∈-NFA of Fig. 2.9 by a ∈-transition to find out the ∈-NFA for given regular expression as shown in Fig. 2.9(*d*).

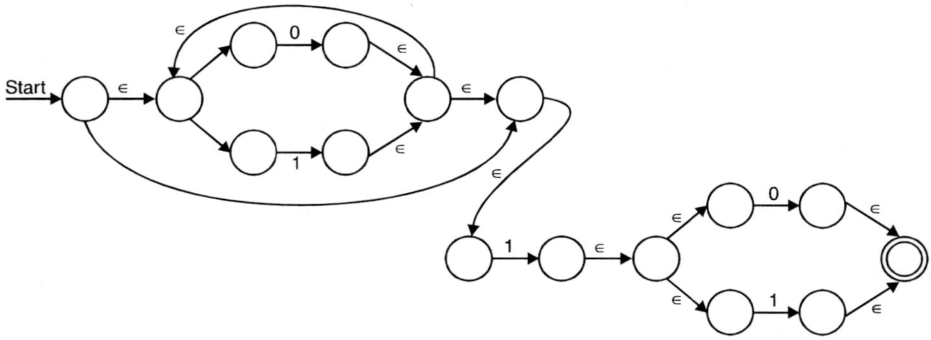

Fig. 2.9 (*d*). Automation for (0 + 1)* 1(0 + 1).

2.4. CONSTRUCTION OF REGULAR EXPRESSION FROM DFA

Now it is clear that regular expression is mathematical expression for a given regular language. We have already know that for every regular language there exist a deterministic finite automate. We can conclude that regular expression, regular language for that regular expression and deterministic finite automata for that regualr language are similar things in different representation. So we can construct regular expression for every deterministic automata.

2.4.1. Ardens's Theorem

Let P and Q be two regular expression over alphabet Σ. If P does not contain null string $\in$, then

$$R = Q + RP$$

has a unique solution that is $R = QP^*$

It can be understand as : $R = Q + RP$

Put the value of R in R.H.S.

$$R = Q + (Q + RP)P = Q + QP + RP^2$$

When we put the value of R again and again we got the following equation.

$$R = Q + QP + QP^2 + QP^3 \ldots$$

$$R = Q(1 + P + P^2 + P^3 \ldots$$

$$R = Q(\in + P + P^2 + P^3 + \ldots$$

$$R = QP^*$$

(By the definition of closure operation for regular expression.)

2.4.2. Use of Arden's Theorem to find Regular Expression of a Deterministic Finite Automata

There are certain assumptions which are made regarding the transition system :

(*i*) The transition diagram should not have $\in$-transitions.

(*ii*) It must have only a single initial state.

(*iii*) It vertices are $q_1 \ldots q_n$.

(*iv*) q_i is final state.

(*v*) w_{ij} denotes the regular expression representing the set of labels of edges from q_i to q_j. We can get the following set of equation in $q_1 \ldots q_n$.

$$q_1 = q_1 w_{11} + q_2 w_{21} + \ldots + q_n w_{n1} + \in \quad \ldots(1)$$

(since q_1 is the initial state)

$$q_2 = q_1 w_{12} + q_2 w_{22} + \ldots q_n w_{n2} \quad \ldots(2)$$

$$\vdots \qquad \vdots$$

$$q_n = q_1 w_{1n} + q_2 w_{2n} + \ldots + q_n w_{nn} \quad \ldots(n)$$

We solve these equation for q_i in terms of w_{ij}'s and it will be required regular expression. One thing should be noted that we add $\in$ (null string) in the equation starts with starting state q_1 and we solve equation to find out q_i (final state) in terms of w_{ij}'s, it is one of the regular expression for given deterministic finite automata.

Example 2.14. Find the regular expression for transition diagram given in Fig. 2.10.

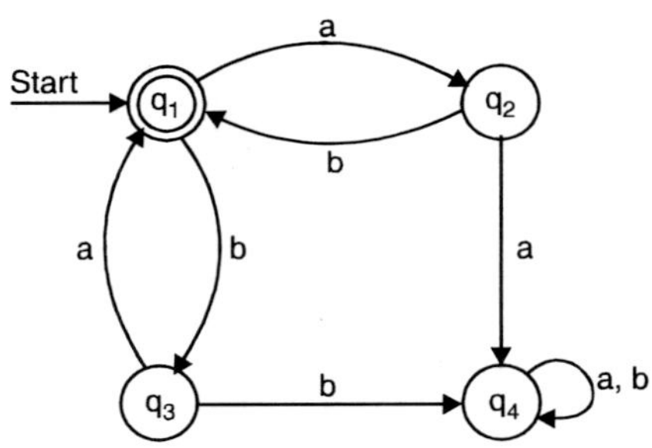

Fig. 2.10.

Solution. Now let us from the equations :

$$q_1 = q_2 b + q_3 a + \in$$
$$q_2 = q_1 a$$
$$q_3 = q_1 b$$
$$q_4 = q_2 a + q_3 b + q_4 a + q_4 b$$

Put q_2 and q_3 in q_1 as

$$q_1 = q_1 ab + q_1 ba + \in$$
$$q_1 = \in + q_1(ab + ba)$$
$$q_1 = \in (ab + ba)^*$$

So required regular expression is $(ab + ba)^*$.

Example 2.15. Construct a regular expression corresponding to the state diagram given in Fig. 2.11.

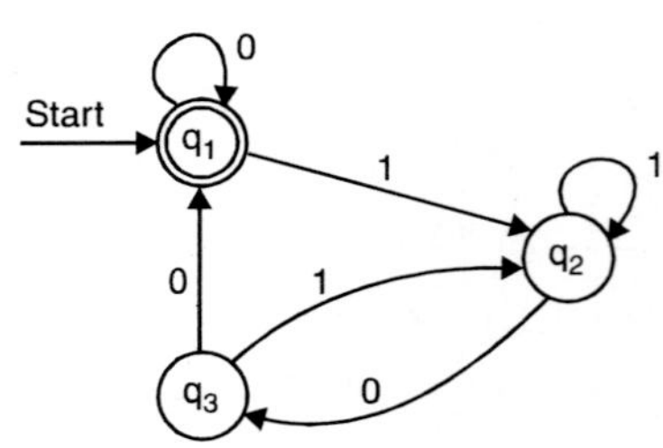

Fig. 2.11.

Solution. Let us form the equations :

$$q_1 = q_1 0 + q_3 0 + \in$$
$$q_2 = q_1 1 + q_2 1 + q_3 1$$
$$q_3 = q_2 0$$
$$q_2 = q_1 1 + q_2 1 + (q_2 0)\, 1 = q_1 1 + q_2 (1 + 01)$$

$$q_2 = q_1 1 (1 + 01)^*$$

So $q_1 = q_1 0 + q_3 0 + \in = q_1 0 + q_2 00 + \in = q_1 0 + (q_1 1\ (1101)^*)00 + \in$

$$q_1 = q_1(0 + 1(1 + 01)^*00) + \in$$

$$q_1 = \in(0 + 1(1 + 01)^*00)^*$$

$$q_1 = (0 + 1(1 + 01)^*\ 00)^*$$

So regular expression is $(0 + 1(1 + 01)^*\ 00)^*$.

Example 2.16. Find the regular expression corresponding to Fig. 2.12.

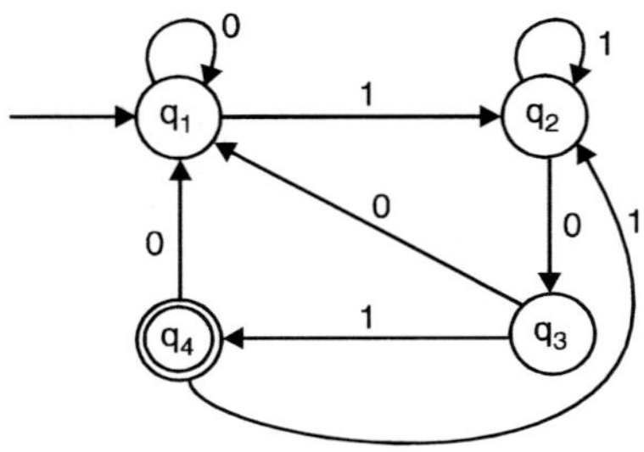

Fig. 2.12.

Solution. Now let us from the equations

$$q_1 = q_1 0 + q_3 0 + q_4 0 + \in$$

$$q_2 = q_1 1 + q_2 1 + q_4 1$$

$$q_3 = q_2 0$$

$$q_4 = q_3 1$$

Now$q_4 = q_3 1 = q_2 01$

Using q_2 equation, we get

$$q_2 = q_1 1 + q_2 1 + q_2 011 = q_1 1 + q_2(1 + 011)$$

$$q_2 = q_1 1(1 + 011)^* = q_1(1(1 + 011)^*)$$

Put q_3 and q_4 in q_1 equation

$$q_1 = q_1 0 + q_2 00 + q_2 010 + \in$$

$$= q_1 0 + q_2(00 + 010) + \in$$

$$= q_1 0 + q_1 1(1 + 011)^*\ (00 + 010) + \in$$

$$q_1 = \in(0 + 1(1 + 011)^*\ (00 + 010))^*$$

Put q_2 in q_4 equation

$$q_4 = q_1 1(1 + 011)^*\ 01 = q_1(1(1 + 01)^*01)$$

$$q_4 = (0 + 1(1 + 011)^*\ (00 + 010))^*\ (1(1 + 011)^*01)$$

So required regular expression is

$(0 + 1(1 + 011)^*\ (00 + 010))^*\ (1(1 + 011)^*01)$

2.5. EQUIVALENCE OF REGULAR EXPRESSIONS AND FINITE AUTOMATA

Regular expressions and finite automata have equivalent expressive power:

(*i*) For every regular expression *r*, there is a corresponding FA that accepts the set of strings generated by *r*.

(*ii*) For every FA, M there is a corresponding regular expression that generates the set of strings accepted by M.

The proof is in two parts:

1. An algorithm that, given a regular expression r, produces an FA, M such that $L(M) == L(r)$.
2. An algorithm that, given an FA, M produces a regular expression r such that $L(r) == L(M)$.

Our construction of FA from regular expressions will allow "epsilon transitions" (a transition from one state to another with epsilon (ε) as the label). Such a transition is always possible, since epsilon (or the empty string) can be said to exist between any two input symbols. We can show that such epsilon transitions are a notational convenience; for every FA with epsilon transitions there is a corresponding FA without them.

2.5.1. Constructing an FA from an Regular Expression

We begin by showing how to construct an FA for the operands in a regular expression.

(*i*) If the operand is a character c, then our FA has two states, q_0 (the start state) and f (the final, accepting state), and a transition from q_0 to f with label c.

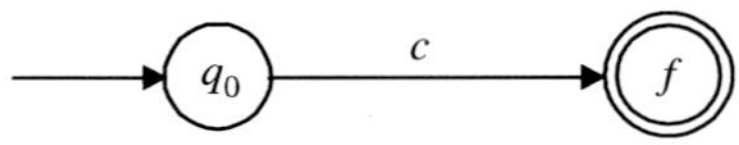

Fig. 2.13.

(*ii*) If the operand is epsilon, then our FA has two states, q_0 (the start state) and f (the final, accepting state), and an epsilon transition from q_0 to f.

Fig. 2.14.

(*iii*) If the operand is null, then our FA has two states, q_0 (the start state) and f (the final, accepting state), and no transitions.

Fig. 2.15.

Given FA for r_1 and r_2, we now show how to build an FA for r_1r_2, $r_1|r_2$, and r_1*. Let M (with start state q_1 and final state f_1) be the machine accepting $L(r_1)$ and B (with start state q_2 and final state f_2) be the machine accepting $L(r_2)$.

(*i*) The machine M_3 accepting $L(r_1r_2)$ includes M_1 and M_2, with start state q_1, final state f_2, and an epsilon transition from f_1 to q_2.

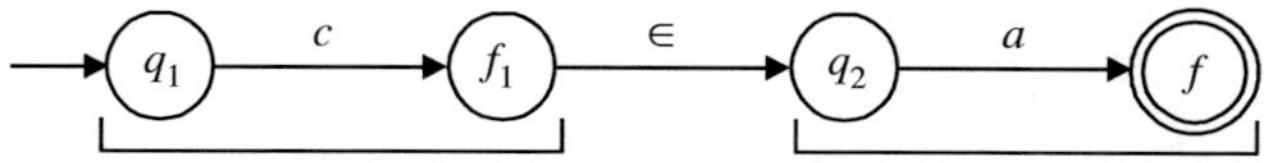

Fig. 2.16.

(*ii*) The machine M_3 accepting $L(r_1|r_2)$ includes M_1 and M_2, with a new start state q_3, a new final state f_3, and epsilon transitions from q_3 to q_1 and q_2, and from f_1 and f_2 to f_3.

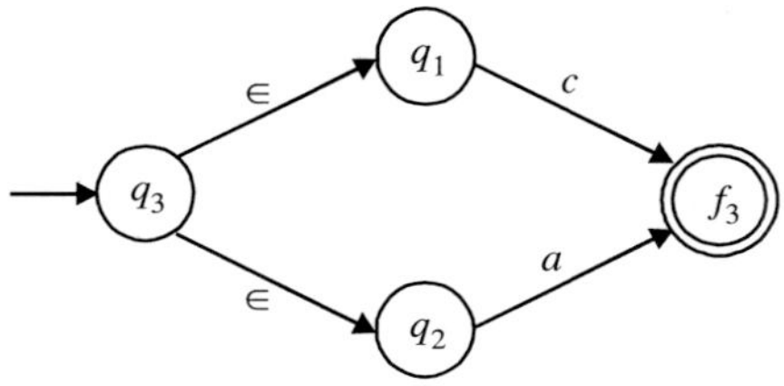

Fig. 2.17.

(*iii*) The machine M_3 accepting $L(r_1{}^*)$ includes M, with a new start state q_4, a new final state f_4, and epsilon transitions from q_4 to q_1 and f_4, and from f_1 to q_1, and from f_1 to f_4.

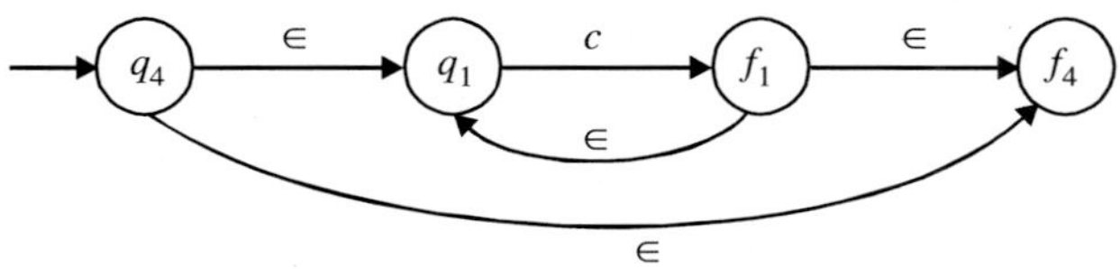

Fig. 2.18.

2.5.2. Eliminating Epsilon Transitions

If we can eliminate epsilon transitions from an FA, then our construction of an FA from a regular expression (which yields an FA with epsilon transitions) can be completed.

Observe that epsilon transitions are similar to nondeterminism in that they offer a choice: an epsilon transition allows us to stay in a state or move to a new state, regardless of the input symbol.

If starting in state q_1, we can reach state q_2 via a series of epsilon transitions followed by a transition on input symbol x, we can replace all of the epsilon transitions with a single transition from q_1 to q_2 on symbol x.

2.5.2.1. Algorithm for Eliminating Epsilon Transitions

We can build a finite automaton M_2 with no epsilon transitions from a finite automaton M_1 containing epsilon transitions as follows:

1. The states of M_2 are all the states of M_1 that have an entering transition labeled by some symbol other than epsilon, plus the start state of M_1, which is also the start state of M_2.

2. For each state in M_1, determine which other states are reachable via epsilon transitions only. If a state of M_1 can reach a final state in M_1 via epsilon transitions, then the corresponding state is a final state in M_2.
3. For each pair of states i and j in M_2, there is a transition from state i to state j on input x if there exists a state k that is reachable from state i via epsilon transitions in M_1, and there is a transition in M_1 from state k to state j on input x.

2.5.3. Constructing an Regular Expression from an Finite Automata

To construct a regular expression from a DFA (and thereby complete the proof that regular expressions and finite automata have the same expressive power), we replace each state in the DFA one by one with a corresponding regular expression.

Just as we built a small FA for each operator and operand in a regular expression, we will now build a small regular expression for each state in the DFA.

The basic idea is to eliminate the states of the FA one by one, replacing each state with a regular expression that generates the portion of the input string that labels the transitions into and out of the state being eliminated.

2.5.3.1. Algorithm for Constructing an RE from an FA

Given a DFA F we construct a regular expression R such that $L(M) == L(r)$.

We preprocess the FA, turning the labels on transitions into regular expressions. If there is a transition with label $\{a, b\}$, then we replace the label with the regular expression $a \mid b$. If there is no transition from a state to itself, we can add one with the label NULL.

For each accepting state f in M, eliminate all states in M except the start state q_0 and f.

To eliminate a state q_1, consider all pairs of states q_2 and q_3 such that there is a transition from q_2 to q_1 with label r_1, a transition from q_1 to q_1 with label r_2 (possibly null, meaning no transition), and a transition from q_1 to q_3 with label r_3. Introduce a transition from q_2 to q_3 with label $r_1r_2{}^*r_3$. If there is already a transition from q_2 to q_3 with label r_4, then replace that label with $r_4|r_1r_2{}^*r_3$.

After eliminating all states except q_0 and f:

(*i*) If $q_0 == f$, then the resulting regular expression is $r_1{}^*$, where r_1 is the label on the transition from q_0 to q_0.

(*ii*) If $q_0 \mathrel{!}= f$, then assume the transition from q_0 to q_0 is labeled r_1, the transition from q_0 to f is labeled r_2, the transition from f to f is labeled r_3, and the transition from f to q_0 is labeled r_4. The resulting regular expression is $r_1{}^*r_2(r_3 \mid r_4r_1{}^*r_2)^*$

Let r_i be the regular expression produced by eliminating all the states except q_0 and f_i. If there are n final states in the DFA, then the regular expression that generates the strings accepted by the original DFA is $r_1 \mid r_2 \mid \ldots r_n$.

2.6. ALGEBRAIC LAWS FOR REGULAR EXPRESSIONS

In this section, we will see a collection of algebraic laws that bring the issue of when two regular laws are equivalent. Instead of examining specific regular expressions, we shall consider pairs of regular expressions with variables as arguments.

TIPS

Two expressions with variables are equivalent if whatever languages we substitute for the variables, the result of the two expressions are the same language.

Like arithmatic expressions, the regular expressions have a number of laws that work for them. Many of these are similar to the laws for arithmatic, if we think of union as addition and concatenation as multiplication. However, there are a few places where the analogy breaks down, and there are also some laws that apply to regular expressions but have no analog for arithmatic, especially when the closure operator is involved. The next sections from a catalog of the major laws.

2.6.1. Associativity and Commutativity

Commutativity is the property of an operator that says that we can switch the order of its operands and get the same result. For example, the associative law of multiplication is $(x \times y) \times z = x \times (y \times z)$. Here are three laws of these types that hold for regular expression :

$$P + Q = Q + P.$$

This law, the commutative law for the union, says that we take union of two languages in either order.

$$(P + Q) + R = P + (Q + R)$$

This law, the associative law for union says that we may take the union of three languages either by taking the union of the first two initially, or taking the union of the last two initially.

Note that, together with the commutative law for union, we conclude that we can take the union of any collection of languages with any order and grouping, and the result will be the same. Intuitively, a string is in $L_1 \cup L_2 \cup ... + L_K$ if and only if it is in one or more of the L_i's.

$$(PQ)R = P(QR)$$

This law, the associative law for concatenation, says that we can concatenate three languages by concatenating either for first two or the last two initially.

2.6.2. Identities

An identity for an operator is a value such that when the operator is applied to identity and some other value, the result is the same value. For instance, 0 is the identity for addition, since $0 + m = m + 0 = m$, and 1 is the identity for the multiplication, since $1 \times x = x \times 1 = x$. There are three laws for regular expression involving these concepts; we list them below.

- $\phi + L = L + \phi = L$. This law assets that ϕ is the identity for union.
- $\in L = L\in = L$. This law assets that $\in$ is the identity for concatenation.
- $\phi L = L\phi = \phi$. This law assets that ϕ is the annihilater for concatenation.

2.6.3. Distributed Law

A distributed law involves two operators, and assets that one operator can be pushed down to be applied to each argument of the operator individually. The most common example is the distributed law of multiplication over addition, that is

$$x \times (y + z) = x \times y + x \times z$$

However, there is an analogous law for regular expression, that we most state in two forms. Since concatenation is not commutative. These laws are :

- $P(Q + R) = PQ + PR$. This law, is the left distributed law of concatenation over union.
- $(Q + R)P = QP + RP$. This law, is the right distributed law of concatenation over union.

2.6.4. The Idempotent Law

An operator is said to be idempotent if the result of applying it to two of the same values as arguments is that value. The common arithmatic operators are not idempotent, $x + x \neq x$ in general and $x \times x \neq x$ in general (although there are some values of x for which the equality holds, such as $0 + 0 = 0$). However, union and intersection are common examples of idempotent operators. Thus for the regular expression, we may assert the following law :

- $L + L = L$. This law, the idempotent law for union, states if we take the union of two identical expression, we can replace them by one copy of the expression.

2.6.5. Laws for Closure

There are number of laws involving the closure operators. We shall list them here, and give some explanation for why they are true.

1. $(L^*)^* = L^*$. This law says that closing an expression that is already closed does not change the language. The language of $(L^*)^*$ is all strings created for concatenating strings in the language L^*. But those strings are themselves composed of strings from L. Thus, the string in $(L^*)^*$ is also a concatenation of strings from L and therefore in the language of L^*.
2. $\phi^* = \in$. The closure of ϕ contains only the string $\in$.
3. $\in^* = \in$. It is easy to check that only string that can be formed by concatenating any number of copies of the empty string is the empty string itself.
4. $L^+ = LL^* = L^*L$. Recall that positive closure of L, L^+ is defined to be $L + LL + LLL + \ldots$. Also, $L^* = \in + L + LL + LLL + \ldots$

   Thus $\qquad LL^* = L + LL + LLL + LLLL + \ldots$

   That proves $\qquad L^+ = LL^*$.
5. $L^* = L^+ + \in$. The proof is easy, since the expansion of L^+ includes every term in the expansion of L^* except $\in$. Note that if the language L contains the string $\in$ then the additional $+\in$ is not needed; that is $L^+ = L^*$ in this special case.
6. $L? = \in + L$. This rule is really the definition of the operator ?

2.6.6. Identities for Regular Expression

On the basis of above discussion we will see some identitites for the regular expression. Let p, q, r are three regular expression. Two regular expression p and q are equivalent (we say $p = q$) if p and q represent the same set of strings.

Some identities are following :

I^1	$\phi + r$	$= r$
I^2	$\phi r = r\phi$	$= \phi$
I^3	$\in r$	$= r\in = r$
I^4	$\in^* = \in$ and ϕ^*	$= \in$
I^5	$r + r$	$= r$
I^6	r^*r^*	$= r^*$
I^7	rr^*	$= r^*r = r^+$
I^8	$(r^*)^*$	$= r^*$
I^9	$\in + rr^*$	$= r^* = \in + r^*r$
I^{10}	$(pq)^*p$	$= p(qp)^*$
I^{11}	$(p + q)^*$	$= (p^*q^*)^* = (p^* + q^*)^*$
I^{12}	$(p + q)r$	$= pr + qr$ and $r(p + q) = rp + rq$

Example 2.17. Prove that (1 + 00*1) + (1 + 00*1) (0 + 10*1)* (0 + 10*1) = 0*1(0 + 10*1)*

Solution. L.H.S. $= (1 + 00^*1)\ (\in + (0 + 10^*1^*(0 + 10^*1)$ using I^{12}

$= (1 + 00^*1)\ (0 + 10^*1)^*$ using I^9

$= (\in + 00^*)1\ (0 + 10^*1)^*$ using I^{12} for $(1 + 00^*1)$

$= 0^*1(0 + 10^*1)^*$ using I^9

= R.H.S.

Hence it is proved.

Example 2.18. Prove $t + RR^* = R^*$ where t is null string and R is a regular expression.

Solution. Two regular expressions P and Q are equivalent (we write $P = Q$). If P and Q are represent the same set of strings.

From Arden's Theorem, we know that, the solution of equation $R' = Q + R'P$ is $R' = QP^*$

Now $t + RR^* = R^*$

L.H.S. $= t + RR^*$ Compare it with Arden's theorem

$\simeq Q + R'P$

$Q = t$

$R' = R$

$P = R^*$

So, the solution of this equation will be $R' = QP^*$

$$R_1 = QP^*$$
$$= tR^*$$
$$= R^*$$

So we can write that $t + RR^* = R^*$

Example 2.19. Prove the following identity :

$$(a^*ab + ba)^* a^* = (a + ab + ba)^*.$$

Solution. $(a^*ab + ba)^*a^* = (a + ab + ba)^*$

By taking L.H.S. $= (a * ab + ba)^* a^*$

$$= (a\ (ab)^* + (ba)^*)^*a^*$$
$$= (ab^* + (ba)^*\ a^*)$$
$$= (ab^* + (ba)^*\ a^*)$$
$$= (a + ab + ba)^*$$
$$= \text{R.H.S.}$$

Example 2.20. Write down regular expression corresponding to each of the following languages over {0, 1}. The langauge of all strings in which every 0 is followed immediately by 11.

Solution. Let us analyse strings, as follows :

1, 1, 111, 011, 011011, 0111, 1011 etc.

So regular expression will be

$$r = (1 + 011)^*$$

Example 2.21. Find a regular expression corresponding to each of the following subset [0, 1] :

(*i*) The language of all strings containing at least two 0's.

(*ii*) The language of all string containing at most two 0's.

(*iii*) The language of all strings ending with 1 and don't contain 00.

(*iv*) The language of all strings in which both the number of 0's and number of 1's are odd.

Solution. (*i*) $r = (0 + 1)^*\ 0\ (0 + 1)^*\ 0\ (0 + 1)^*$

(*ii*) $r = 1^* + 1^*01^* + 1^*01^*01^*$

It means string may contain all 1's, at most one 1 or atmost two 0's.

(*iii*) $r = (1 + 01)^*\ 1$

(*iv*) $r = (1(11)^* + 0(00)^*)$

Example 2.22. Find the regular expression representing the following sets :

(*i*) The set of all strings over [0, 1] having at most one pair of 0's or at most one pair of 1's.

(*ii*) The set of all strings over [*a*, *b*] in which there are at least two occurence of *b* between any two occurences of *a*.

Solution.

(*i*) Given $t = [0, 1]$ at most means upto one pair of 0's and 1's.

$$0^* = \{\in, 0, 00, 000, 0000, ...\} \quad ...(i)$$

$$1^* = \{\in, 1, 11, 111, 1111, ...\} \quad ...(ii)$$

$$01^* = \{0, 01, 011, 0111, 01111, ...\} \quad ...(iii)$$

$$10^* = \{1, 10, 100, 1000, 10000, ...\} \quad ...(iv)$$

Now we can find that which regular expression does not have a pair of 0's *i.e.* 00.

Hence the set 1* and 01* (*i.e.* from 2 and 3) not containing pair of 0's.

Now we can write as :

$$1^*+01^*$$

$$(1 + 01)^* \quad ...(v)$$

Thus (1 + 01)* is regular expression that contain no pair of 0's. To find the string that contain exactly one pair of zero's, we simply write the string which have not a pair of 0's on both side of the pair of 0's *i.e.* 00. Thus, the string which have exactly one pair 0's can be shown by

$$(1 + 01)^*\ 00\ (1 + 01)^* \quad ...(vi)$$

Similarly, regular expression does not have pair of 1's *i.e.* 0* and 10* (*i.e.* from (*i*) and (*iv*)) we can write these two as

$$0^* + 10^*$$

$$(0 + 10)^* \quad ...(vii)$$

Similarly, as above, the string that not have pair of 1's on the both side of the pair of 1's *i.e.* 11. Thus this string which have exactly one pair of 1's :

$$(0 + 10)^*\ 11\ (0 + 10)^* \quad ...(viii)$$

From (*v*), (*vi*), (*vii*) and (*viii*), we get a set of strings over {0, 1} having at most one pair of 0's or 1's *i.e.*

$$r = (1 + 01)^* + (1 + 01)^*\ 00\ (1 + 01)^* + (0 + 10)^* + (0 + 10)^*\ 11\ (0 + 10)^*$$

(*ii*) When we say $(bb)^*\ a\ (bb)^*$ it means that one *a* between even *b*. Then if we do $(a(bb)^*\ a\ b(bb)^*\ a)^*$ then obviously there will be at least two *b* between two '*a*' hence corresponding regular expression (RE) will be :

$$r = (a(bb)^*\ a(bb)^*a)^*$$

Example 2.23. Let Σ = {*a*, *b*} be an alphabet. Show that

(*i*) $a^*(a \cup b)^*$

(*ii*) $aa^*(a \cup b)^*a$

(*iii*) $(a \cup b)^*(a \cup \phi)$

are regular expression over Σ.

Solution.

(*i*) $\Sigma = \{a, b\}$ so a^* and a are regular expression over Σ. $a \cup b$ is also regular expression by the definition of regular expression, therefore $(a \cup b)^*$ is regular expression over Σ. Hence by concatenation of two regular expressions $a^* (a \cup b)^*$ is a regular expression.

(*ii*) a, b and a^* are regular expression and hence aa^* is r.e. (regular expression) by concatenation $aa^* (a \cup b)$ is regular expression. This regular expression concatenates with the regular expression a produces regular expression $aa^* (a \cup b)^* a$ on Σ.

(*iii*) Since $a, b \in \Sigma$, ϕ, a, b are regular expressions. Hence $a \cup \phi$ and therefore $(a \cup \phi)^*$ is a regular expression. Again $(a \cup b)$ and $(a \cup b)^*$ is a regular expression. By concatenation $(a \cup b)^* (a \cup \phi)$ is a regular expression over Σ. Note that ϕ represents the empty set, that is the set with no strings.

Example 2.24. Find regular expression '*r*' over $\Sigma = \{a, b\}$ such that $L(r) = \{a^m b^m : m > 0\}$.

Solution. Given language is $L = \{a^m b\, a^n : m > 0 \,;\, n > 0\}$

Regular expression, $r = aa^*\, baa^*$

Example 2.25. Exhibit the language *L* (a^*, $(a + b)$ in set notaion.

Solution.

$$\begin{aligned} L(a^* . (a + b)) &= L(a^*) . L(a + b) \\ &= (L(a))^* . (L(a) + L(b)) \\ &= \{\in, a, aa, aaa, ...\} . \{a + b\} \\ &= \{a, aa, aaa, ..., b, ab, aab, ...\} \end{aligned}$$

Example 2.26. Find a regular expression for the language $L = \{w \in (0, 1)^* : w$ has no pair of consecutive zero's}.

Solution. If language is set of strings which does not have any consecutive zero's as substring we can analyse that *L* as the repetition of string 1 and 01 only. So as per this logic regular expression is :

$$r = (1 + 01)^* \underline{(0 + \in)}$$

String may end in 0 or 1.

Example 2.27. Let $L = \{uvu : u, v \in (a, b)^*, |\, w \,| = 2\}$, find regular expression for the same.

Solution. We can have following case for $|\, w \,| = 2$.

(*i*) $u = aa$ (*ii*) $u = ab$

(*iii*) $u = ba$ (*iv*) $u = bb$

So regular expression is

$$r = aa(a + b)^*\, aa + ab(a + b)^*\, ab + ba(a + b)^*\, ba + bb(a + b)^*\, bb$$

EXERCISES

1. Give the English description for the following regular expression :

(*a*) $r = (aa)^* (bb)^* b$

(*b*) $r = (0 + 1)^*00 (0 + 1)^*$

(*c*) $r = ((0 + 1) (0 + 1)^*)^* 00(0 + 1)^*$

2. Find the regular expression for the following :

(*a*) $L = \{a^n b^m, n \geq 1, m \geq 1, nm \geq 3\}$

(*b*) $L = \{ab^n w : n \geq 3, w \in \{ab\}^+\}$

(*c*) $L = \{a^n : n \geq 0, n \neq 3\}$

3. Write regular expression for the following language over alphabet {0, 1} :

(*a*) All string ending in 01

(*b*) All strings not ending in 01

(*c*) All strings containing an even number of 0's

(*d*) All strings with at most two occurrences of the substring 00.

(*e*) All strings not containing the substring 10.

4. Prove that

(*a*) $(r_1^*)^* = r_1^*$

(*b*) $r_1^*(r_1 + r_2)^* \equiv (r_1 + r_2)^*$

(*c*) $(r_1 + r_2)^* \equiv (r_1^* r_2^*)^*$

For all regular expression r_1 and r_2. Here ≡ stands for equivalence in the sense of the language generated.

5. Give the English description for the following regular expression :

(*a*) $r = a(a + b)^* ab$

(*b*) $r = a^* b + b^* a$

(*c*) $r = (aa + b)^* (bb + a)^*$

6. Prove the following identity :

$(a^* ab + ba)^* a^* = (a + ab + ba)^*$

7. Construct the transition system for the following regular expressions :

(*a*) $r = 1 + 00 + 010^*1$

(*b*) $r = (ab + c^*)^* b$

8. Construct a finite automata for the regular expression

$$r = (a + b)^* abb.$$

9. Find the regular expression for the following transition diagram.

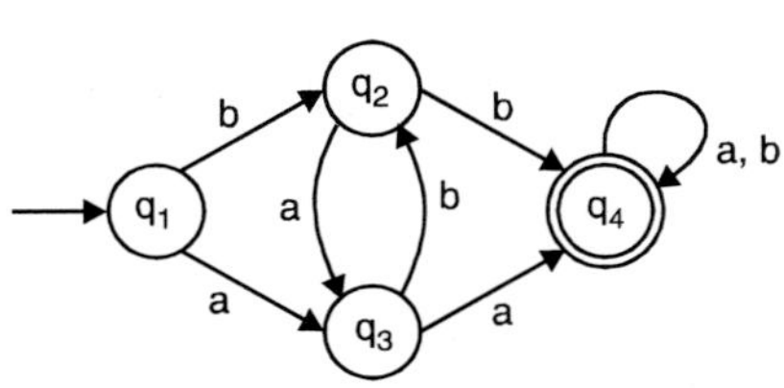

Fig. 2.19.

10. Find the regular expression corresponding to the automation given in Fig. 2.14.

11. Find the regular expression representing the following sets :

    (*i*) The set of all strings over {0, 1}, having at most one pair of 0's or at most 1 pair of 1's.

    (*ii*) The set of all strings over {0, 1} in which there are atleast two occurrence of 1 between any two occurance of *a*.

    (*iii*) The set of all strings over {*a*, *b*} ending with *bb* and begining with *a*.

    (*iv*) The set of all strings over {0, 1} not containing 101 as substring.

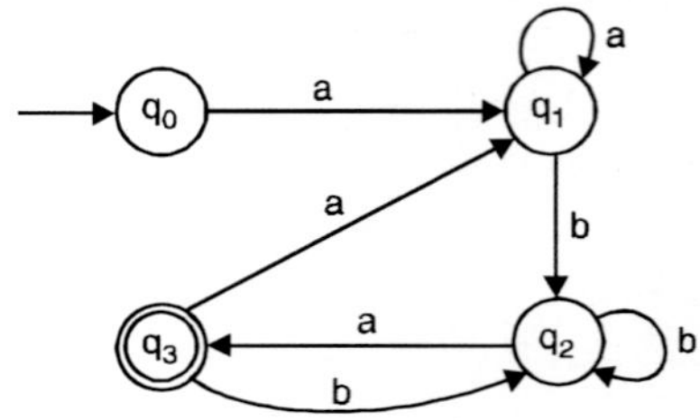

Fig. 2.20.

12. Construct a regular expression defining each of the following language over the alphabet Σ = {*a*, *b*}

    (*a*) All strings that do not end in a double letter.

    (*b*) All strings in which letter *b* never triples.

    (*c*) All strings in which *a* is tripled or *b* is tripled.

    (*d*) All strings that have an even number of *a*'s and an odd number of *b*'s.

    (*e*) All strings that have an odd number of *a*'s and an odd number of *b*'s.

*****

CHAPTER 3

Introduction to Machines

INSIDE THIS CHAPTER

3.1. INTRODUCTION

One limitation of the finite automation as we have defined it in previous chapter is that output is limited to a binary signal : accept (1)/not accepted (0). The only explicit task a machine has done so far is to recognize a language. Computers as we know, often have the more useful function of performing calculations and conveying results. In this chapter, we expand the notion of machine task.

If we assume that all the printing of output is to be done at the end of the program run, at which time we have an instruction that dumps a buffer that has been assembled, then we have a maximum on the number of charaters that dumps a buffer that has been assembled, then we have a maximum on the number of characters that the program can print, namely, the size of the buffer. However, theoritically we should be able to have outputs of any finite length. For example, we might simply want to print out a copy of the input string, which could itself be arbitrarily long. These questions that have to be faced if we are to claim that our mathematical nodels of FAs and Transition graphs represent actual physical machines. In this chapter we will discuss two different model of FAs with output capabilities. These were created by G.H. mealy in 1955 and, independently, by E.F. Moore in 1956. We will present these two models, prove that they are equivalent and give some examples of how they arise in the "logic" section of a computer.

3.2. MOORE MACHINE

It is the FA in which output is associated with each state. In moore machine every state of this finite machine has a fix output.

Definition Mathematically moore machine is a six-tuple machine and defined as

$$M_0 = (Q, \Sigma, \Delta, \delta, \lambda', q_0)$$

M_0 is the six-tuple moore machine in which

Q : A nonempty finite set of state in M_0.

Σ : A nonempty finite set of input symbols.

Δ : A nonempty finite set of outputs.

δ : It is transition function which takes two arguments as in finite automata, one is input state and another input symbol. The output of this function is a single state, so clearly δ is the function which is responsible for the transitions in M_0.

q_0 : is the initial state of M_0 and $q_0 \in Q$.

λ' : is a mapping function which maps Q to Δ, giving the output associated with each state.

Let M_0 be a moor machine and Σ is $a_1, a_2, a_3, ..., a_n$, $n \geq 0$ then output of M_0 for Σ is $\lambda'(q_0), \lambda'(q_1), \lambda'(q_2)\ldots. \lambda'(q_n)$, where $q_0, q_1, q_2, \ldots q_n$ is the sequence of states such that

$$\delta(q_{i-1}, a_i) = q_i \text{ for } 1 \leq i \leq n$$

It is to be noticed that moore machine gives output $\lambda'(q_0)$ for input $\in$, *i.e.* in case of moore machine length of output sequence is $n + 1$.

3.2.1. Representation of Moore Machine

Moore machine can be represented by transition table as well as transition diagram same as finite automata.

Let M_0 be moore machine having following transition table $\Sigma = \{0, 1\}$

Present state	*Next state at input*		*Output*
	a = 0	*a = 1*	
$\rightarrow q_0$	q_3	q_1	0
q_1	q_1	q_2	1
q_2	q_2	q_3	0
q_3	q_3	q_0	0

$$M_0 = (Q, \Sigma, \Delta, \delta, \lambda', q_0)$$
$$Q = \{q_0, q_1, q_2, q_3\}$$
$$\Sigma = \{0, 1\}$$
$$\Delta = \{0, 1\}$$

$\lambda'(q_0) = 0$, $\lambda'(q_1) = 1$, $\lambda'(q_2) = 0$ and $\lambda'(q_3) = 0$

q_0 is the initial state. There is no concept of final state in moore machine since we are considering output for the each corresponding state.

We can also viewed it by transition diagram as

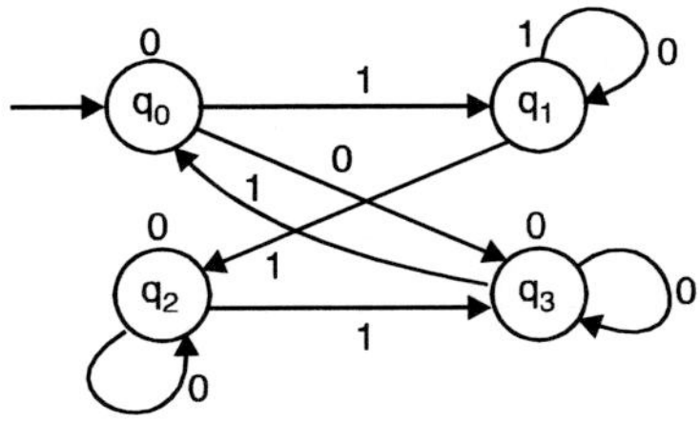

Fig. 3.1.

By the analysing moore machine transition diagram of Fig. 3.1. We can easily noticed that it is FA without any final state and output is there for every corresponding state. Output for every state is written just above the corresponding state.

3.2.2. Processing of the String through Moore Machine

Let us process the string (input) $w = 0111$ by the moore machine discussed in Fig. 3.1.

$$\xrightarrow{\text{Start}} \overset{0}{q_0} \xrightarrow[\text{Input}]{0} \overset{0}{q_3} \xrightarrow[\text{Input}]{1} \overset{0}{q_0} \xrightarrow[\text{Input}]{1} \overset{1}{q_1} \xrightarrow[\text{Input}]{1} \overset{0}{q_2}$$

The output string will be
$$w' = \lambda'(q_0)\ \lambda'(q_3)\ \lambda'(q_0)\ \lambda'(q_1)\ \lambda'(q_2)$$
$$= 00010$$
$$|w| = 4$$
$$|w'| = 5$$

So clearly if the length of input string is n then the length of output state will be $(n + 1)$. Since for the input $\in$ (null string), the output is $\lambda'(q_0) = 0$.

Let us discuss some example for better understand.

Example 3.1. Design a moore machine which determine the residue mod-3 for each binary string treated as binary integer.

Solution. We have to analyse the problem. First of all you must understand the meaning of residue mod-3 for each binary string treated as binary integer. Binary string means $\Sigma = \{0, 1\}$ is given, so binary string is combination of 0's and 1's. Binary integer means, you must find decimal equivalent of input string as

Binary	*Decimal equivalent*
0000	0
0001	1
0010	2
0011	3
0100	4
0101	5
0110	6
0111	7
1000	8

residue mod-3 means reminder, when decimal number divided by the 3.

So now things are clear, we have to design a moore machine which prints the reminder, when decimal equivalent of input string is divided by the 3.

So clearly $\Delta = \{0, 1, 2\}$

Since 0, 1 and 2 are the possible reminder when a decimal number is divided by the three. So we need three state in moore machine.

Let moore machine is M_0

$$M_0 = (Q, \Sigma, \Delta, \delta, \lambda', q_0)$$
$$Q = \{q_0, q_1, q_2\}$$
$$\lambda'(q_0) = 0$$
$$\lambda'(q_1) = 1$$
$$\lambda'(q_2) = 2$$
$$\Sigma = \{0, 1\}$$

and q_0 is initial state.

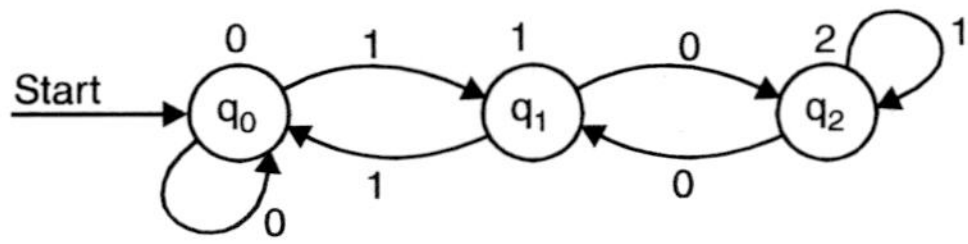

Fig. 3.2. Transition diagram of M_0.

Let us assume input string is 111, decimal equivalent is 7, 7/3 = 1(reminder). When we pass 111 by M_0 it ends at q_1 and $\lambda'(q_1) = 1$ which reminder of 7/3.

Example 3.2. Design a Moore machine which counts the occurrence of substring *aab* in input string.

Solution. Let Moore machine be

$$M_0 = (Q, \Sigma, \Delta, \delta, \lambda', q_0)$$
$$Q = \{q_0, q_1, q_2, q_3\}$$
$$\Sigma = \{a, b\}$$

We will design this machine in such a way that, machine prints out the character 0 except for state q_3, which prints a 1. To get q_3, we must come from state q_2 and have just read '*b*'. To get state q_2, we most read at least two *a*'s in a row, having started in any state.

$$\Delta = \{0, 1\}$$
$$\lambda'(q_0) = 0, \lambda'(q_1) = 0, \lambda'(q_2) = 0, \lambda'(q_3) = 1.$$

The following machine will counts *aab* occurrence for us.

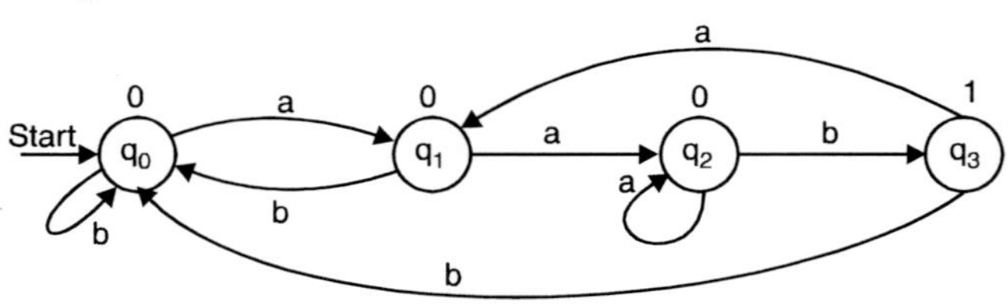

Fig. 3.3.

Let us pass string $w = a\underline{aab}\ b\underline{aab}aa$

$$q_0 \xrightarrow{a} q_1 \xrightarrow{a} q_2 \xrightarrow{a} q_2 \xrightarrow{b} q_3 \xrightarrow{b} q_0 \xrightarrow{b} q_0 \xrightarrow{a} q_1 \xrightarrow{a} q_2 \xrightarrow{b} q_3 \xrightarrow{a} q_1 \xrightarrow{a} q_2$$

Output string $w' = \lambda'(q_0)\ \lambda'(q_1)\ \lambda'(q_2)\ \lambda'(q_2)\ \lambda'(q_3)\ \lambda'(q_0)\ \lambda'(q_0)$
$\lambda'(q_1)\ \lambda'(q_2)\ \lambda'(q_3)\ \lambda'(q_1)\ \lambda'(q_2)$
$= 000010000100$

There are number of 1's are two so occurrence of *aab* will be 2.

Example 3.3. Construct a moore machine which calculate residue mod-4 for each binary string treated as binary integer.

Solution. This problem can be solved as we solved example 3.1. When we divide any number by 4 then reminder can be 0, 1, 2 and 3, so clearly moore machine will have four state.

Let moore machine is M_0

$$M_0 = (Q, \Sigma, \Delta, \delta, \lambda', q_0)$$
$$Q = \{q_0, q_1, q_2, q_3\}$$
$$\Sigma = \{0, 1\}$$
$$\Delta = \{0, 1, 2, 3\}$$
$$\lambda'(q_0) = 0, \lambda'(q_1) = 1, \lambda'(q_2) = 2 \text{ and } \lambda'(q_3) = 3$$

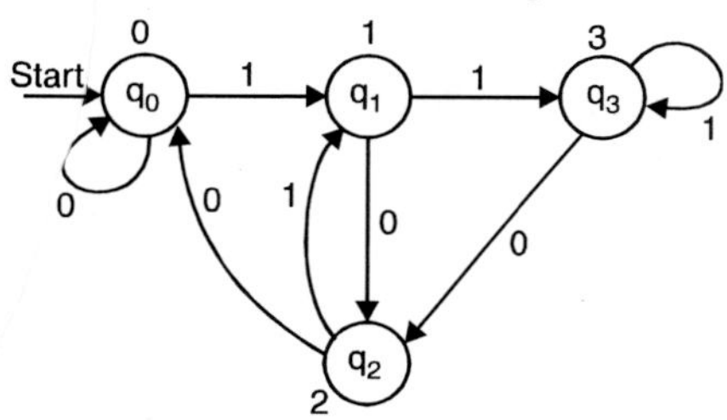

Fig. 3.4.

Let us pass the string 11, decimal equivalent is 3 and reminder of 3/4 is 3.

$$q_0 \xrightarrow{1} q_1 \xrightarrow{1} q_3$$

and $\lambda'(q_3) = 3$, which is reminder of 3/4.

Example 3.4. Give the moore machine for the input from (0 + 1 + 2)* which prints the residue module 5 of the input treated as a ternary (base 3, with digits 0, 1, 2) number.

Solution. First of all, let us analyse the problem, as input is from (0 + 1 + 2)* that input string will be combination of 0, 1 and 2. Like 012, 210, 11102, ... residue module 5 means reminder when a decimal divided by 5. Clearly reminder may be 0, 1, 2, 3, 4. Another important pointed is that conversion of string in two decimal number as

For example input string is 0 1 2 1, then decimal equivalent of it is w(say)
↓ ↓ ↓ ↓
3 2 1 0

$$w = 0 \times 3^3 + 1 \times 3^2 + 2 \times 3^1 + 1 \times 3^0$$
$$= 0 + 9 + 6 + 1$$
$$= 16$$

According problem 16/5 = 1 (reminder). So far input 0121 moore machine should print as output, when last character that is 1 is passed through machine.

It should be clear that moore machine will have 5 states since there may be five possible reminder when any number is divided by the 5 possibilities are 0, 1, 2, 3 and 4 transition diagram of this moore machine will be following :

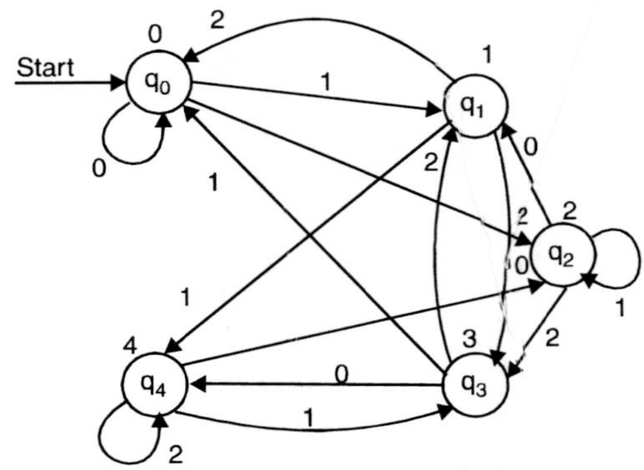

Fig. 3.5.

$$M_0 = (Q, \Sigma, \Delta, \delta, \lambda', q_0)$$

q_0 is initial state

$$\Sigma = \{0, 1, 2\}$$

$$Q = \{q_0, q_1, q_2, q_3, q_4\}$$

$$\lambda'(q_0) = 0, \lambda'(q_1) = 1, \lambda'(q_2) = 2, \lambda'(q_3) = 3, \lambda'(q_4) = 4.$$

Let us pass 0121

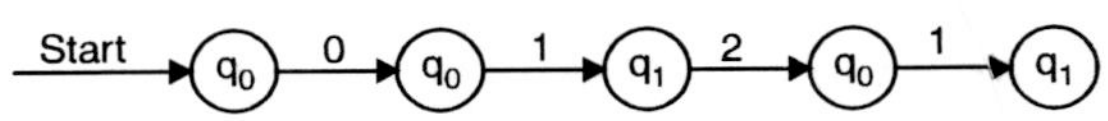

and $\lambda'(q_1) = 1$ which reminder of 16/5 = 1 (reminder)

3.3. MEALY MACHINE

Definition It is finite automata in which output is associated with each transition. In mealy machine every transition for a particular input symbol has a fix output.

Mathematically mealy machine is a six-tuple machine and defined as

$$M_e = (Q, \Sigma, \Delta, \delta, \lambda', q_0)$$

Q : A nonempty finite set of state in M_e.

Σ : A nonempty finite set of input symbols.

Δ : A nonempty finite set of outputs.

q_0 : initial state of M_e.

δ : It is a transition function which takes two arguments as in moore machine, one is input state and another is input symbol. The output of this function is a single state, so clearly δ is the function which is responsible for the transition in M_e.

λ' : is a mapping function which maps $Q \times \Sigma$ to Δ, giving the output associated with each transition.

So λ' maps $Q \times \Sigma \to \Delta$, that is $\lambda'(q, a)$ gives the output associated with the transition from state q_0 on input a. The output of M_e in response to input $a_1, a_2, a_3 \ldots a_n$ is $\lambda'(q_0, a_1), \lambda'(q_1, a_2) \ldots \lambda'(q_{n-1}, a_n)$ where $q_0, q_1, q_2, \ldots q_n$ is the sequence of states such that $\delta(q_i - 1, a_i) = q_i$ for $1 \le i \le n$. Clearly in this case length of output sequence is n.

3.3.1. Representation of Mealy Machine

Mealy machine can also be represented by transition table, as well as transition diagram, same as moore machine and finite automata.

Let M_e be mealy machine having following transition table.

Present state	*For input a = 0*		*For input a = 1*	
	State	*Output*	*State*	*Output*
$\to q_0$	q_3	0	q_2	0
q_2	q_1	1	q_4	0
q_3	q_2	1	q_1	1
q_4	q_4	1	q_3	0

$$M_e = (Q, \Sigma, \Delta, \delta, \lambda', q_0)$$
$$Q = (q_1, q_2, q_3, q_4)$$
$$\Sigma = \{0, 1\}$$
$$\Delta = \{0, 1\}$$

q_0 is initial state and δ is transition function.

$$\lambda'(q_1, 0) = 0 \qquad \lambda'(q_1, 1) = 0$$
$$\lambda'(q_2, 0) = 1 \qquad \lambda'(q_2, 1) = 0$$
$$\lambda'(q_3, 0) = 1 \qquad \lambda'(q_3, 1) = 1$$
$$\lambda'(q_4, 0) = 1 \qquad \lambda'(q_4, 1) = 0$$

Transition diagram for M_e is following :

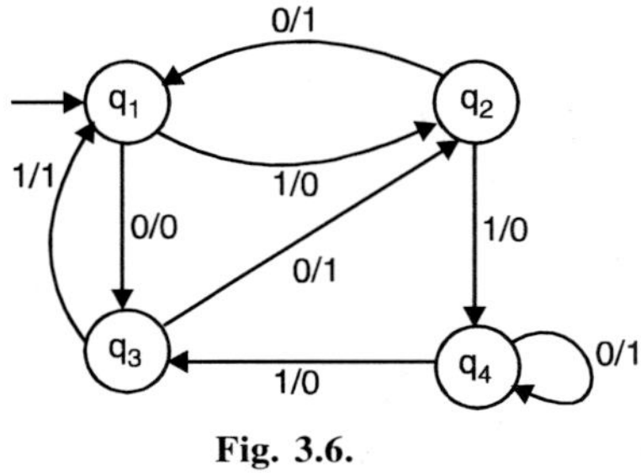

Fig. 3.6.

By the analysing mealy machine transition diagram of Fig. 3.6. We can easily noticed that, it is also finite automata without any final state and output is there for input-symbol on corresponding transition. Output for every input symbol is represented as level on each transition in the form input symbol/corresponding output. For example in Fig. 3.6 arc between q_1 and q_2 is labelled by 0/1, 0 is input symbol and 1 is output symbol.

3.3.2. Processing of the String through Mealy Machine

Let us process the string $w = 0111$ by the mealy machine, describe in Fig. 3.6.

$$\xrightarrow{\text{Start}} q_1 \xrightarrow{0/1} q_2 \xrightarrow{1/0} q_4 \xrightarrow{1/0} q_3 \xrightarrow{1/1} q_1$$

So output string will be 1001

$$w_1 = \underline{1001}$$
$$|w| = 4$$
$$|w_1| = 4$$

So the length of input string is same as length of output string. It is difference with moore machine.

Example 3.5. Design a melay machine, which prints 1's complement of input bit string over alphabet $\Sigma = \{0, 1\}$.

Solution. If input string is 101110 then 1's complements of 101110 will be 010001, so we have to design a mealy machine which will point output 010001 for input string 101110.

Let $$M_e = (Q, \Sigma, \Delta, \delta, \lambda', q_0)$$

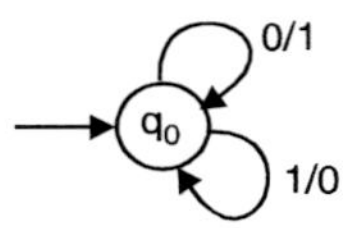

Fig. 3.7.

$$Q = \{q_0\}$$
$$\Sigma = \{0, 1\} \text{ given}$$

q_0 is also initial state.

Let us pass input $w = 011$ to M_e

$$q_0 \xrightarrow{0/1} q_0 \xrightarrow{1/0} q_0 \xrightarrow{1/0} q_0$$

output string is $w' = 100$ which 1's complement of $w = 011$.

Example 3.6. Design mealy machine, which marks as a increment machine that assumes that its input is a binary number and prints out the binary number that is one larger.

Solution. We assume that the input bit string is a binary number fed is backward, that is, unit digit first.

If input string is 100 then first input symbol will be 0 not 1.

The output string will be the binary representation of the number one greater and also will be generated right to left.

The machine will have three states : initial, carry, no carry. The carry state represents the overflow when two bits equal to 1 are added — we print 0 and we carry a 1. From the start state, we read the first bit. If we read a 0, we print a 1 and we do not owe a carry bit. If we read a 1, we print 0 and we do owe a carry bit. If at any point in the process we are in no-carry (which means that do not owe a carry), we print the next bit just as we read it and remain in no-carry. However if at some points in the process we are in owe-carry, the situation is different. If we read a 0, we print a 1 and go to the no-carry state. If we are in owe-carry and we read a 1, we print 0 and we loop back to owe-carry. The complete picture for this machine is following

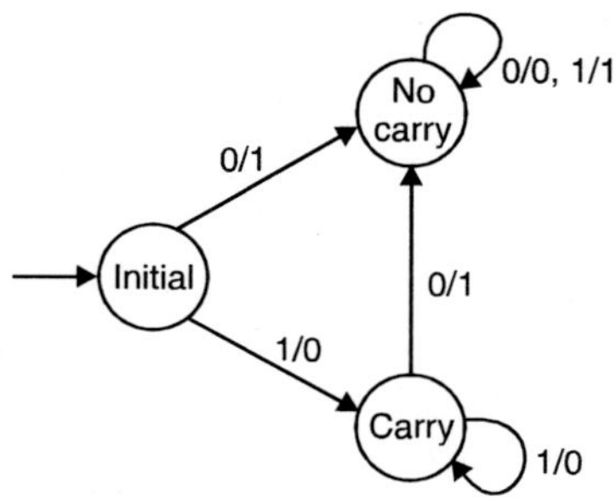

Fig. 3.8.

Let us pass 100 *i.e.* 4, string will be passed as 001 *i.e.* in reverse order.

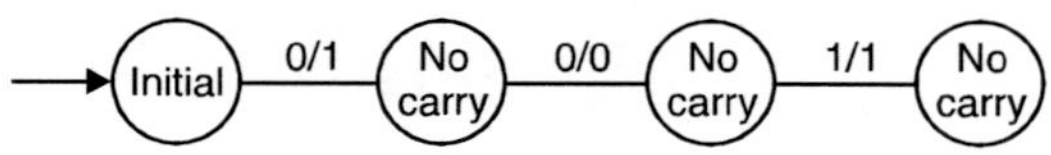

Output string = 101 *i.e.* 5 which 4 + 1

Now input string is w = 11 *i.e.* 3

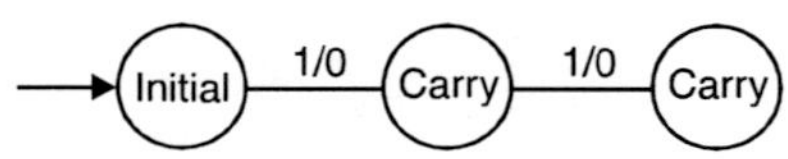

Output string will be = 100

Since process end in "**carry state**" so we put 1 infront 00 and number becomes 100.

Example 3.7. Construct a mealy machine for regular expression (0 + 1)* (00 + 11).

Solution. Let us first analyse the problem, check the string set that is language of (0 + 1)* (00 + 11). It is set of strings either end with 00 or end with 11, like 00, 11, 1011, 010100, ...

Here we define a three state mealy machine that use it's state to remember the last symbol read, emits output y whenever the current input matches the previous one, and emits 'n' otherwise. Here y is for the string belongs to regular expression and 'n' otherwise.

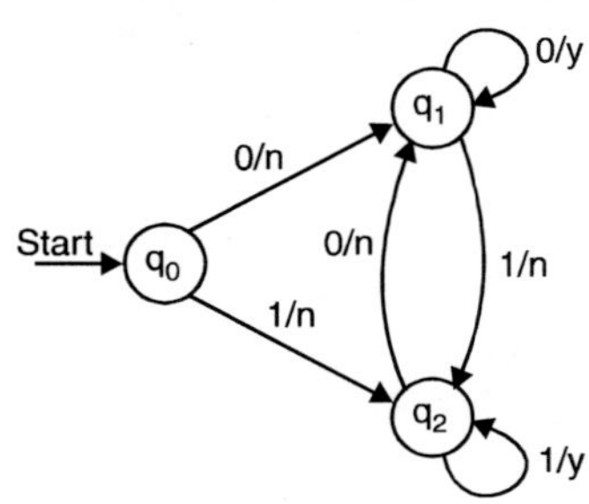

Fig. 3.9.

Let us pass input string w through mealy machine construct in Fig. 3.9.

$$w = 011\underline{00}$$

$$q_0 \xrightarrow{0/n} q_1 \xrightarrow{1/n} q_2 \xrightarrow{1/y} q_2 \xrightarrow{0/n} q_1 \xrightarrow{0/y} q_1$$

When we pass last input symbol '0' from q_1 we get y as output so string belongs to set of strings of regular expressions.

Example 3.8. Construct a mealy machine which calculate residue mod-4 for each binary string treated as binary integer.

Solution. This problem is same as example 3.3 but there we have to design moore machine now for the same logic we want to design mealy machine.

Let mealy machine is

$$M_e = (Q, \Sigma, \Delta, \delta, \lambda', q_0)$$
$$Q = \{q_0, q_1, q_2, q_3\}$$
$$\Sigma = \{0, 1\}$$
$$\Delta = \{0, 1, 2, 3\}$$

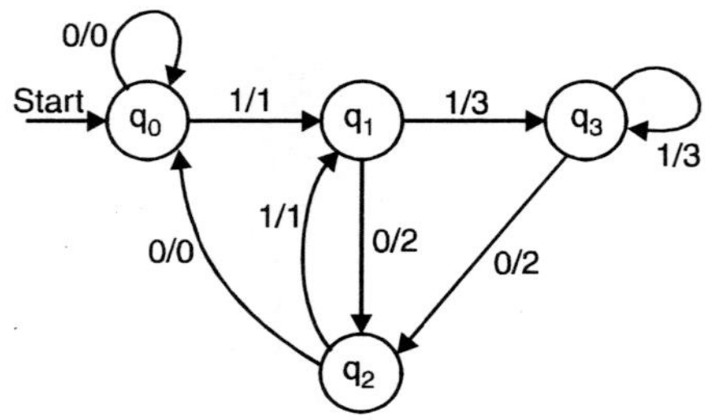

Fig. 3.10.

Let us pass the string 11, decimal equivalent of 11 is 3 and reminder of 3/4 is 3.

$$q_0 \xrightarrow{1/1} q_1 \xrightarrow{1/3} q_3$$

So when we pass last then output is 3 which is reminder of 3/4.

3.4. EQUIVALENCE OF MEALY AND MOORE MACHINE

Given the mealy machine M_e and the moore machine M_0, which prints the automatic start state character '*a*', we will say that these two machines are equivalent if for every input string the output string from M_0 is exactly '*a*' concatenated with the output from M_e.

3.4.1. Mealy Equivalent of Given Moore Machine

For every given moore machine M_0, there exist a equivalent mealy machine M_e.

We will learn to convert M_0 into M_e in various steps.

Step 1. Let M_0 is given moore machine, select any particular state of M_0, call it q_6. It gives instruction to print a certain character, call it *P*. Let us consider all the edges that enter this state. Each of them is labelled with an input letter. Let us change this, now relabel all the edges coming into q_6. If they were previously labelled *a* or *b* or *c* ..., let them now be labelled *a*/*p* or *b*/*p* or *c*/*p* ... and let us erase *P* from above the state q_6. This means that we shall be printing a *P* on the incoming edges before they enter q_6.

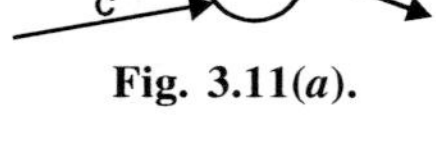

Fig. 3.11(*a*).

now becomes

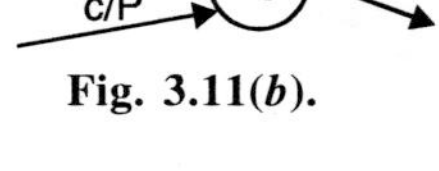

Fig. 3.11(*b*).

Step 2. We leave the outgoing edge from q_6 alone. They will be relabelled to print the character as associated with state to which they lead. The outgoing edged in Fig. 3.11(*a*), which is labelled δ is not change in Fig. 3.11(*b*).

Step 3. Repeat step 2 and step 3 for every state q_0, q_1, q_2 ... of M_0, we turn M_0 into melay machine M_e.

Definition The symbol that used to printed automatically when the machine started in state q_0 is no longer the first output character, but this does not stop the rest of the output string from being the same.

Example 3.9. Convert given moore machine into equivalent mealy machine.

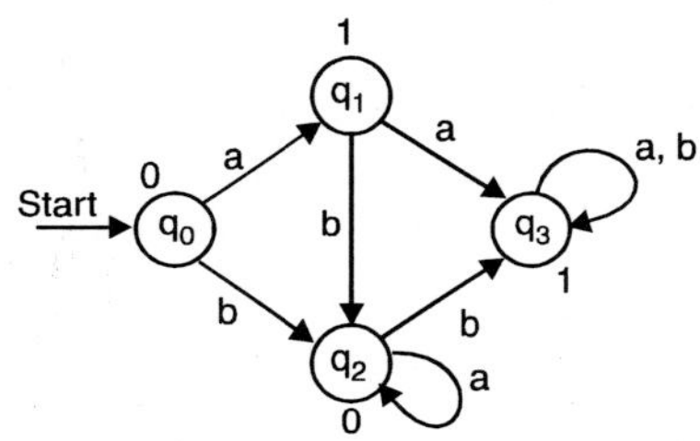

Fig. 3.12.

Solution. Now we will convert moore machine given in Fig. 3.12 into equivalent mealy machine by using steps discussed in 3.4.1, as follows :

(*i*) at q_0 there is no incoming arc, so output at q_0 becomes useless

(*ii*) at q_1 one arc labelled by a is coming, now it will be labelled by *a*/1, since output for q_1 was 1.

(*iii*) at q_2 three arcs labelled by *b*, *a* and *b* are incoming, no they becomes *b*/0, *a*/0 and *b*/0 as output for q_2 was 0.

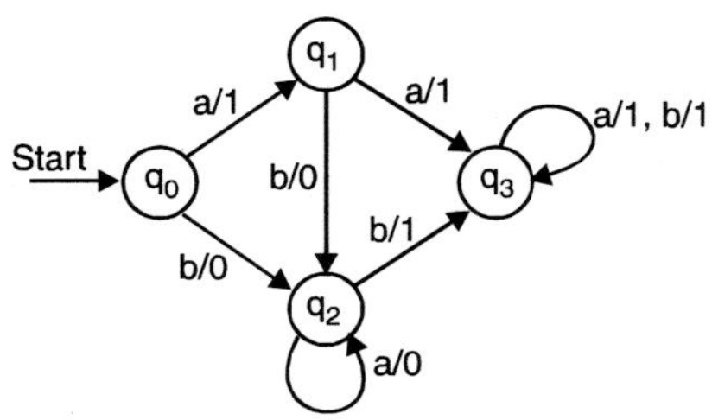

Fig. 3.13. Equivalent mealy machine.

(*iv*) at q_3 arcs was labelled by *a, b, a* and *b*, now the becomes *a*/1, *b*/1, *a*/1 and *b*/1 since output for q_3 was 1.

3.4.2. Moore Equivalent of given Mealy Machine

For every given mealy machine M_e, there exist a equivalent moore machine M_0.

We cannot just do the reverse of previous procedure. If we were try to push the printing instruction from the edge as it is in M_e to the inside of the state as it should be for a moore machine, we might end up with a conflict. We will handle these conflicts in steps, as follows :

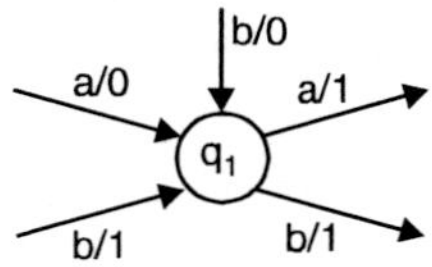

Fig. 3.14.

Step 1. Let us consider the situation in which two edges might come into the same state but have different printing instructions, as follows

We need then are two copies of the same state. The edge $a/0$ will go into q_1^1 (q_1 copy 1, a state that prints a '0') and edge $b/1$ will go into q_1^2 (q_1 copy 2, a state that print a '1'). The edge labelled $b/0$ will also go into q_1^1. Above these states, we include the printing instructions 0 and 1. The arrow coming out of each of these copies of what used to be q_1 must be the same as the edges coming out of q_1 originally. We get two sets of the output edges each equal to the original outgoing edges, but one set of the original incoming edges is divided between the two copies.

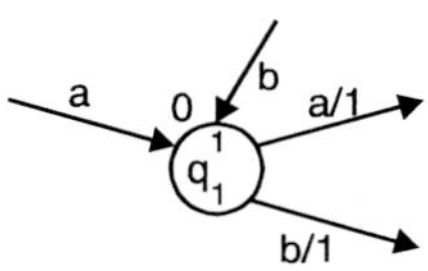

Fig. 3.15.

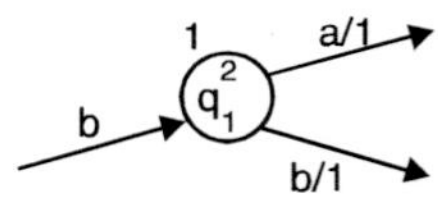

Fig. 3.16.

It can be noticed from Fig. 3.15 and Fig. 3.16 that instruction 0 is printed above q_1^1 and instruction 1 is printed above the q_1^2.

Repeat the this step for all the states in given mealy machine.

Step 2. Another conflict may come on the states, those are having loop. For example

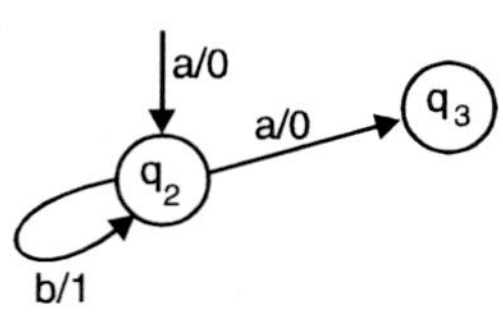

Fig. 3.17.

What happens in the preceding example is that the edge labelled $a/0$ has to enter a version of q_2 that prints 0. Let us call q_2^1. The loop labelled $b/1$ at q_3 has to enter a version of q_2 that prints a 1. We call this q_2^2. When we enter q_2 from edge $a/0$, we enter q_2', but we must also be able to loop with b's while staying in a q_2-like state. Therefore, an edge labelled b must correct q_2^1 to q_2^2. Because we must be allowed to repeat as many b's as we want, there must be a b-loop at state q_2^2. Each b-loop we go around prints another 1 when it reenters q_3^2. As with all such clones of an original state, they must both be corrected to q_3 by $a/0$ as follows :

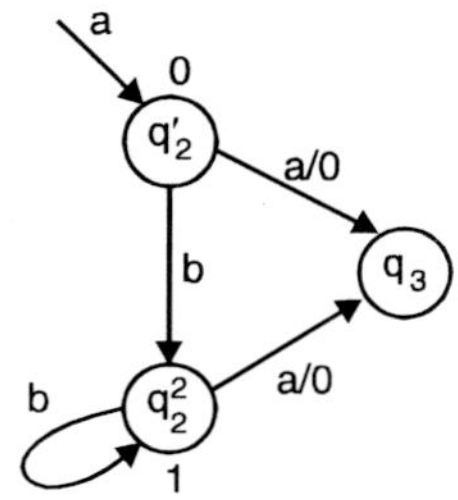

Fig. 3.18.

Step 3. If there is ever a state that has no edges entering it, we can assign if any printing instruction we want, even if this state is the start state.

Step 4. Let us repeat this process *i.e.* from step 1 to step 3 for each state of M_e, q_0, q_1, q_2, q_3, ... This will produced M_0. If we have to make copies of the start state in M_e, we can let any one of them be the start state in M_0 because they all give the identical directions for proceeding to other states. Having a choice of start states means that the conversion of M_e into M_0 is not unique. We should except this because may M_e is equivalent to more then one M_0.

Example 3.10. Convert the following mealy machine into moore machine.

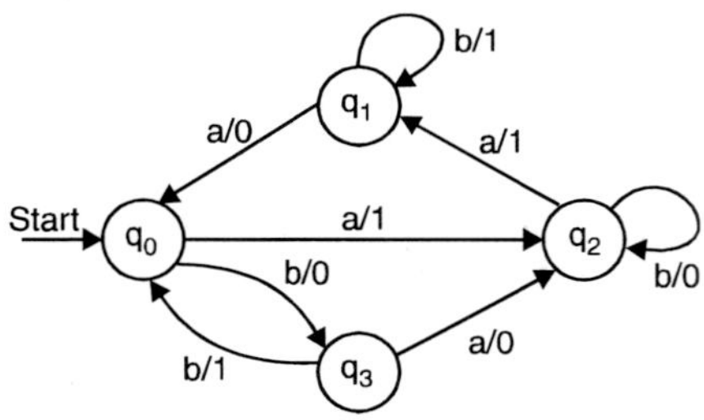

Fig. 3.19.

Solution. We can begin the conversion process anywhere because the algorithm does not specify the order of replacing states, so let us start with q_0.

Two edges comes into this state, one labelled *a*/0 and one labelled *b*/1. Therefore, we need two copies of this state, one that prints a '0' called q_0' and one that print a 1 (called q_0^2). Both these states must be connected to q_2 through an edge labelled *a*/1 and to q_3 through an edge labelled *b*/0. There is no loop at q_0, so these two clones of q_0 are not connected to each other. The machine becomes,

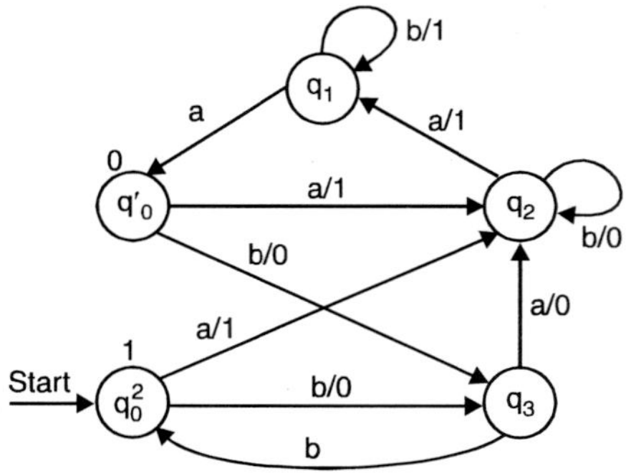

Fig. 3.20 (*a*).

Let us select arbitraily q_0^2 as start state.

Now consider state q_1, has only two edges coming into it : one from q_2 labelled $a/1$ and a loop labelled $b/1$. So whenever we enter q_1, we are always printing a 1. We have no trouble hour transfering the print instructions from the edges into state. Now machine looks like this :

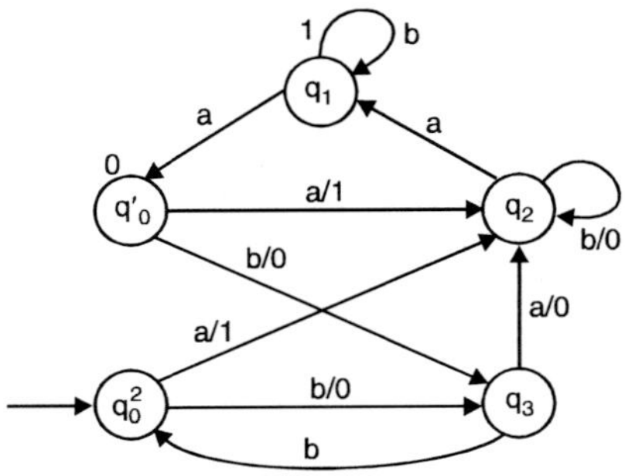

Fig. 3.20 (*b*).

Now handle the state q_3. Two edges come into it, both labelled $b/0$, so we change the state to q_3 (output is 0) and simply edges labels to b alone :

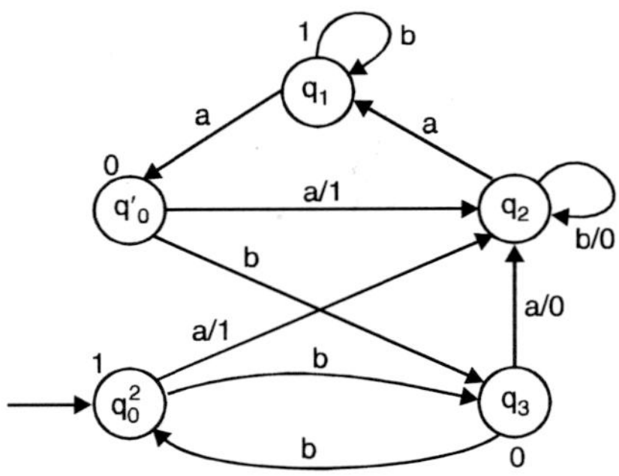

Fig. 3.20 (*c*).

Now consider last state q_2. It has some 0-printing edges entering it and some 1-printing edges. Therefore, we must split it into two copies, q_2^1 and q_2^2. Let us first print a 0 and the second print a_1. The two copies will be connected by a b-edge going from q_2^2 to q_2^1 (to print a 0). There will also be a loop at q_2^1. The final M_0 will be following :

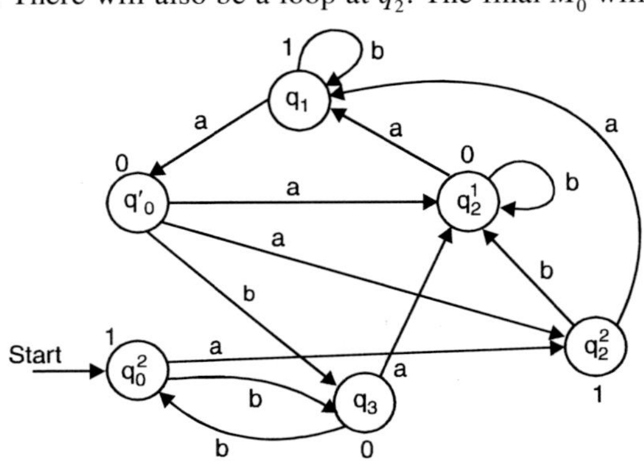

Fig. 3.20 (*d*). Equivalent moore machine.

Example 3.11. Differentiate Between Moore Machine and Melay Machine. **[UPTU 2004-05]**

Solution.

S. No.	*Moore Machine*	*Melay Machine*
1.	Its output depends only on present state.	Its output depends on the transition (input) and present state.
2.	Its transition function is excess into Q (finite set of states).	Its transition function is excess into Δ (output alphabet).
3.	If input string is of length n the output string is of length $n + 1$.	It input string is of length n then the output string is of length n.
4.	At the different input on the same state, its output is same.	At the different input on the same state, its output is also different.

EXERCISES

1. Each of the following is a moore machine with alphabet $\Sigma = \{a, b\}$ and output alphabet $\Delta = \{0, 1\}$. Given the transition and output tables. Draw the machines.

(*i*)

	a	*b*	*output*
q_0	q_1	q_2	1
q_1	q_1	q_1	0
q_2	q_1	q_0	1

(*iii*)

	a	*b*	*output*
q_0	q_3	q_2	0
q_1	q_1	q_0	0
q_2	q_2	q_3	1
q_3	q_0	q_1	0

(*ii*)

	a	*b*	*output*
q_0	q_0	q_2	0
q_1	q_1	q_0	1
q_2	q_2	q_1	1

(*iv*)

	a	*b*	*output*
q_0	q_1	q_2	0
q_1	q_2	q_3	0
q_2	q_3	q_4	1
q_3	q_4	q_4	0
q_4	q_0	q_0	0

2. Convert the following moore machine into mealy machines :

(*i*)
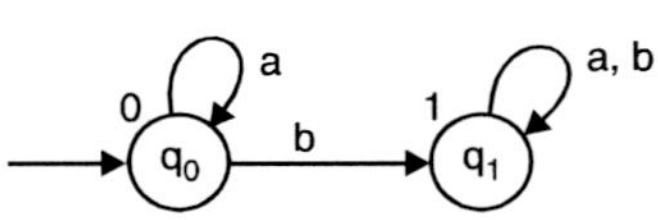

(*ii*)
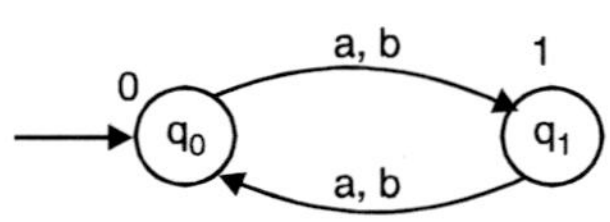

(*iii*)
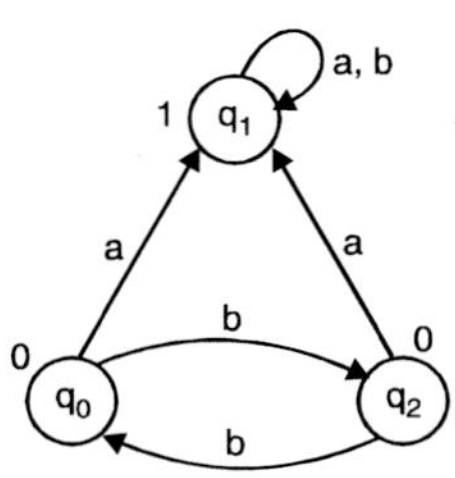

(*iv*)
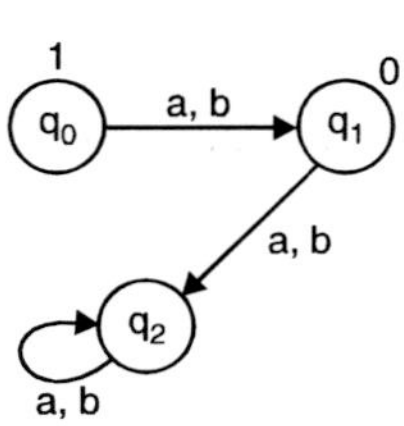

(*v*)
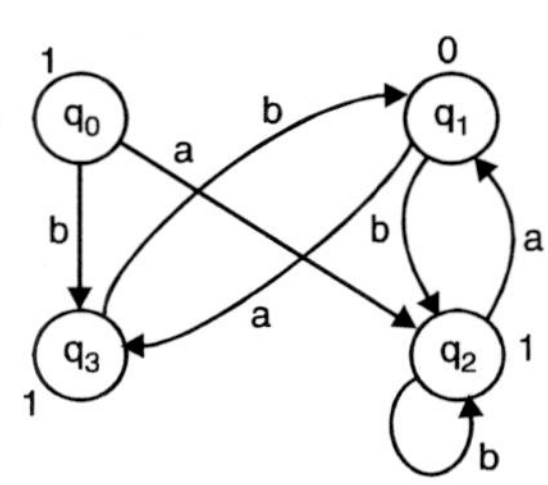

3. One each of the moore machine in problem 1 and 2, run the input sequence *aabab*. What are their respective outputs ?
4. Convert the following mealy machine into moore machines.

(*i*)

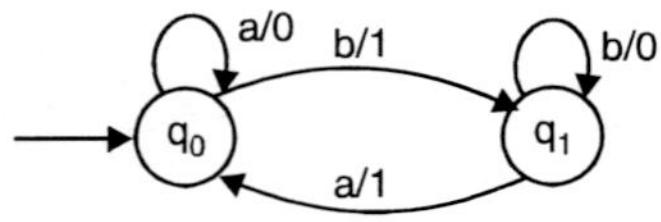

(*iii*)

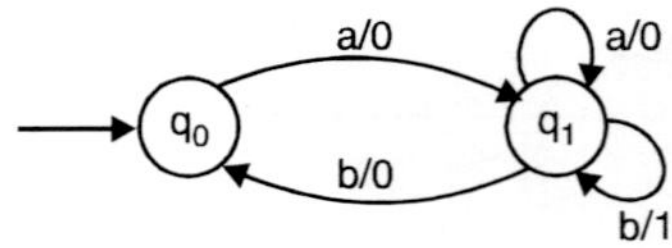

(*ii*)

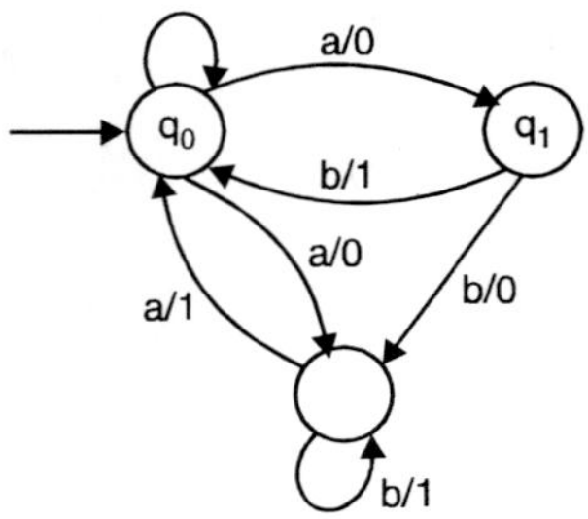

(*iv*)

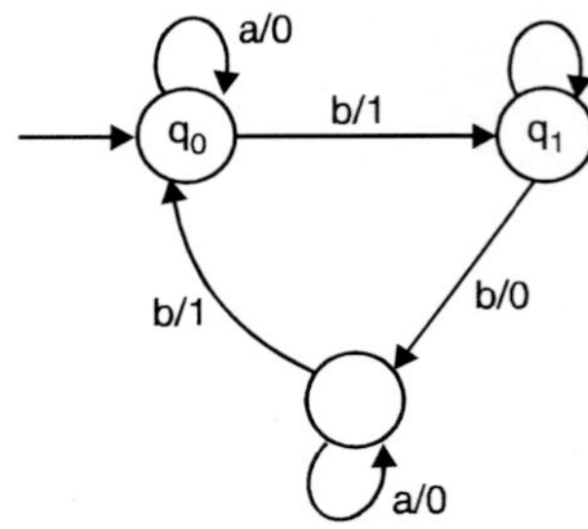

5. Design a machine to perform a parity check on the input string; that is, the output string ends in 1 if the total number of 1-bits in the input string is odd so if the total number of 1-bits in the string is even (the front part of the output string is ignored).
6. Construct a moore machine which converts binary number to octal number.
7. Give the mealys moore machine for the following processes. For input from $(0 + 1)^*$, if the input ends in 101, output *A*; if the input ends in 110 output *B*, otherwise output *C*.
8. Construct a melay machine which can output EVEN(E) ODD(0) according as the total number of 1's encountered is even or odd. The input symbols are 0 and 1.
9. Given a bit string of length '*n*', the shift-left-cyclic operation places the first bit at the end, leaving the rest of the bits unchanged. For example, SLC (100110) = 001101, build a mealy machine with input and output alphabet {0, 1, \$} such that for any string *x* when we input the *n* + 1 bits as *x*\$, we get output the *n* + 1 bit string \$SLC(*x*).

*****

SECTION-B

- ➤ Properties of Regular Sets
- ➤ Context-Free Grammars and Languages
- ➤ Simplified Context-Free Grammars and Its Normal Form

SYLLABUS

Properties of Regular Sets : The Pumping Lemma for Regular Sets, Applications of the Pumping Lemma, Closure Properties of Regular Sets, Myhill-Nerode Theorem and Minimization of Finite Automata, Minimization Algorithm.

Grammars : Definition, Context Free and Context Sensitive Grammar, Ambiguity Regular Grammar, Reduced Forms, Removal of Useless Symbols and Unit Production, Chomsky Normal From (CNF), Griebach Normal From (GNF).

4

CHAPTER

Properties of Regular Sets

4.1. INTRODUCTION

This chapter explores the properties of regular languages. Our first tool for this exploration is a way to prove that certain languages are not regular. One important kind of fact about the regular languages is called a "closure property." These properties let us build recognize for languages that are constructed from other languages by certain operations. As an example, the intersection of two regular languages is also regular. Thus, given automata that recognize two different regular languages, we can construct mechanically or automation that recognizes exactly the intersection of these two languages.

Some other important facts about regular languages are called "decision properties." Our study of these properties gives us algorithms for answering important questions about automata.

4.2. PROVING LANGUAGES NOT TO BE REGULAR

We have established that the class of Languages known as the regular languages for at least four different descriptions. They are the languages accepted by DFA's, NFA's, and by $\in$-NFA's; they are also the languages defined by regular expression.

Not every language is regular. In this section, we shall introduce a powerful technique, known as the "Pumping Lemma", for showing certain language not to be regular.

4.2.1. The Pumping Lemma for Regular Languages

Pumping Lemma is a powerful tool for proving certain language non-regular. It's also useful in the development of algorithm to answer certain questions concerning, finite automata, such as weather the language accepted by a given FA is finite or infinite.

Statement : Let L be a regular language. Then there exist a constant n (which depends on L) such that for every string w in L, such that $| w | \geq n$, we can break w into three substrings, $w = xyz$, such that :

(1) $y \neq \in$

(2) $| xy | \leq n$

(3) For all $i \geq 0$, the string xy^iz is also in L.

That is we always find a nonempty string y not too far from the beginning of w that can be "pumped"; that is, repeating y any number of times, keeps the resulting string in the language.

Proof. Suppose L is regular. Then $L = L(M)$ for some DFA, M. Suppose M has n states. Now, consider any string w of length n or more, say $w = a_1 a_2, ... a_m$, where $m \geq n$ and each a_i is an input symbol, for $i = 0, 1, 2, ..., m$ define state p_i to be $\hat{\delta}(q_0, a_1a_2 ... a_i)$, where δ is the transition function of M, and q_0 is the start state of M. That is p_i is the state, M is in after reading the first i symbols of w.

By the pigeonhole principle, it is not possible for the $n + 1$ different p_i's for $i = 0, 1, ..., n$ to be distinct, since there are only n different states. Thus we can find two different integers i and j, with $0 \leq i < j \leq n$, such that $p_i = p_j$. Now we can break $w = xyz$ as follows :

1. $x = a_1a_2 ... a_i$.
2. $y = a_{i+1} a_{i+2} ... a_j$.
3. $z = a_{j+1} a_{j+2} ... a_m$.

That is x takes us to p_i once; y takes us from p_i back to p_i (since p_i is also p_j), and z is balance of w. The relationship among the strings and states are discussed in Fig. 4.1.

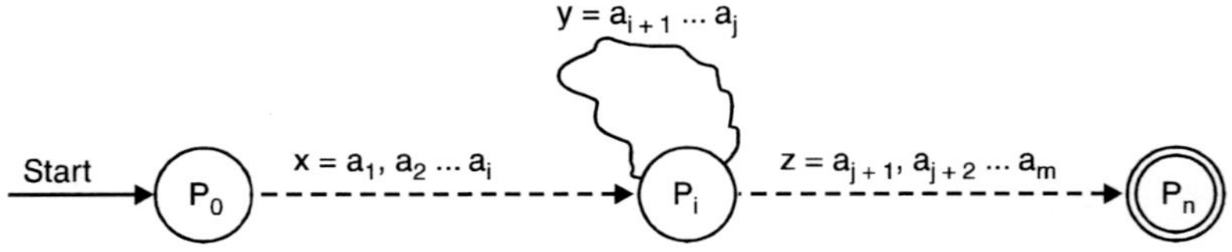

Fig. 4.1.

Note that x may be empty, in the case that $i = 0$. Also, z may be empty if $j = n = m$. However y cannot be empty, since i is strictly less than j.

Now consider what happens if the automation M receives xy^iz for any $i \geq 0$. If $i = 0$, then automation M goes from the start state q_0 to P_i on input x. Since p_i is also p_j, it must be that M goes from p_i to the accepting state in Fig. 4.1 on input z. Thus, in accepts xz.

If $i > 0$, then M goes from p_0 to p_i on input string x, circles from p_i to p_i, i times on input y_i, and then goes to the accepting state on input z. Thus, for any $i \geq 0$, xy^iz is also accepted by M; that is, xy^iz is in L.

We can also understand these transactions as follows :

For $i = 0$

$$\hat{\delta}(p_0, x) = \hat{\delta}(p_i, x) = p_i$$

$$\hat{\delta}(p_i, z) = P_n$$

For $i > 0$

$$\hat{\delta}(q_0, x) = \hat{\delta}(p_0, x) = p_i$$

$$\hat{\delta}(p_i, y) = p_i$$

In general $\hat{\delta}(p_i, y^i_{i\geq 0}) = p_i$

$$\hat{\delta}(p_i, z) = p_n$$

4.2.2. Application of the Pumping Lemma

The pumping lemma is extremely useful in proving that certain sets are non-regular. The general methodology in its application is an "adversary argument" of the following :

1. Select the language L, you wish to prove non-regular.
2. The "adversary" picks n, the constant mentioned in the pumping lemma. Once the adversary has picked n, he may not change it.
3. Select a string z in L. Your choice may depend implicitly on the value of n chosen.
4. The adversary breaks z into u, v and w, subject to the constants that $| uv | \leq n$ and $| v | \geq 1$.
5. You achieve a contradiction to the pumping lemma by showing, for any u, v and w determined by the adversary that there exit i for which uv^iw is not in L. It may then concluded and L is not regular. Your selection of i may depend on, n, u, v and w.

Now let us discuss some example to see the application of pumping lemma.

Example 4.1. Prove that Language $L = \{a^nb^n$ for $n = 0, 1, 2, 3, ... \}$ is not regular.

Solution. Let us see how we could apply the pumping lemma directly to this case.

The pumping lemma says that there must be strings x, y and z such that all words of the form xy^nz are in L.

How do we break this into three pieces x, y and z ?

Case 1. If middle part y is made off entirely of a's, as

$$x\ aaaa\ ...\ z$$

How if we pump it as $xyyz$, $xyyyz$ then number of a's increases, but in language $L = \{a^nb^n$ for $n = 0, 1, 2, 3, ...\}$ a's and b's are equal so it is not allowed.

Case 2. If middle part y is made off entirely of b's as

$$x\ bbbbb\ ...\ z$$

For the same reason, it is also not allowed.

Case 3. y part is made of some positive number of a's and some positive number of b's. This would mean that y contain the substring ab.

$$x \ldots aaababbb \ldots z$$

Then $xyyz$ would have two copies of the substring ab. But every word in L contains substring ab exactly once. Therefore $xyyz$ cannot be a word in L.

This proves that the pumping lemma cannot apply to L and therefore L is not regular.

Example 4.2. Prove that $L = \{a^n b\, a^n$ for $n = 0, 1, 2, \ldots\}$ is not regular.

Solution. If this language were regular then there would exist three strings x, y and z such that xyz and $xyyz$ were both words in this language. We can show that this is impossible.

Observation 1. If the y string contained the b, then $xyyz$ would contain two b's which is not allowed, according to the language L.

Observation 2. If the y string is all a's than the b in the middle of the word xyz is in the x-side or z-side. In either case, $xyyz$ has increased the number of a's either infront of the b or after b, but not both.

Conclusion. Therefore, $xyyz$ does not have its b in the middle and is not in the form $a^n ba^n$. This language cannot be pumped and is therefore not regular.

Example 4.3. Prove that $L = \{a^n b^n\, ab^{n+1}$ for $n = 1, 2, 3, \ldots\}$ is not regular.

Solution. We are going to show that this language too is not regular by showing that if xyz is in this language for any three strings x, y and z, then $xyyz$ is not in this language :

Observation 1. For every word in this language, if we know the total number of a's, we can calculate the exact number of b's (twice the total number of a's – 1).

And conversely, if we know the total number of b's, we can uniquely calculate the number of a's (add 1 and divide by 2). So no two different words have the same number of a's and b's.

Observation 2. All words in this language have exactly two substrings two substrings equal to ab and one equal to ba.

Observation 3. If xyz and $xyyz$ are both in this language, then y cannot contain either substring ab or the substring ba because then $xyyz$ would have too many.

Conclusion 1. It must be a solid dump of a's or solid dump of b's.

Conclusion 2. If y is solid a's, then xyz and $xyyz$ are different words with same total b's, violating observation 1. If y is solid b's, then xyz and $xyyz$ are different words with the same number of a's violating observation 1.

Conclusion 3. It is impossible for both xyz and $xyyz$ to be in this language for any string x, y and z. Therefore, the language is unpumpable and not regular.

Example 4.4. Prove that language which contains set of strings of balanced parentheses is not regular.

Solution. Let us suppose language is regular, and n be a constant according to pumping lemma. We can divide each string in three parts xyz as $|xy| \leq n$ and $y \neq \in$. So xy^iz must be in L for $i \geq 0$.

Observation. It is clear that language will contains the strings like (( ) ( ) ( )), ( ) ( ) ( ), (((( )))) ....

Let us select a string w = (((... ( ) ....))), where there are n left parenthesis by n right parenthesis. This string is in L, and is of length at least n. So we can divide w in three parts $w \simeq xyz$ as $|xy| \leq n$ and $y \neq \in$.

Conclusion. Not let us choose $i = 0$. The resulting string xz is not in L, reason is very clear.

Since $w \simeq xyz$ $|xy| < n$.

It means y will be made of only left parenthesis. So xz is not in L since it has fewer left parenthesis then right parenthesis. This contradict the pumping lemma, so our original assumption, that L was regular, must have been incorrect.

Example 4.5. Prove that $L = \{0^n 1^m 2^n, n, m \geq 0\}$ is not regular.

Solution. We assume that L is regular and try to obtain a contradiction.

Observation. Suppose L is regular, let n be constant provided by the pumping lemma. Let w be the string $0^n 2^n$. (Here $m = 0$ which is allowed). This string is in L, and is of length at least $2n$.

So we can write $w \simeq xyz$, $|xy| \leq n$
$y \neq \in$

Clearly xy will contain only 0's since in $w = 0^n 2^n$ first n placed are occupied by the 0's only. So y will also made of 0's only.

Conclusion. According to pumping lemma $xy^i z$ is also in L for $i \geq 0$. Let us choose $i = 0$. The resulting string xz, is not in L since it has fewer 0's then 1's. This contradicts the pumping lemma, so our original assumption that L was regular, must have been incorrect.

Example 4.6. Prove that language $L = \{0^n 1^m \mid n \leq m\}$ is not regular.

Solution. Let us analyse the language first, $L = \{0^n 1^m \mid n \leq m\}$ means that it contains set of strings, like 01, 011, 0111, 0011, 00111, ...

We will adopt same approach as we assume L is regular then try to obtain a contradiction to pumping lemma.

Let L be regular, and p be the constant provided by the pumping lemma. Let w be the string of L

$$w = 0^p 1^p.$$

The string w is in L and is of length at least p.

Let us divide w in three parts $w = xyz$, $|xy| \leq p$ and y never empty.

Observation. Since $|xy| < p$, y consists entirely of 0's and since y is not $\in$ there is at least one 0 in y. If $xyz \in L$ then according to pumping lemma $xy^k z$ must be in L for all $k \geq 0$.

Conclusion. So let us choose k to be 2. The resulting string $xyyz$, is not in L, since it has more 0's than 1's. This contradicts the pumping lemma, so our original assumption, that L was regular, must have been incorrect.

Example 4.7. Prove that $L = \{0^n 1^{2n} \mid n \geq 1\}$ is not regular.

Solution. Let us assume that L is regular and p be the constant provided by the pumping lemma. Let w be the string of L

$$w = 0^p 1^{2p}$$

Observation. The string w is in L, and its length at least p. So w can be written as xyz with $| xy | \leq p$ and y not $\in$ according to pumping lemma if xyz is in L then $xy^k z$ is in L for $k \geq 0$.

Conclusion. Since $| xy | \leq p$, y consists entirely of 0's, and since y is not $\in$ there is at least one 0 in y. So now let us choose k to be $2p + 1$. (Note that we are allowed to have our k depend on the p. The value p was given to us by pumping lemma.)

$$xyz \in L$$

$\Rightarrow$ $$xy^k z \in L$$

Now $$xy^{2p+1} z \notin L$$

Since the number of 0's is at least $2p + 1$ but the number of 1's is exactly $2p$.

This contradicts the pumping lemma, so our original assumption, that L was regular, must have been incorrect.

Example 4.8. Prove that language $L = \{0^n \mid n$ is a perfect square$\}$ is not regular.

Solution. Let us suppose L be regular, let p be a constant provided by the pumping lemma. Let w be the string O^{P2}. This string is in L and is of length at least p. So w can be written as xyz with $| xy | \leq p$ and y not $\in$, and each $xy^k z$ also in L. Since $| xy | \leq p$, y consists of no more than p 0's, and since y is not $\in$ so there is atleast one '0' in y.

So now let us choose k to be 2. We claim that the resulting string $xyyz$ is not in L.

Now let us see the proof of this claim :

We want to show that the length of $xyyz$ is not a perfect square. In fact what we do is show that it must lie strictly in between 2 consecutive perfect squares, namely, p^2 and $(p - 1)^2$.

So let q stand for the length of y. Then the length of $xyyz$ is $p^2 + q$. So it suffices to show that $p^2 < p^2 + q < (p + 1)^2$. The first in equality holds since $q > 0$ (since y was not $\in$). For second inequality, we compute : $(p + 1)^2 = p^2 + 2p + 1$, and this greater than $p^2 + q$ since $q < p$ (since $| xy | \leq p$). The claim is established, so $xyyz$ is not in L.

This contradicts the pumping lemma, so our original assumption, that L was regular, must have been incorrect.

Example 4.9. Prove that $L = \{0^n \mid n$ is perfect cube$\}$ is not regular.

Solution. We will apply same approach that is we assume that L is regular and try to obtain contradiction by applying pumping lemma.

Observation. Let us assume L is regular, let p be a constant provided by the pumping lemma. Let w be the string O^{P3}. This string is in L, and is of length at least p. So w can be written as xyz with $| xy | \leq p$ and y not $\in$.

Conclusion. Pumping lemma say that if $xyz \in L$ then $xy^k z$ also in L. y contains at least one 0 in y since y is not $\in$.

Let us choose $K = 2$. We claim that resulting string $xyyz$ is not a perfect cube.

Now let us see the proof for this claim :

We want to show that the length of $xyyz$ is not a perfect cube. Let q be the length of y, then the length of $xyyz$ is $p^3 + q$. So it suffices to show that $p^3 < p^3 + q < (p + 1)^3$. The first inequality holds since $q > 0$ (since y was not $\in$). For second inequality, we compute $(p + 1)^3 = p^3 + 3p^2 + 3p + 1$, and this is greater than $p^3 + q$ since $q < p$ (since $|xy| \leq p$). The claim is established, so $xyyz$ is not in L.

This contradicts the pumping lemma, so our original assumption, that L was regular, must have been incorrect.

Example 4.10. Let

$$\Sigma = \{a, b\}. \text{ Show that}$$

$$L = \{w \in \Sigma^* : n_a(w) < n_b(w)\}$$

is not regular.

Solution. Let us assume that L is regular and p is a constant provided by the pumping lemma. Let $w = a^p b^{p+1}$ be a string in L.

Observation. Let us write w as xyz where $|xy| \leq p$ and y not $\in$. w is in L and y consists of no more than p a's since $|xy| \leq p$. According to pumping lemma xy^kz will be in L for $k \geq 0$.

Conclusion. Let us choose $k = 2$ then $xyyz$ is not in L, since as we pump y by two then number of a's increases which violates the condition $n_a(w) < n_b(w)$.

So we got a contradiction for the pumping lemma so our assumption that L is regular must have been incorrect.

Example 4.11. Prove that set of all strings begining with a nonempty string of the form ww, is not regular. It is given that $\Sigma = \{0, 1\}$.

Solution. Let us assume that given language L is regular. Because the regular language over any alphabet are closed under intersection (we will discuss this theorem in Section 4.3).

So clearly language $L' = L \cap L$ is regular.

Now let us assume that n be constant given by the pumping lemma for language L'.

Let $$w = 10^n\,10^n \in L'$$

By the pumping lemma, we can write $w = xyz$ such that $xy \leq n$, $y \neq \in$, and for all $k \in N$, $xy^kz \in L'$. Now let us consider xy^2z (that is choose $k = 2$).

Conclusion. If y contains a 1, then xy^2z certain more than two 1's and hence is not in L'.

If y contains no 1's then $xy^2z = 10^{n+1}10^n$ for $i > 1$, and hence is not in L' (since $y \neq \in$).

In either case, we obtain a contradiction. We conclude that L is not regular.

Example 4.12. Prove that language L, which is the set of all strings beginning with a string w of length at least 3 such that $w = w^R$, is not regular. It is given that alphabet $\Sigma = \{0, 1\}$.

Solution. Let us assume L is not regular. Because the set of regular languages over any alphabet is closed under intersection, the language $L' = L \cap L$ is regular, (let say 001* 01*00).

Let n be the constant guaranteed by the pumping lemma for L'. Let $w = 001^{n+1}01^{n+1}00 \in L'$. By the pumping lemma, we can write $w = xyz$ such that $xy \leq n$, $y \neq \in$ and for all $k \in N$, $xy^kz \in L'$.

Conclusion. Because every string in L' has exactly five 0s, y can contain no 0's, for otherwise, xy^0z would not be in L'. Therefore $xy^iz = 001^{n+i+1} + 01^{n+1}00$ for some $i > 0$. Clearly xy^2z contains no prefix of length at least three that is palindrome so $xy^2z \notin L'$—a contraction. Therefore, L is not regular.

Example 4.13. Prove that language $L = \{0^{2^K} \mid K \in N\}$ is not regular.

Solution. By contradiction.

Let us assume that L is regular. Let n be constant provided by the pumping lemma.

Let $w = 0^{2^i}$ for some $i \in N$ such that $2^i > n$. Clearly $w \in L$, by the pumping lemma we can write $w = xyz$ such that $\mid xy \mid \leq n$, $y \neq \in$, and for all k, $k \in N$, $xy^kz \in L$.

Clearly, $0 < \mid y \mid < 2^i$, by adding 2^i to each term, we obtained

$$2^i < \mid y \mid + 2^i < 2^{i+1}$$

Thus

$$xy^2z = 0^{2^{i+\mid y \mid}}$$

$$\notin L$$

a contradiction. Therefore L is not regular.

Example 4.14. Prove that language $L = \{0^K\}$ K is prime number $\}$ is not regular.

Solution. By contradiction. Let us assume that L is regular, let n be constant guaranteed by pumping lemma. Let $w = 0^P$ where p is some prime number so smaller then n. Clearly w is in L.

By pumping lemma we can write $w = xyz$ such that $\mid xy \mid < n$, $y \neq \in$ and for all $k \in N$ $xy^kz \in L$. Then

$$xy^{p+1}z = 0^{p+p\mid y \mid}$$

$$= 0^{P(1+\mid y \mid)}$$

$$\notin L \text{ because } \mid y \mid > 0$$

a contradiction. Therefore, L is not regular.

Example 4.15. Prove that $L = \{x \in (0, 1)^* \mid x = x^R\}$ is not regular.

Solution. We will apply same approach of proving by contradiction.

Let us assume that L is regular and n be a constant guaranteed by pumping lemma.

Let $w = 0^n 10^n \in L$.

By the pumping lemma we can write

$w = xyz$ such that $\mid xy \mid \leq n$, $y \notin \in$ and for all $k \in N$ $xy^kz \in L$.

Because $\mid xy \mid \leq n$ we have

$$xy^2z = 0^{n+\mid y \mid}10^n \quad \text{For } k = 2$$

$$\notin L \text{ because } \mid y \mid > 0$$

So it is contradiction, hence L is not regular.

4.3. CLOSURE PROPERTIES OF REGULAR LANGUAGES

In this section, we shall prove several theorems of the form "if certain languages are regular, and a language L is formed from them by certain operations, then L is also regular. These theorems are often called closure properties of the regular languages, since they show that the class of regular languages is closed under the operation mentioned. Here is a summary of the principle closure properties of regular languages.

1. The union of two regular languages is regular.
2. The intersection of two regular languages is regular.
3. The complement of a regular language is regular.
4. The difference of two regular languages is regular.
5. The reversal of a regular language is regular.
6. The closure of a regular language is regular.
7. The concatenation of regular language is regular.
8. A homomorphism (substitution of strings for symbols) of a regular language is regular.
9. The inverse homomorphism of a regular language is regular.

Theorem 4.1. If L_1 and L_2 are regular languages, then so are $L_1 \cup L_2$, L_1L_2 and L_1^*.

We can say that family of regular language is closed under union, concatenation, and star-closure.

Proof. If L_1 and L_2 are regular, then there exist regular expression r_1 and r_2 such that $L_1 = L(r_1)$ and $L_2 = L(r_2)$.

By the definition, $r_1 + r_2$, r_1r_2 and r_1^* are regular expression denoting the languages $L_1 \cup L_2$, L_1L_2, and L_1^*, respectively. Thus, closure under union, concatenation and star-closure is immediate.

Theorem 4.2. If L is regular then complement of L that is $\bar{L}$ is also regular. In other words, the set of regular languages is closed under complementation.

Proof. To show closure under complementation, let $M = (Q, \Sigma, \delta, q_0, F)$ be a *dfa* that accepts L, then *dfa*

$$\bar{M} = (Q, \Sigma, \delta, q_0, Q - F)$$

accepts $\bar{L}$.

We can justify it as if M accepts the language L then some of the states of this FA are final states and, must likely, some are not. Let us reverse the final status of each state, that is if it was a final state, make it a non-final state, and if it was a non-final state, make it a final state. If an input string formally ended in a non-final state, it now ends in a final state and vice versa. This new machine we have built accepts all input strings that were not accepted by the original FA (all the words in $\bar{L}$) and rejects all the input strings that the FA used to accept (the words in L). Therefore in machine accepts exactly the language $\bar{L}$. So $\bar{L}$ is also regular language.

Example 4.16. An FA that accepts only the strings *aba* and *abb* is shown adjoining figure, find FA for the complement of *L i.e.* accepted every string other then *aba* and *abb*.

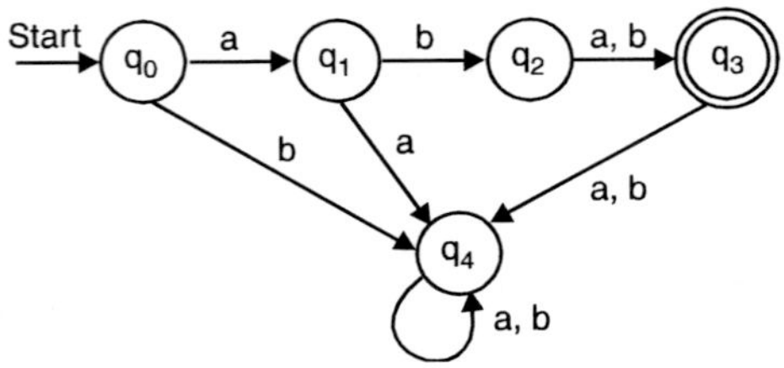

Fig. 4.2.

Solution. Apply Theorem 4.2. to find out the required FA which accepts strings other then *aba* and *abb* that is every non-final state will be final state and every final state will be non-final.

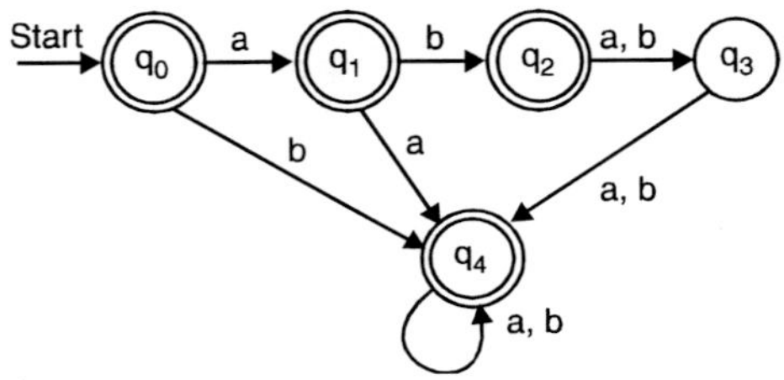

Fig. 4.3. DFA for the complements of *dfa* in figure 4.2.

Theorem 4.3. If L_1 and L_2 are regular languages, then $L_1 \cap L_2$ is also regular language. In other words the set of the regular language is closed under intersection.

Proof. By Demorgon's law for sets of any kind (regular languages or not) :

$$L_1 \cap L_2 = \overline{(\bar{L}_1 + \bar{L}_2)}$$

This can be justify by the Venn diagrams as follows :

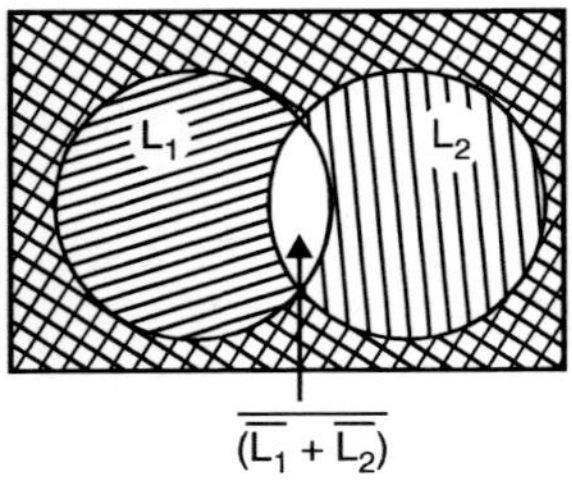

Fig. 4.4.

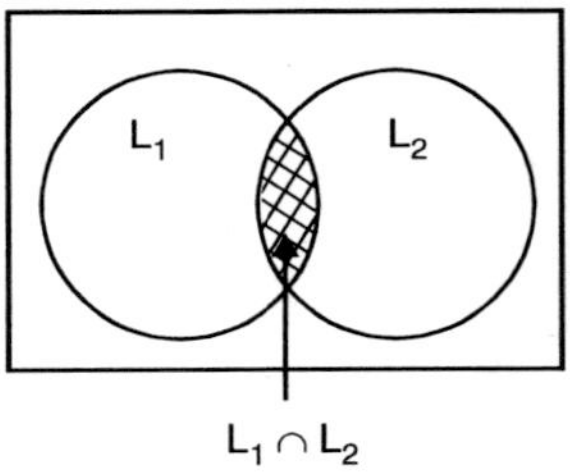

Fig. 4.5

So it is clear from Fig. 4.4 and Fig. 4.5 that

$$L_1 \cap L_2 = \overline{(\bar{L}_1 + \bar{L}_2)}$$

Because L_1 and L_2 are regular so $\bar{L}_1$ and $\bar{L}_2$ are regular (from Theorem 4.2). So is $\bar{L}_1 + \bar{L}_2$ (by Theorem 4.1). And because $\bar{L}_1 + \bar{L}_2$ is regular then so is $\overline{(\bar{L}_1 + \bar{L}_2)}$ which means $L_1 \cap L_2$ is regular.

Theorem 4.4. If L and m are regular languages, then so is $L - m$.

Proof. Observe that $L - m = L \cap \bar{m}$. Here $\bar{m}$ is also regular since m is regular (according to Theorem 4.2) and $L \cap \bar{m}$ is regular according to the Theorem 4.3. Therefore $L - m$ is regular.

4.3.1. Reversal

The reversal of a string $a_1\, a_2\, a_3 \ldots a_n$ is the string written backwards, that is $a^n\, a^{n-1} \ldots a_1$. We use w^R for the reversal of string w. Thus, 1101^R is 1011 and $\in^R = \in$.

TIPS

The reversal of a language L, written L^R, is the language consisting of the reversal of all its strings. For instance, if $L = \{001, 10, 111\}$, then $L^R = \{100, 01, 111\}$. Reversal is another operation that preserve regular languages; that is if L is regular language, so is L^R. There are two simple proofs, one based on automata and one based on regular expression.

Given a language L that is $L(A)$ for some finite automation, perhaps with non-determinism and $\in$-transitions, we may construct automation for L^R by :

1. Reverse all the arcs in the transition diagram for A.
2. Make the start of A be the only accepting state for the new automation.
3. Create a new start state p_0 with transition on $\in$ to all the accepting states of A.

The result is an automation that simulates A "in reverse", and therefore accepts a string w if and only if A accepts w^R.

Theorem 4.5. If L is a regular language, so is L^R.

Proof. Let us assume that L is defined by the regular expression r. The proof is a structural induction on the size of r. We show that there is another regular expression r^R such that $L(r^R) = (L(r))^R$; that is the language of r^R is the reversal of the language of r.

Basis. If r is $\in$, ϕ, or a, for some symbol a, then r^R is the same as r. That is, we know $\{\in\}^R = \{\in\}$, $\phi^R = \phi$ and $\{a\}^R = \{a\}$.

Induction. There are three cases, depending on the form of r.

1. $r = r_1 + r_2$. Then $r^R = r_1{}^R + r_2{}^R$. The justification is that the reversal of the union of two languages is obtained by computing the reversal of the two languages and taking the union of those languages.
2. $r = r_1 \,.\, r_2$. Then $r^R = r_2{}^R r_1{}^R$. Note that we reverse the order of the two languages, as reversing the language themselves.
3. $r = r_1{}^*$. Then $r^R = (r_1{}^R)^*$. The justification is that any string w in $L(r)$ can be written as $w_1, w_2, \ldots, w_n$, where each w_i is in $L(r)$. But

$$w^R = w_n{}^R w_{n-1}{}^R \ldots w_1{}^R.$$

Each w_i^R is in $L(r)^R$, so w^R is in $(r_1^R)^*$.

Example 4.17. Let L be defined by the regular expression $(0 + 1)0^*$, then find L^R.

Solution.

$$L = (0 + 1)\, 0^*$$

$L^R = (0^*)^R\ (0 + 1)^R$, by the rule for the concatenation.

4.3.2. Homomorphism

Definition A string homomorphism is a function on strings that works by substituting a particular string for each symbol.

Suppose Σ and Σ' are alphabets. Then function

$$h : \Sigma \longrightarrow \Sigma'$$

is called a "homomorphism".

In other words, a homomorphism is a substitution in which a single letter is replaced with a string. The domain of the function h is extended to strings in an obvious fashion, if

$$w = x_1 x_2 \ldots x_n,$$

then

$$h(w) = h(x_1)\, h(x_2)\, h(x_3) \ldots h(x_n)$$

If L is a language on Σ, then its homomorphic image is defined as

$$h(L) = \{h(w) : w \in L\}.$$

Example 4.18. Let $\Sigma = \{0, 1\}$ and $\Sigma' = \{0, 1, 2\}$ and defined n by

$$h(0) = 01$$

$$h(1) = 112$$

Find $h(010)$ and homomorphic image of $L = \{00, 010\}$.

Solution. Given $\Sigma = \{0, 1\}, \Sigma' = \{0, 1, 2\}$

h is defined as :

$$h(0) = 01$$

$$h(1) = 112$$

So $h(010) = 0111201$.

The homomorphic image of $L = \{00, 010\}$ is the language $h(L) = \{0101, 0111201\}$.

If we have a regular expression r for a language L, then a regular expression for $h(L)$ can be obtained by simply applying the homomorphism to each Σ symbol of r.

Example 4.19. $\Sigma = \{0, 1\}$, and $\Sigma' = \{1, 2, 3\}$. Define h by

$$h(0) = 3122$$

$$h(1)\ 132$$

if L is regular language denoted by

$$r = (0 + 1^*)\,(00)^*$$

the find out the regular expression for language $h(L)$.

Solution. Given

$$\Sigma = \{0, 1\}$$

$$\Sigma' = \{1, 2, 3\}$$

and h is defined as $\quad h(0) = 3122$

$$h(1) = 132$$

$r = (0 + 1^*)\ (00)^*$; where r is regular expression for language L.

Let regular expression for the language $h(L)$ be r', then

$$r' = (h(0) + h(1)^*)\ (h(0)\ h(0))^*$$

Let us put $h(0)$ and $h(1)$ in r'.

$$r' = (3122 + (132)^*)\ (3122\ 3122)^*$$

denotes then regular language $h(L)$.

Theorem 4.6. Let h be a homomorphism. If L is a regular, then its homomorphic image $h(L)$ is also regular. The family of regular languages is therefore closed under arbitrary homomorphisms.

Proof. Let L be regular language denoted by some regular expression r. We find $h(r)$ by substituting $h(a)$ for each symbol $a \in \Sigma$ or r. It can be shown directly by an appeal to the definition of a regular expression that the result is a regular expression. It is equally easy to see that expression denotes $h(L)$. All we need to do is to show that for every $w \in L(r)$. The corresponding $h(w)$ is in $L(h(r))$ and conversely that for every x in $L(h(r))$ there is a w in L, such that $x = h(w)$. Leaving the detail as an exercise, we claim that $h(L)$ is regular.

4.3.3. Inverse Homomorphism

Definition Homomorphism may also be applied "backwards," and in this made they also preserve regular language. That is, suppose h is a homomorphism from some alphabet Σ to strings in another (possible the same) alphabet Σ'. Let L be a language over alphabet Σ'. Then $h^{-1}(L)$, read "h inverse of L," is the set of strings w in Σ^* such that $h(w)$ is in L.

Figure 4.6 suggests the effect of a homomorphism on a language L in part (a), and the effect of an inverse homomorphism in part (b).

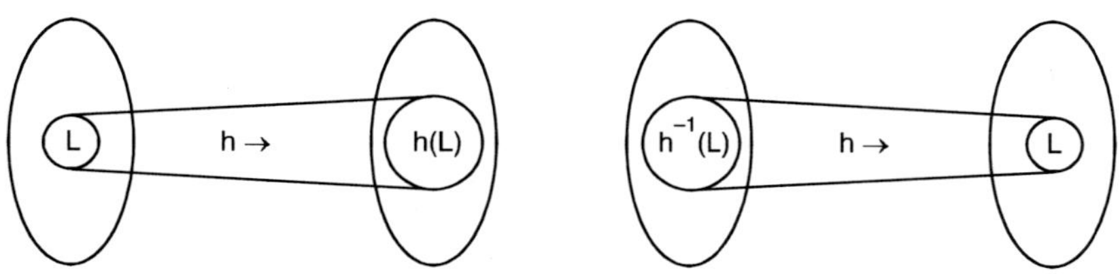

(*a*) *A homomorphism applied in forward direction.* (*b*) *A homomorphism applied into inverse direction.*

Fig. 4.6.

Theorem 4.7. If n is a homomorphism from alphabet Σ to alphabet Σ', and L is a regular over Σ', then $h^{-1}(L)$ is also regular language.

Proof. The proof starts with a DFA A for L. We construct from A and h a DFA for $h^{-1}(L)$ using the plan suggested by Fig. 4.7. This DFA uses the states of A but translates the input symbol according to h before deciding on the next state.

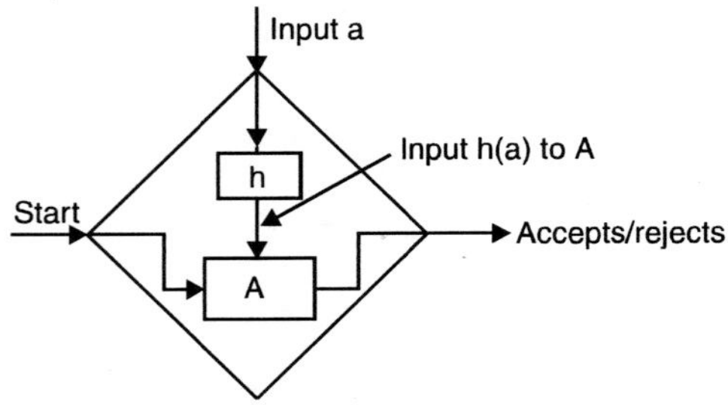

Fig. 4.7.

The DFA for $h^{-1}(L)$ applies to its input, and then simulates the DFA for L.

Formally, let L be $L(A)$, where DFA A = $(Q, \Sigma', \delta, q_0, F)$. Define a DFA $B = (Q, \Sigma, \delta', q_0, F)$, where transition function δ' is constructed by the rule $\delta'(q, a) = \hat{\delta}(q, h(a))$. That is, the transition B makes on input a is the result of the sequence of transitions that a makes on the string of symbols $h(a)$. Remember that $h(a)$ could be $\in$, it could be one symbol, or it could be many symbols, but δ is properly defined to take care of all these cases.

It is an easy induction on $| w |$ to show that $\hat{\delta}'(q_0, w) = \hat{\delta}(q_0, h(w))$. Since the accepting states of A and B are the same, B accepts w if and only if A accepts $h(w)$. Put another way, B accepts exactly those strings what are in $h^{-1}(L)$.

4.4. DECISION PROPERTIES OF REGULAR LANGUAGES

In this section we consider how one answers important questions about regular languages. First, we must consider what it means to ask a question about a language. The typical language is infinite, so you cannot present the strings of the language to someone and ask a question that requires them to inspect the infinite set of strings. Rather, we present a language by giving one of the finite representations for it that we have developed : a DFA, an NFA, an $\in$-NFA or a regular expression.

Of course the language so described will be regular and in fact there is no way at all to represent completely arbitrary languages. Let us consider some of the fundamental questions about languages :

1. Is the language described empty ?
2. Is a particular string w is in the described language ?
3. Do two descriptions of a language actually describe the same language ? This question is often called "Equivalence" of languages.

4.4.1. Testing Emptiness of Regular Languages

TIPS

If our representation is any kind of finite automation, the emptiness question is whether there is any path whatsoever from the start state to some accepting state, if so, language is nonempty, while if the accepting states are all separated from the start state, then the language is empty.

The start state is surely reachable from the start state, if there is an arc from q to p with any label (an input symbol, or $\in$ if the automation is an $\in$-NFA) then p is reachable. In that manner we can compute the set of reachable states. If any accepting state is among them, we answer "no" (the language of the automation is not empty), and otherwise we answer "yes". Note that the reachability calculation takes no more time that $0(n^2)$ if the automation has n states, and in fact it is no worse than proportional to the number of arcs in the automation is transition diagram, which could be less than n^2 and cannot be more than $0(n^2)$.

If we are given a regular expression representing the language L, rather than an automation, we could convert the expression to an $\in$-NFA and proceed as above. Since the automation that results from a regular expression of length n has at most $0(n)$ states and transitions, the algorithm takes $0(n)$ time.

4.4.2. Testing Membership in a Regular Language

The next question of importance is, given a string w and a regular language L, is w in L. While w is represented explicitly, L is represented by an automation or regular expression.

If L is represented by a DFA, the algorithm is simple. Simulate the DFA processing the string of input symbol w, beginining in the start state. If the DFA ends in an accepting state, the answer is "yes"; otherwise the answer is "no". If $| w | = n$, and the DFA is represented by a suitable data structure, such as two-dimensional array that is the transition table, then each transition requires constant time, and the entire test takes $0(n)$ time.

Theorem 4.8. Whether two regular sets are identical is solvable.

Proof. Let us take two finite automate (M_1 and M_2) and examine a picture of the sets they accept.

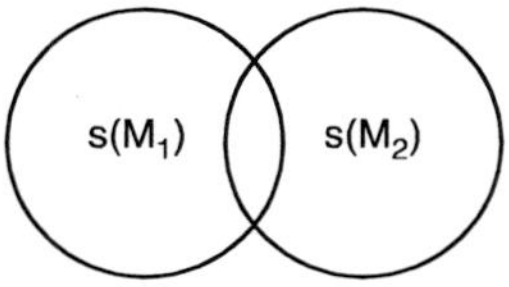

Fig. 4.8.

If the intersection is the same as both sets then indeed they are identical. Or, on the other hand, if the areas outside the intersection are empty then both sets ($s(M_1)$ and $s(M_2)$) are identical.

Let us examine these outside areas.

The picture in Fig. 4.9 (*a*) represents the set accepted by M_1 and rejected by M_2 while that in Fig. 4.9 (*b*) is the set which M_2 accepts and M_1 rejects.

$s(M_1) \cap \overline{s(M_2)}$ $\qquad$ $s(M_2) \cap \overline{s(M_1)}$

(*a*) $\qquad$ (*b*)

Fig. 4.9.

If these two areas (or sets) are empty then sets accepted by M_1 and M_2 are exactly the same. This means that the equivalence problem for $s(M_1)$ and $s(M_2)$ is exactly same as the emptiness problem for :

$$\left| s(M_1) \cap s(\bar{M}_2) \right| \cup \left| s(M_2) \cap s(\bar{M}_1) \right|$$

So, if we can solve the emptiness problem for the above set, then we can solve the equivalence problem for $s(M_1)$ and $s(M_2)$. Since the regular sets are closed under union, complement and intersection; the above set is a regular set. And we know that emptiness for the class of regular sets is solvable. Hence it proved.

Theorem 4.9. Whether a regular set is finite is solvable.

Proof. We know that if a finite automation accepts any strings at all then some will be of length less than the size of the state set. (This does not help directly but it gives a hint as to what we need for this theorem.)

If we were to find a string accepted by finite automation m which was longer than or equal to the size of it's state set, we could use the inflation aspect of the pumping lemma to show that machine m must accept on infinite number of strings. This means that :

"a finite automation accepts only strings of length less than the size of its set of states, if and only if it accepts a finite set."

Thus, to solve the finiteness problem for $M = (Q, \Sigma, \delta, q_0, F)$, we need to determine whether or not :

$L(M)$ – | strings of length $< | Q | | = \phi$

A question now arises as to how many inputs we must examine in order to tell it M will accept on input longer than or equal to the size of its state set. The answer is that we only need to consider input strings upto twice the size of the state set.

Theorem 4.10. If a finite automation accepts any strings, it will accepts one of length less than the size of its state set.

Proof. Let m be a finite automation which accepts the string x and that the length of x is no less than the size of M's state set. Assume further that M accepts no string shorter than x. This is the opposite of our theorem.)

Immediately the pumping lemma asserts that there are strings u, v and w such that $uvw = x$, $v \neq \in$ and $uw \in L(M)$. Since $v \neq \in$, uw is shorter than $uvw = x$. Thus M accepts shorter strings than x.

Example 4.20. Show that $L = \{a^i b^j c^k : 0 \leq k < j < k\}$ is not regular.

Solution. Assume that given language L is regular, n be an constant. L must follow pumping lemma as follows :

Let $\quad w = a^n b^{n+1} c^{n+2} \in L$

$\quad \simeq xyz, |y| > 0, |xy| \leq n.$

Observation. xy is a string of a's, so that the string $xy^2z = a^{n+|y|} b^{n+1} c^{n+2}$.

Since $|y| > 0$, the number of a's is not at least equal to the number of b's so that $xy^2z \notin L$. Hence L is not regular.

Example 4.21. The language $L = \{(ab)^n a^k / n > k, k \geq 0\}$ is not regular.

Solution. Again we can prove L non regular by the method of contradiction. Let L be regular then it must follow pumping lemma for regular language.

Let $\quad w = (ab)^{m+1} a^m \in L$, m is constant

$\simeq xyz, |y| > 0, [xy] \leq m$

w can be written as follows :

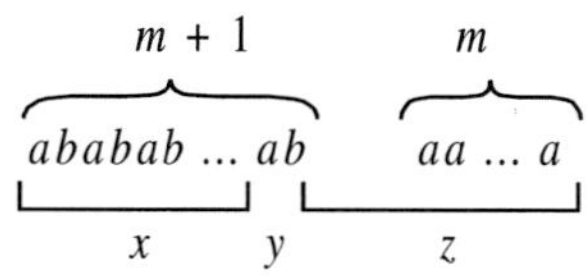

Now $|xy| \leq m$, $y = a$, $z = ba^m$. Choose $i = 0$. Then $w_i = xy^iz \Rightarrow w_0 = xz = (ab)^m ba^m \notin L$. Also if we pick $y = ab$, then we get the string $(ab)^m a^m \notin L$. In same way, we can deal with any possible choice for y, to prove L is not regular.

Example 4.22. The language $L = \{0^i w : i \geq 0, w \in (0, 1)^*, |w| \leq i\}$ is not regular.

Solution. L can be proved non regular by contradiction. Let L be regular then it must follow the pumping Lemma. Let string $w' = xyz = 0^i 1^{i-1} \in L$, $|xy| \leq i$, $|y| > 0$. Choose $x = 0^{i-2}$, $y = 0^2$, then $|xy| \leq i$. Now pumping Lemma says tha $xy^0z \in L$, this is a contradiction, because $xy^0z = 0^{i-2} 1^{i-1} \notin L$. Therefore L is not regular.

Example 4.23. Show that the language $L = \{a^m b^n : m \neq n\}$ is not regular.

Solution. Here choosing a string with $m = n + 1$, or $m = n + 2$, will not serve the purpose since y can always choose a decomposition that will make it impossible to pump the string out of the language (that is pump it so that if has an equal number of a's and b's). So let take $(m = l)!$, $n = (l + 1)!$. If we choose y of length $k < m$, we pump i times to generate a string with $l! + (i - 1)\, k$ a's. We get a contradiction of pumping Lemma if we can pick i such that $l! + (i - 1)k = (l + 1)!$. This is always possible since $i = 1 + \dfrac{l\,.\,l!}{k}$, and $k \leq l$. The right side is therefore an integer and hence violating the condition of the pumping lemma.

Example 4.24. Show that $L = \{a^{n!} : n \geq 0\}$, is not regular.

Solution. Let $w = a^{m!}$. The various decompositions of w obiously differ only in the lengths of the substrings. Suppose, if we choose y such that $|y| = k \leq m$, we look at xz which has length $m! - k$. This string is in L only if there exists a j such that $m! - k = j!$. But this is not possible, since for $m > 2$ and $k \leq m$, we have $m! - k > (m - 1)!$. Hence L is not regular.

4.5. MINIMIZATION OF DFA'S

Any DFA defines a unique language, but the converse is not true. For a given language, there may be many DFA's that accepts it. To under stand the minimization first of all one should be very clear about the dead state, inaccessible states and indistinguishable states, so let us discuss them first.

Dead state. All those non final states which transit to itself for all input symbols in Σ, are called Dead state.

Let $\Sigma = \{a, b\}$

then q is said to be dead state.

If q is not the member of F

and $\delta(q, a) = q$

i.e., $\delta(q, b) = q$

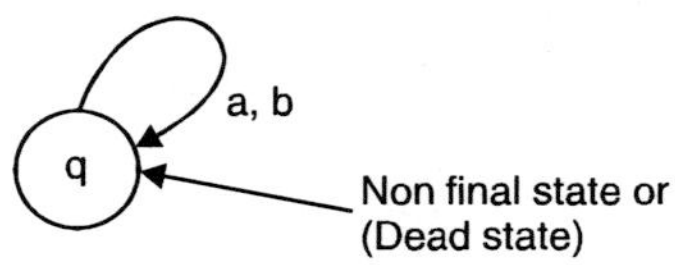

Fig. 4.10.

in the following Fig. 4.11 q_4 is dead state.

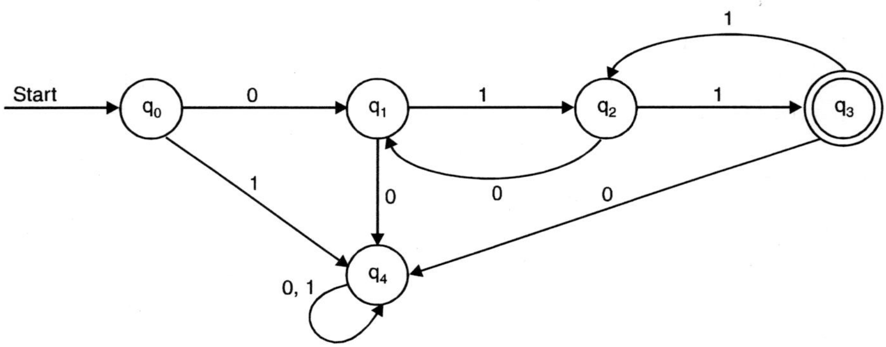

Fig. 4.11.

Inaccessible states. All those states which can never be reached from initial state are called inaccessible states. For example :

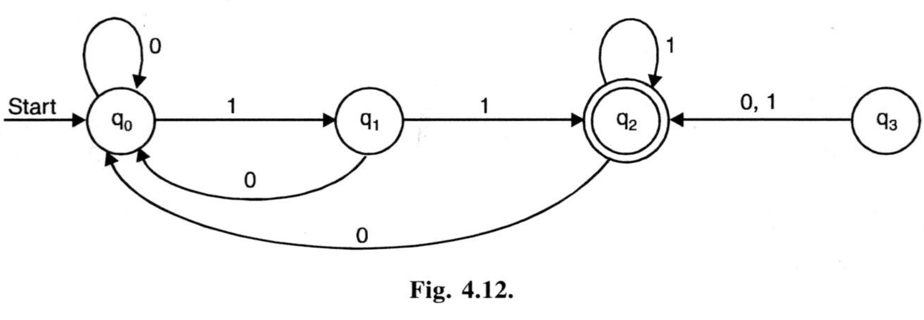

Fig. 4.12.

In Fig. 2.12, q_3 is the state which is not reachable from q_0 (initial state). So q_3 is the unreachable state.

Indistinguishable states. Two states q and r of DFA are called indistinguishable if

$$\hat{\delta}(q, w) \in F \Rightarrow \hat{\delta}(r, w) \in F$$

and $$\hat{\delta}(q, w) \notin F \Rightarrow \hat{\delta}(r, w) \notin F$$

for all $$w \in \Sigma^*.$$

Distinguishable states. Two states q_1 and q_2 of DFA are called distinguishable if

$$\hat{\delta}(q_1, w) \in F \text{ and } \hat{\delta}(q_2, w) \notin F \text{ or vice versa; for all } w \in \Sigma^*.$$

4.5.1. Minimization Algorithm for DFA

One important result of finite automata, both theoretically and practically, is that for any regular language there is a unique DFA having the smallest number of states that accepts it. Let $M = (Q, \Sigma, \delta, q_0, F)$ be a DFA that accepts a language L. Then the following algorithm produces the DFA, denote it by M_1, that has the smallest number of states among the DFAs that accept L.

Minimization Algorithm for DFA :

Step 1. All states that are unreachable from the initial state via any set of transitions of the DFA M are removed.

Step 2. Draw the transition table for rest states, after removing the unreachable states.

Step 3. Split the transition table in two tables T_1 and T_2, where

T_1 = {contains all rows which starts with states from Q-F}

T_2 = {contains all rows which starts with states from F}

Step 4. Find the similar rows from T_1 sauch that :

$$\delta(q, a) = P$$

$$\delta(r, a) = P$$

Where $q, r \in Q$-F, $a \in \Sigma$ and $P \in Q$, then rows starts with q and r are similar and we can remove one of them, let us assume we are removing row, which starts with r then we must replace r with q in other rows.

Step 5. Repeat step 4 till there is no similar rows are available in T_1.

Step 6. Repeat the step 4 and step 5 for Table T_2 also.

Step 7. Now combined the reduced T_1 and T_2. The combined transition table is transition table of minimized DFA.

Note : Basically by applying above algorithm, we are removing non-reachable states, dead states and indistinguishable states.

Example 4.25. Minimize the DFA given below.

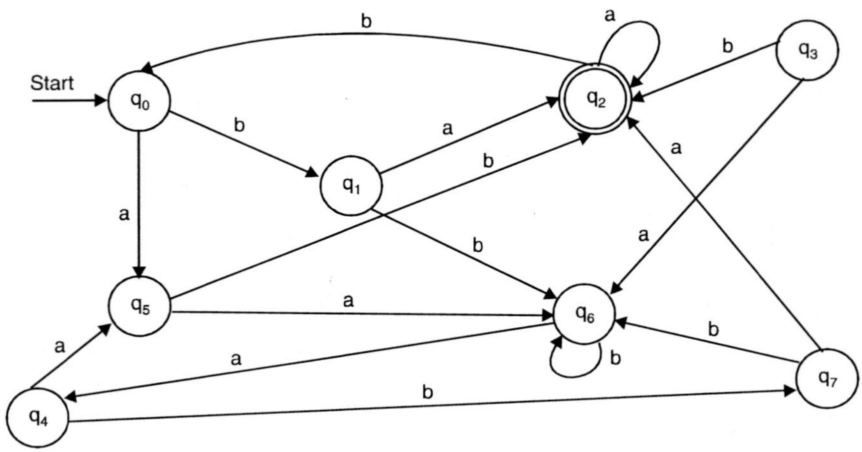

Fig. 4.13.

Solution. Let us discuss the solution step by step.

Step 1. Remove the unreachable states from the given DFA. The DFA given in Fig. 4.13, q_3 is unreachable state so remove it.

Step 2. Draw the transition table for rest states, after removing q_3. (unreachable states).

δ/Σ	a	b
$\rightarrow q_0$	q_5	q_1
q_1	q_2	q_6
$* q_2$	q_2	q_0
q_4	q_5	q_7
q_5	q_6	q_2
q_6	q_4	q_6
q_7	q_2	q_6

Step 3. Now divide rows of transition table in two sets as :

(*i*) One set contains only those rows, which starts from non-final states *i.e.*

q_0	q_5	q_1
q_1	q_2	q_6
q_4	q_5	q_7
q_5	q_6	q_2
q_6	q_4	q_6
q_7	q_2	q_6

Set-1 (say)

(*ii*) Other set contains only those rows, which starts from final states.

$q_2 \quad q_2 \quad q_0$ **Set-2 (say)**

Step 4. Apply step 4 on both the sets individually. Now first consider set-1

q_0	q_5	q_1	$\to$	row-1
q_1	q_2	q_6	$\to$	row-2
q_4	q_5	~~q_7~~ q_1	$\to$	row-3
q_5	q_6	q_2	$\to$	row-4
q_6	q_4	q_6	$\to$	row-5
~~q_7~~	~~q_2~~	~~q_6~~	$\to$	row-6

row 2 and row 6, are similar since q_1 and q_7 transit to same states on a and b. So skip one of them let skip row 6, then q_7 is replaced by q_1 in the rest. Write the rest as follows :

q_0	q_5	q_1	$\to$	row-1
q_1	q_2	q_6	$\to$	row-2
q_4	q_5	q_1	$\to$	row-3
q_5	q_6	q_2	$\to$	row-4
q_6	q_4	q_6	$\to$	row-5

Step 5. Repeat step 4 on the rest of set 1 till you are finding similar rows and write rest, now this set is minimised.

q_0	q_5	q_1
q_1	q_2	q_6
~~q_4~~	~~q_5~~	~~q_1~~
q_5	q_6	q_2
q_6	~~q_4~~ q_0	q_6

as row-1 and row-3 are similar

Now minimised set is

q_0	q_5	q_1
q_1	q_2	q_6
q_5	q_6	q_2
q_6	q_0	q_6

See, now there are no two rows are similar.

Step 6. Do same process for set 2 as :

$$q_2 \quad q_2 \quad q_0$$

So only single row, already minimised.

Step 7. Now combine set 1 and set 2 as :

δ/Σ	a	b
$\to q_0$	q_5	q_1
q_1	q_2	q_6
q_5	q_6	q_2
q_6	q_0	q_6
$*q_2$	q_2	q_6

Now it is the transition table of minimised DFA.

Transition diagram of minimised DFA :

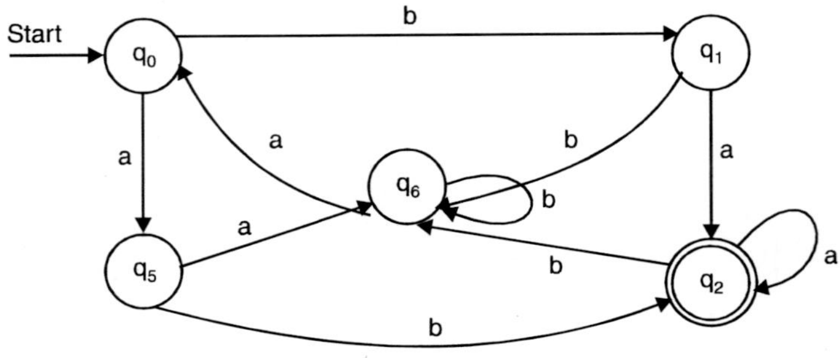

Fig. 4.14.

Example 4.26. Minimise the given DFA.

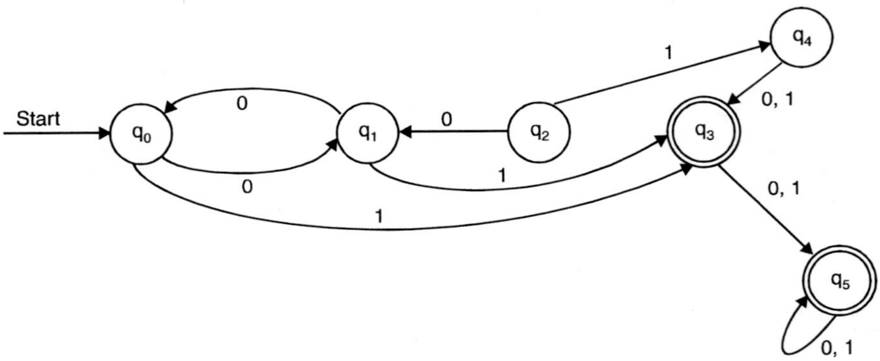

Fig. 4.15.

Solution. DFA given in Fig. 4.15 is having q_2 and q_4 as unreachable states, so remove them and draw the transition table for rest of the states.

	δ/Σ	0	1
set 1	→ q_0	q_1	q_3
	q_1	q_0	q_3
set 2	*q_3	q_5	q_5
	~~*q_5~~	~~q_5~~	~~q_5~~

Set 1 has no, two similar rows so we have to minimise only set 2. Now transition table for minimised DFA is as follows :

δ/Σ	0	1
→ q_0	q_1	q_3
q_1	q_0	q_3
* q_3	q_3	q_3

Transition diagram is as follows :

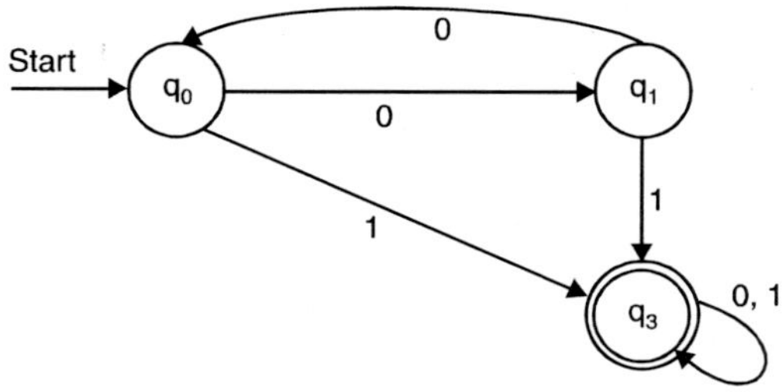

Fig. 4.16.

4.6. MYHILL-NERODE THEOREM

Myhill-Nerode theorem is used to eliminate useless states from a DFA. Before discussing the theorem, let us see some basics first.

Think of an equivalence relation as being true or false for a specific pair of strings x and y. Thus xRy is true for some set of pairs x and y. We will use a relation R such that $xRy \Leftrightarrow yRx$, x has a relation to y if and only if y has the same relation to x. This is known as symmetric.

xRy and $yRz \Rightarrow xRz$. This is known as transitive.

xRx is true. This is known as reflexive.

The notation R_L means an equivalence relation 'R' over the language L. The notation R_M means an equivalence relation R over machine M. We know for every regular language L there is a machine M that exactly accepts the strings in L.

Our R_L is defined $\quad xR_Ly \Leftrightarrow$ for all z in Σ^*

(xz in $L \Leftrightarrow yz$ in L)

Our R_M is defined $\quad xR_My \Leftrightarrow xzR_Myz$ for z in Σ^*.

In other words

$$\begin{aligned}\delta(q_0, xz) &= \delta(\delta(q_0, x), z)\\ &= \delta(\delta(q_0, y), z)\\ &= \delta(q_0, yz)\end{aligned}$$

for x, y and z strings in Σ^*.

R_M divides the set Σ^* into equivalence classes, one class for each state reachable in that particular state of M from the starting state q_0. To get R_L from this we have to consider only the final reachable states of M.

Statement of the Myhill-Nerode Theorem.

The Myhill-Nerode theorem says the following three statements are equivalent :

1. The set L, a subset of Σ^*, is accepted by a DFA. (We know this means L is regular language).
2. L is the union of some of the equivalence classes of a right invariant (with respect to concatenation) equivalence relation of finite index.
3. Let equivalence relation R_L be defined by : xR_Ly if and only if for all z in Σ^*, xz is in L exactly when yz is in L. Then R_L is of finite index.

4.7. IMPLEMENTATION OF MYHILL-NERODE THEOREM

Step 1. Start with a FA = $(Q, \Sigma, \delta, q_0, F)$ as usual remove from Q, F and δ all states that cannot be reached from q_0.

Step 2. Build a two dimensional matrix labelling the right side $q_0, q_1, q_2, ...$ running down and denote this as the "P" first subscript. Label the top as $q_0, q_1, q_2, ...$ and denote this as "q" second subscript.

Put dashes in the major diagonal and the lower triangular part of the matrix (every thing below the diagonal) we will always use the upper triangular part because $xR_My = yR_Mx$ is symmetric. We will also use (p, q) to index into the matrix with the subscript of the state called "p" always less than the subscript of the state called "q".

Step 3. We can have one of three things in a matrix location where there is no dash.

A "X" indicates a distinct state from our initialization, "x", at the location (p, q) indicate that state p is distinguishable from q.

We will level all empty matrix location with "0". The "0" location means the p and q are equivalent and will be the same state in the minimum machine.

Step 4. Now put at (p, q) in $F \times (Q - F)$ such that subscript of p < subscript of q. Do not write over dashes. If (p, q) has a dash, put X in (q, p).

Step 5. Begin for every pair of distinct states (p, q) in $F \times F$ and for every pair of distinct state (p, q) in $(Q - F) \times (Q - F)$ do

{

if for any input symbol, (r, s) has X or x then put x at (p, q), check (s, r) if (r, s) has a dash.

(where : $r = \delta(p, a)$ and $s = \delta(q, a)$: Σ is the input symbol in Σ.)

- If (p, q) has a dash then mark x at (q, p)
- You do not have to write another 'x' if one is there already.

Step 6. Mark all the empty position '0' in upper triangle.

Step 7. To find out the states in minimised DFA, check all the rows of final matrix. The minimised machine will contain the states equal to the number of rows contains "0" entries (including the initial state if row starts with initial state does not contain any '0' entry).

If a row start with q_x is having "0" at q_y and q_z positions then $\{q_x, q_y, q_z\}$ will a state in minimum machine.

Initial state will be that set which contains initial state as member state. (If there is no "0" entry at the row, starts with the initial state, then initial state of new minimised machine will be same as in old machine). Final states of minimised machine are all those sets which contains the any accepting state of old machine as member state.

Step 8. For finding the transition between the states of new machine, we will adopt the same strategy as we adopt in the minimization process discussed earlier, start with the initial state of M and check transition of it for all input symbols in Σ. Repeat this process till you are getting any new state.

EXERCISES

1. Use the pumping lemma to show that each of these languages is non-regular :

(*i*) $\{a^n\ b^{n+1}\} = \{abb,\ aabbb,\ aaabbbb\ ...\}$

(*ii*) $L = \{a^n b^n a^n \mid n > 0\}$

(*iii*) $L = \{a^n b^{2n} \mid n > 0\}$

(*iv*) $L = \{a^n b a^n \mid n > 0\}$

(*v*) $L = \{a^n b^n a^m$ where $n = 0, 1, 2, ...$ and $m = 0, 1, 2, ...\}$

2. Define the language square as follows :

$$\text{Square} = \{a^n \text{ where } n \text{ is a square, } n > 0\}$$
$$= \{a,\ aaaa,\ aaaaaaaaa\ ...\}$$

Using the pumping lemma to prove that square is non-regular.

3. Prove that $L = \{a^n b^n$ where n is square and $n > 0\}$ is non-regular.

4. Define the language double prime as follows

Double prime = $\{a^p b^p$ where p is any prime$\}$

prove that double prime is non-regular.

5. Define the language double factorial as follows :

$$\text{DF} = \{a^{n!}\ b^{n!} \mid n > 0\}$$

prove that DF is non-regular.

6. Let L_1, L_2, L_3 ... be an infinite sequence of regular languages.

(*i*) Let L be the infinite union of all these language taken together. Is L necessarily regular ?

(*ii*) Is the infinite intersection of all these languages necessarily regular ?

7. Consider the following language :

$$\overline{p} = \{a^n \text{ where } n \text{ is not prime}\}$$
$$= \{\in,\ a,\ aaaa,\ aaaaaa,\ aaaaaaaaa,\ ....\}$$

prove that $\overline{p}$ is non-regular.

8. Consider the language L defined as

$L = \{a^n$ where n is any integer with an even number of digits in base 10$\}$

$= \{\in,\ a^{10},\ a^{11},\ a^{12},\ ...\}$

prove that L is non-regular.

9. (*i*) Show that if we add a finite set of words to a regular language, the result is a regular language.

(*ii*) Show that if we subtract a finite set of words from a regular language, the result is a regular language.

10. Prove that following languages are not regular.

(*i*) $L = \{a^n b^l a^k : k \geq n + 1\}$

(*ii*) $L = \{a^i b^j a^k : k \neq i + j\}$

(*iii*) $L = \{a^i b^j a^k : i = j \text{ or } j = k\}$

(*iv*) $L = \{a^i b^j : i \leq 1\}$

(*v*) $L = \{w : n_a(w) \neq n_b(w)\}$

11. Prove that following language regular ?

(*i*) $L = \{xyy^R z : x, z, y \in \{0, 1\}^+\}$

(*ii*) $L = \{xyy^R z : x, z, y \in \{0, 1\}^+, | x | \geq | z |\}$

12. Suppose that we know that $L_1 \cup L_2$ and L_1 are regular. Can we conclude from this that L_2 is regular.

13. Prove that following languages over alphabet {0, 1} are not regular.

(*i*) The set of strings of 0's and 1's, beginning with *a* 1, such that when interpreted as an integer, that integer is prime.

(*ii*) The set of strings of the form *w* 1*n*, where *w* is a string of 0's and 1's of length *n*.

(*iii*) The set of strings of 0's and 1's of the form $w\bar{w}$, where $\bar{w}$ is formed from *w* by replacing all 0's by 1's, and vice-versa that is, $\overline{011} = 100$, and 011 100 is an example of string in the language.

14. Given an algorithm to tell whether a regular language *L* is infinite.

15. Let *A* be any FA with *N* states. Then :

(*i*) If *A* accepts an input string *w* such that

$N \leq$ length $(w) < 2N$, then *A* accepts an infinte language.

(*ii*) If *A* accepts many words, then *A* concepts some word *w* such that $N \leq$ length $(w) < 2N$.

*****

5

CHAPTER

Context-Free Grammars and Languages

5.1. GRAMMARS

INSIDE THIS CHAPTER

All of us know that a grammar is nothing but a set of rules to define valid sentences in any languages. In this chapter we introduce the context-free grammars, which generates context-free languages. Context-free languages have great practical significance in defining programming languages and in simplifying the translation for programming languages.

Initially Linguists were trying to define precisely valid sentences and to give structural description for these sentences. They tried to define rules for natural languages like Hindi, English etc. Noam Chomsky gave a mathematical model for the grammars in 1956. Although it was useless to describe natural languages but it becomes very useful for computer languages.

The original motivation for grammars was the description of natural languages. We can write rules for the grammar of natural languages as follows :

<sentence> → <noun phrase> <verb phrase>

<noun phrase> → <article> <noun>

According to above set of rules "The girl eats" is valid sentence.

TIPS

It can be easily observed that, some words in the grammar works as terminator for the sentence, these symbols are called terminal symbol. Rest of the symbols of the vocabulary are called non-terminals. So it is clear that terminals and non-terminals forms vocabulary for any language.

Let us now formalise the idea of a grammar and how it is used. There are four important components in a grammatical description of a language :

1. There is a finite set of symbols that form the strings of the language being defined. We call these alphabets terminal symbols, represented by V_t.
2. There is a finite set of variables, also called sometimes non-terminals or syntactic categories. Each variable represents a language; *i.e.*, a set of strings, represented by V_n.
3. One of the variable represents the language being defined; it is called the start symbol. Generally it is denoted by S.
4. There is a finite set of rules or productions that represents the recursive definition of a langauge. Each production consists of :
   (*a*) A variable that is being (partially) defined by the production. This variable is often called the head of the production.
   (*b*) The production symbol $\rightarrow$
   (*c*) A string of zero or more terminals and variables. This string, called the body of the production, represents one way to form strings in the language of the variable of the head. So, we leave terminals unchanged and substitute for each variable of the body any string that is known to be in the language of that variable.

After this discussion, we are able to define context free grammar.

5.2. CONTEXT-FREE GRAMMARS

Now let us formalize the concept of context free grammar, as we discussed intuitive notions for context free grammar in previous section.

Definition

Mathmatically context-free grammar is defined as follows :

"A grammar $G = (V_n, V_t, P, S)$ is said to be context-free"

where V_n : A finite set of non-terminals, generally represented by capital letters, $A, B, C, D, \ldots$

V_t : A finite set of terminals, generally represented by small letters, like, $a, b, c, d, e, f, \ldots$

S : Starting non-terminal, called start symbol of the grammar. S belongs to V_n.

P : Set of rules or productions in CFG.

G is context-free and all productions in P have the form

$$\alpha \rightarrow \beta$$

where $\alpha \in V_n$ and $\beta \in (V_t \cup V_n)^*$

Every regular grammar is context-free, so a regular language is also a context-free one. It is already proved by pumping lemma that language $\{a^n b^n / n > 0\}$ is not regular, but it is possible to design a context-free-grammar for these languages (we will design it in next section). **So now it is very clear that "the family of regular language is a proper subset of the family of context-free language".**

> **TIPS**
>
> Context-free grammars derive their name from the fact that the substitution of the variable on the left of a production can be made any time, such a variable appears in the sentential form. It does not depend on the symbols in the rest of the sentential form (the context). This feature is the consequence of allowing only a single variable on the left side of the production.

5.2.1. Derivations

We now define the notations to represent a derivation. First we define two notations $\underset{G}{\Rightarrow}$ and $\overset{*}{\underset{G}{\Rightarrow}}$. If $\alpha \to \beta$ is a production of P in CFG and a and b are strings in $(V_n \cup V_t)^*$, then

$$a \, \alpha \, b \underset{G}{\Rightarrow} a \, \beta \, b.$$

We say that the production $\alpha \to \beta$ is applied to the string $a \, \alpha \, b$ to obtain $a \, \beta \, b$ or we say that $a \, \alpha \, b$ directly drives $a \, \beta \, b$.

Now suppose $\alpha_1, \alpha_2, \alpha_3, ..., \alpha_m$ are string in $(V_n \cup V_t)^*$,

$$m \geq 1 \text{ and } \alpha_1 \underset{G}{\Rightarrow} \alpha_2, \alpha_2 \underset{G}{\Rightarrow} \alpha_3, \alpha_3 \underset{G}{\Rightarrow} \alpha_4, ..., \alpha_{m-1} \underset{G}{\Rightarrow} \alpha_m$$

Then we say that $\alpha_1 \overset{*}{\underset{G}{\Rightarrow}} \alpha_m$, *i.e.*, we say α_1, drives α_m in grammar G. If α drives β by exactly i steps, we say $\alpha \overset{i}{\underset{G}{\Rightarrow}} \beta$.

5.2.2. Language of Context-Free Grammar

If $G = (V_n, V_t, P, S)$ is a CFG, the language of G, denoted by $L(G)$, is the set of terminal strings that have derivations from the start symbol. That is

$$L(G) = \left\{ w \text{ in } V_t \,/\, S \overset{*}{\underset{G}{\Rightarrow}} w \right\}$$

If a language L is the language of some context-free grammar, then L is said to be context-free language, or CFL.

5.2.3. Sentential Forms

Derivations from the start symbol produce strings that have a special rule. We call these "Sentential forms". That is, if $G = (V_n, V_t, P, S)$ is a CFG, then any string α in $(V_n \cup V_t)^*$ such that $S \overset{*}{\Rightarrow} \alpha$ is a sentential form.

Note that the language $L(G)$ is those sentential forms that are in V_t^*, *i.e.*, they consists solely of terminals.

Now let us discuss some examples.

Example 5.1. Consider a grammar $G = (V_n, V_t, P, S)$ where

$V_n = \{S\}$, $V_t = \{a, b\}$ and set of productions P is given by

$$\begin{aligned} P &= \{S \to aSb \\ S &\to ab \\ \}. \end{aligned}$$

Solution. Here S is the only non-terminal which is the starting symbol for the grammar; 'a' and 'b' are terminals. There are two productions $S \to aSb$ and $S \to ab$. Now we will show how the strings a^2b^2 can be derived.

$$\begin{aligned} S &\Rightarrow aSb \\ &\Rightarrow aabb \\ &\Rightarrow a^2b^2 \end{aligned}$$

Here we need to apply the first production then second production.

By applying first production $n - 1$ times, followed by an application of second production, we get

$$\begin{aligned} S &\Rightarrow aSb \\ &\Rightarrow aaSbb \\ &\Rightarrow \,: \\ &\quad\; : \\ &\Rightarrow a^{n-1}Sb^{n-1} \\ &\Rightarrow a^nb^n \end{aligned}$$

Hence we can say that language for the above grammar is

$$L(G) = \{a^n\, b^n /\, n >=1\}.$$

Example 5.2. Following is a CFG for the language

$$L = \{wcw^R / w \in (a, b)^*\}$$

Solution. Let G be CFG for language $L = \{wcw^R/w \in (a, b)^*\}$

$$G = (V_n, V_t, P, S)$$

Here $V_n = \{S\}$

$V_t = \{a, b, c\}$

and P is given by

$$\begin{aligned} S &\to aSa \\ S &\to bSb \\ S &\to c \end{aligned}$$

Let us check that *abbcbba* can be drived from the given CFG.

$$\begin{aligned} S &\Rightarrow aSa && \text{(use the } S \to aSa) \\ &\Rightarrow abSba && \text{(use the } S \to bSb) \\ &\Rightarrow abbSbba && \text{(use the } S \to bSb) \\ &\Rightarrow abbcbba && \text{(use the } S \to c) \end{aligned}$$

So string *abbcbba* can be derived from given CFG.

5.2.4. Bacos Naur Form (BNF)

While linguists were studying CFG's computer scientists began to describe programming languages by notation called Bacus-Naur form or Bacus Normal Form, which is the CFG notation with minor changes in format and some shorthand. For example, the above grammar can be rewritten in BNF as

$$S \rightarrow bA/aB$$
$$A \rightarrow bAA/aS/a$$
$$B \rightarrow aBB/bS/a$$

Hence BNF is a shorthand notation for context free grammar.

Example 5.3. Write a CFG, which generates string of balanced parenthesis.

Solution. This grammar will accept the balanced right and left parenthesis. For example ( ) ( ) is acceptable, ((( ))) is also accepted.

Let us design the CFG for this

Let CFG be $G = (V_n, V_t, P, S)$ where

V_n = set of non-terminals = $\{S\}$,

V_t = set of terminals = $\{(,)\}$

and set of production P is given by

$$S \rightarrow SS$$
$$S \rightarrow (S)$$
$$S \rightarrow \in$$

Now we see what this grammar generates :

$$\begin{aligned} S &\Rightarrow SS \\ &\Rightarrow (S)\, S \\ &\Rightarrow (S)\, SS \\ &\Rightarrow (S)\,(S)\, S \\ &\Rightarrow (S)\,(S)\,(S) \\ &\Rightarrow (\,)\,(S)\,(S) \\ &\Rightarrow (\,)\,(\,)\,(S) \\ &\Rightarrow (\,)\,(\,)\,(\,) \end{aligned}$$

Thus the above grammar always generates balanced pairs of parenthesis.

Example 5.4. Write a CFG, which generates palindrome for binary numbers.

Solution. Grammar will generate palindrome for binary numbers, that is 00, 010, 11, 101, 11100111 ...

Let CFG be $G = (V_n, V_t, P, S)$

where V_n = set of non-terminal = $\{S\}$

V_t = set of terminals = $\{0, 1\}$

and production rule P is defined as

$$S \rightarrow 0S0/1S1$$
$$S \rightarrow 0/1/\in$$

Obviously this grammar generates palindrome for binary numbers, it can be seen by the following derivation

$$\begin{aligned} S &\Rightarrow 0S0 \\ &\Rightarrow 01S10 \\ &\Rightarrow 010S010 \\ &\Rightarrow 0101010 \end{aligned}$$

which is a palindrome.

Example 5.5. Write a CFG for the regular expression

$$\mathbf{r = 0^*\,1(0+1)^*}$$

Solution. Let us analyse regular expression

$$r = 0^*\,1\,(0+1)^*$$

Clearly regular expression is the set of strings which starts with any number of 0's, followed by a one and end with any combination of 0's and 1's.

Let CFG be $G = (V_n, V_t, P, S)$

where $V_n = \{S, A, B\}$

$V_t = \{0, 1\}$

and productions P are defined as

$$\begin{aligned} S &\to A1B \\ A &\to 0A/\in \\ B &\to 0B/1B/\in \end{aligned}$$

Let us see the derivation of the string 00101

$$\begin{aligned} S &\Rightarrow A1B \\ &\Rightarrow 0A10B \\ &\Rightarrow 00A101B \\ &\Rightarrow 00101 \end{aligned}$$

So clearly G is CFG for regular expression r.

Example 5.6. Write a CFG which generates strings having equal number of *a*'s an *b*'s.

Solution. Let CFG be $G = (V_n, V_t, P, S)$

$V_n = \{S\}$

$V_t = \{a.\ b\}$

where P is defined as

$$S \to aSbS/bSaS/\in$$

Let us derive a string

$$w = \text{bbabaabbaa}$$

$$\begin{aligned} S &\Rightarrow b\underline{S}aS \\ &\Rightarrow bb\underline{S}aSaS \\ &\Rightarrow bbaSb\underline{S}aSaS \\ &\Rightarrow bbaSba\underline{S}bSaSaS \\ &\Rightarrow bba\underline{S}baaSbSbSaSaS \end{aligned}$$

$$\Rightarrow bbabaa\underline{S}bSbSaSaS$$
$$\Rightarrow bbabaaba\underline{S}bSaSaS$$
$$\Rightarrow bbabaabb\underline{S}aSaS$$
$$\Rightarrow bbabaabba\underline{S}aS$$
$$\Rightarrow bbabaabbaa\underline{S}$$
$$\Rightarrow bbabaabbaa$$

Example 5.7. Design a CFG, which can generate string, having any combination of *a*'s and *b*'s, except null string.

Solution. Let CFG be $G = (V_n, V_t, P, S)$

$$V_n = \{S\}$$
$$V_t = \{a, b\}$$

Productions P are defined as

$$(P_1, \text{say})\ S \to aS$$
$$(P_2, \text{say})\ S \to bS$$
$$(P_3, \text{say})\ S \to a$$
$$(P_4, \text{say})\ S \to b$$

We can produce the word *baab* as follows :

$$S \Rightarrow bS \text{ (by } P_2)$$
$$\Rightarrow baS \text{ (by } P_1)$$
$$\Rightarrow baaS \text{ (by } P_1)$$
$$\Rightarrow baab \text{ (by } P_4)$$

The language generated by this CFG is the set of all possible strings of the letters *a* and *b* except for the null string, which we cannot generate.

We can generate any word by the following algorthim :

At the beginning, the working string is the start symbol S select a word to be generated. Read the letters of the desired word from left to right, one at a time. If an *a* is read that is not the last letter of the word, apply P_1 to the working string. If '*b*' is read that is not the last letter of the word, apply P_2 to the working string.

If the last letter is read and it is an *a*, apply P_3 to the working string. If the last letter is read and it is *b*, apply P_4 to the working string.

Productions 3 and 4 can be used only once and one of them can be used. For example, to generated *babb*, we apply in order productions P_2, P_1, P_2 and P_4 as below

$$S \Rightarrow bS \Rightarrow baS \Rightarrow babS \Rightarrow babb.$$

Example 5.8. Design CFG for regular expression

$$\boldsymbol{r = (a + b)^* \, aa \, (a + b)^*}.$$

Solution. Let CFG be

$$G = (V_n, V_t, P, S)$$
$$V_n = \{S, T\}$$
$$V_t = \{a, b\}$$

Productions P are defined as

$$S \rightarrow TaaT \qquad \text{(say } P_1\text{)}$$
$$T \rightarrow aT \qquad \text{(say } P_2\text{)}$$
$$T \rightarrow bT \qquad \text{(say } P_3\text{)}$$
$$T \rightarrow \in \qquad \text{(say } P_4\text{)}$$

Last three productions, *i.e.*, P_2, P_3 and P_4 allows us to generate any word we want from terminal T. If the non-terminal T appears in any working string, we can apply productions to turn in to any string we want. Therefore, the words generated from S have the form any substring *aa* any substring

or $(a + b)^* \, aa \, (a + b)^*$

which is the language of the all words with a double '*a*' in them somewhere.

For example to generate baabaab, we can proceed as follows :

$$\begin{aligned} S &\Rightarrow TaaT \\ &\Rightarrow bTaaT \\ &\Rightarrow baTaaT \\ &\Rightarrow baaTaaT \\ &\Rightarrow baabTaaT \\ &\Rightarrow baab\in aaT \\ &\Rightarrow baabaaT \\ &\Rightarrow baabaabT \\ &\Rightarrow baabaab\in \\ &\Rightarrow baabaab \end{aligned}$$

There are other sequences that can also derive the word *baabaab*.

Example 5.9. Design a CFG for the regular expression $r = (a + b)^*$.

Solution. Let G be CFG

$$\begin{aligned} G &= (V_n, V_t, P, S) \\ V_n &= \{S\} \\ V_t &= \{a, b\} \end{aligned}$$

P are desined as

$$\begin{aligned} S &\rightarrow aS \\ S &\rightarrow bS \\ S &\rightarrow a \\ S &\rightarrow b \\ S &\rightarrow \in \end{aligned}$$

The word *ab* can be generated by the derivation

$$\begin{aligned} S &\Rightarrow aS \\ &\Rightarrow abS \\ &\Rightarrow ab\in \\ &\Rightarrow ab \end{aligned}$$

or by the derivation

$$\begin{aligned} S &\Rightarrow aS \\ &\Rightarrow ab. \end{aligned}$$

Example 5.10. The difference between even 'palindrome' and 'odd palindrome' (whose definitions is obvious) is that when we are finally ready to get rid of *S* in the even palindrome working string, we must replace it with $\in$. If we were force to replace it with an *a* or *b* instead, we would create a central letter and the result would be a grammar for odd palindrome as follows :

$$S \to aSa$$
$$S \to bSb$$
$$S \to a$$
$$S \to b$$

Solution. If we allow the option of turning the central *S* into either $\in$ or a letter, we would have a grammar for the entire language palindrome :

$$S \to aSa$$
$$S \to bSb$$
$$S \to a$$
$$S \to b$$
$$S \to \in$$

The language $\{a^n b^n\}$ and palindrome are amazingly similar in grammartical structure while the first is nearly a regular expression and the other is for from it.

Example 5.11. Write a CFG for the language $L(G) = \{ww^r : w \in \{0, 1\}^*\}$.

Solution. Let grammar be *G*.

$$G = (V_n, V_t, P, S)$$
$$V_n = \{S\}$$
$$V_t = \{0, 1\}$$

Productions are defined as

$$S \to 0S0 \qquad (P_1)$$
$$S \to 1S1 \qquad (P_2)$$
$$S \to \in \qquad (P_3)$$

Let us derive a string $\underset{w}{\underline{001}}\,\underset{w^r}{\underline{100}}$ by above CFG.

$$S \Rightarrow 0S0 \qquad (\text{by } P_1)$$
$$\Rightarrow 00S00 \qquad (\text{by } P_1)$$
$$\Rightarrow 001S100 \qquad (\text{by } P_2)$$
$$\Rightarrow 001100 \qquad (\text{by } P_3)$$

Example 5.12. Design a CFG for the language

$$L(G) = \{ab\,(bbaa)^n\,bba\,(ba)^n : n \geq 0\}.$$

Solution. Let CFG be *G*

$$G = (V_n, V_t, P, S)$$
$$V_n = (S, X, Y)$$
$$V_t = \{a, b\}$$

Productions are defined as $S \to abX$

$$X \to bbYa$$

$$Y \to aaXb$$
$$Y \to \in$$

or we can also defined production as :

$$S \to abX$$
$$X \to bbaaXba$$
$$X \to bba$$

Let us derive a string $ab(bbaa)^2\ bba(ba)^2$

$$\begin{aligned} S &\Rightarrow ab\underline{X} \\ &\Rightarrow abbb\underline{Y}a \\ &\Rightarrow abbbaa\underline{X}ba \\ &\Rightarrow abbbaabb\underline{Y}aba \\ &\Rightarrow abbbaabbaa\underline{X}baba \\ &\Rightarrow ab(bbaa)^2bbYa(ba)^2 \\ &\Rightarrow ab(bbaa)^2bb\in a(ba)^2 \\ &\Rightarrow ab(bbaa)^2bba(ba)^2 \end{aligned}$$

or

$$\begin{aligned} S &\Rightarrow abX \\ &\Rightarrow abbbaaXba \\ &\Rightarrow abbbaabbaaXbaba \\ &\Rightarrow ab(bbaa)^2bba(ba)^2. \end{aligned}$$

Example 5.13. Design a CFG for the language

$$L = \{a^n\ b^m : n \neq m\}.$$

Solution. If $n \neq m$ then there are only two cases are possible.

Case 1. $n > m$

Let us say langauge L on condition $n > m$ is

$$L_1 \text{ and } L_1 = \{a^n\ b^m : n > m\}.$$

Let us say G_1 be the CFG for the language L_1,

$$G_1 = (V_n^1, V_t^1, P^1, S^1)$$
$$V_n^1 = \{S_1, A, S^1\}$$
$$V_t^1 = \{a, b\}$$

then productions P^1 are defined as

$$S^1 \to AS_1$$
$$S_1 \to aS_1b/\in$$
$$A \to aA/a$$

Case 2. $n < m$

Let us say L on condition $n < m$ is L_2

and $$L_2 = \{0^n1^m : n < m\}$$

Let CFG for G_2 be G_2

$$G_2 = (V_n^2, V_t^2, P^2, S^2)$$
$$V_n^2 = \{S^2, S_2, B\}$$
$$V_t^2 = \{a, b\}$$

P^2 are defined as follows :

$$S^2 \rightarrow S_2B$$
$$S_2 \rightarrow aS_2b/\in$$
$$B \rightarrow bB/b$$

by the help (by the combining G_1 and G_2) of G_1 and G_2 we can write the CFG for L as

$$S \rightarrow S^1/S^2$$

where S is start symbol of CFG for L.

Example 5.14. Design a CFG for the language

$$L = \{0^n\ 1^n/n \geq 0) \cup (1^n\ 0^n/n \geq 0)\}.$$

Solution. We can assure L as

$$L = L_1 \cup L_2$$
$$L_1 = \{0^n\ 1^n/n \geq 0\}$$

Let say G_1 be CFG for the language L_1

$$G_1 = (V_n, V_t, P, S)$$
$$V_n = \{S_1\}$$
$$S = \{S_1\}$$
$$V_t = \{0, 1\}$$

and P is defined as

$$S_1 \rightarrow 0S_11/\in$$

and $$L_2 = \{1^n0^1/n \geq 0\}$$

Let say G_2 be CFG for language L_2

$$G_2 = (V_n, V_t, P, S)$$
$$V_n = \{S_2\}$$
$$S = \{S_2\}$$
$$V_t = \{0, 1\}$$

P is defined as

$$S_2 \rightarrow 1S_20/\in$$

Since $L = L_1 \cup L_2$, suppose G be CFG for language L, with starting non-terminal S. Then productions in G will be

$$S \rightarrow S_1/S_2$$
$$S_1 \rightarrow 0S_11/\in$$
$$S_2 \rightarrow 1S_20/\in.$$

Example 5.15. Design CFG for $\Sigma = \{a, b\}$ that generates the set of

(*a*) all strings with exactly one *a*.

(*b*) all strings with at least one *a*.

(*c*) all strings with at least 3 *a*'s.

Solution. (*a*) Let CFG be X G_1

$$G_1 = (V_n, V_t, P, S)$$
$$V_n = \{S, A\}$$
$$V_t = \{a, b\}$$

P is defined as
$$S \rightarrow AaA$$
$$A \rightarrow bA/\in$$

Let us derive a string *bbaab*

$$\begin{aligned} S &\Rightarrow AaA \\ &\Rightarrow bAaA \\ &\Rightarrow bbAaA \\ &\Rightarrow bbAabA \\ &\Rightarrow bbAab\in \\ &\Rightarrow bb\underline{A}ab \end{aligned}$$

How we don't have any choice for derving second G so string *bbaab* cannot be derive from G_1.

(*b*) Let CFG be G_2

$$\begin{aligned} G_2 &= (V_n, V_t, P, S) \\ V_n &= \{S, A\} \\ V_t &= \{a, b\} \end{aligned}$$

P is defined as
$$S \rightarrow AaA$$
$$A \rightarrow aA/bA/\in$$

Let us derive string *baab*

$$\begin{aligned} S &\Rightarrow AaA \\ &\Rightarrow bAaA \\ &\Rightarrow baAaA \\ &\Rightarrow ba\in AaA \\ &\Rightarrow baaA \\ &\Rightarrow baabA \\ &\Rightarrow baab\in \\ &\Rightarrow baab \end{aligned}$$

(*c*) Let CFG be G_3

and
$$\begin{aligned} G_3 &= (V_n, V_t, P, S) \\ V_n &= \{S, A\} \\ V_t &= \{a, b\} \end{aligned}$$

P is defined as

$$S \rightarrow AaAaAaA$$
$$A \rightarrow aA/bA/\in.$$

Example 5.16. Give the simple description of the language generated by the grammar with production

$$\boldsymbol{S \rightarrow aA}$$
$$\boldsymbol{A \rightarrow bS}$$
$$\boldsymbol{S \rightarrow \in.}$$

Solution. Given CFG is
$$S \rightarrow aA \quad \text{(say } P_1\text{)}$$
$$A \rightarrow bS \quad \text{(say } P_1\text{)}$$
$$S \rightarrow \in \quad \text{(say } P_3\text{)}$$

From the production P_3, CFG can generate null string. Now let us see production P_1 (*i.e.* $S \to aA$), once we use it then we have to $A \to bS$. Now again we have to use either production P_1 ($S \to aA$) or production P_3 ($S \to \in$), so clearly the langauge of above CFG will be as follows :

$$L = \{(ab)^n / n \geq 0\}$$

Clearly language L is the set of strings. Starts with *ab* followed by any number of *ab*'s. Null string ($\in$) also included in language L.

Example 5.17. Give the simple description of the language generated by the grammar with productions.

$$S \to SS$$
$$S \to (S)$$
$$S \to \in.$$

Solution. Given CFG is

$$S \to SS$$
$$S \to (S)$$
$$S \to (\in)$$

The language generated by the CFG is set of strings, contains balanced set of paranthesis that is for every left paranthesis there is a right paranthesis, and strings like ( ) ( ) ( ) ... or ((( ))) ( ) ( ) ((( ))) ... are the part of the language of CFG.

Example 5.18. Write the CFG for the language

$$L = \{a^n\, b^n\, c^m\, d^m / n >=1, m >=1\}.$$

Solution. Let CFG for the language L be G

$$G = (V_n, V_t, P, S)$$
$$V_n = \{S, X, Y\}$$
$$V_t = \{a, b\}$$

Productions P is defined as follows :

$$S \to XY \qquad \text{(Production } P_1\text{)}$$
$$X \to aXb / ab \qquad \text{(Production } P_2\text{)}$$
$$Y \to cYd / cd \qquad \text{(Production } P_3\text{).}$$

When we use P_1 (*i.e.* $S \to XY$) then X can generate equal number of a's and b's with no a's follows when any b is encountered. Similarly Y can generate strings of equal number of c's and d's with no c's follows when any d is encountered.

Let us derive a string aaabbbcd that $n = 3$, $m = 1$

$$S \Rightarrow XY$$
$$\Rightarrow aXbY$$
$$\Rightarrow aaXbbY$$
$$\Rightarrow aaabbbY$$
$$\Rightarrow aaabbbcd$$

which required string.

Example 5.19. Design a CFG for the language

$$L = \{a^n\, b^m\, c^m\, d^n / n \geq 1, m >= 1\}.$$

Solution. Let us analyse the language L first, language L is the set of strings which contain equal number of a's and d's and equal number of b's and c's in between a's and d's. There is no 'a' followed by b, c and, there is no 'b' followed by c and, there is no c followed by d. Similarly there is no b, c, and d before any 'a', there is no 'c' and d before any 'b', finally there is no d before any c.

On the base of above description let us design a CFG for the language.

Let us assume CFG be G

$$G = (V_n, V_t, P, S)$$
$$V_n = \{S, A\}$$
$$V_t = \{a, b, c, d\}$$

and P is defined as

$$S \rightarrow aSd \quad \text{(say } P_1\text{)}$$
$$S \rightarrow aAd \quad \text{(say } P_2\text{)}$$
$$A \rightarrow bAc \quad \text{(say } P_3\text{)}$$
$$A \rightarrow bc \quad \text{(say } P_4\text{)}$$

Let us derive a string *aabbbcccdd* from grammar G.

$$S \Rightarrow aSd \quad \text{(by } P_1\text{)}$$
$$\Rightarrow aaAdd \quad \text{(by } P_2\text{)}$$
$$\Rightarrow aabAcdd \quad \text{(by } P_3\text{)}$$
$$\Rightarrow aabbAccdd \quad \text{(by } P_3\text{)}$$
$$\Rightarrow aabbbcccdd \quad \text{(by } P_4\text{)}$$

which is required string.

Example 5.20. Design a CFG for the language $L = \{a^n\, b^{2n} / n >= 0\}$.

Solution. Let CFG be G for the language L

$$G = (V_n, V_t, P, S)$$
$$V_n = \{S\}$$
$$V_t = \{a, b\}$$

and P is defined as

$$S \rightarrow aSbb / \in$$

Let us derive a string *aaabbbbbb*

$$S \Rightarrow aSbb$$
$$\Rightarrow aaSbbbb$$
$$\Rightarrow aaaSbbbbbb$$
$$\Rightarrow aaa\in bbbbbb$$
$$\Rightarrow aaabbbbbb$$

which is required string.

Now let us derive another string *abbb*.

$$S \Rightarrow aSbb$$

From this derivation only it is clear that third b is not possible against a single 'a'.

Example 5.21. Write the CFG for the language $L = \{a^{2n}\, b^m / n > 0, m >= 0\}$.

Solution. Let us analyse the language first language L is the set of strings, if string starts with a then number of a's are even, followed by any number of times b's.

There is no 'a' in any string after first 'b' is encountred similarly there is no 'b' before any 'a' in the string.

Let suppose CFG for the L is G

$$G = (V_n, V_t, P, S)$$
$$V_n = \{S, A, B\}$$
$$V_t = \{a, b\}$$

Productions P is defined as

$$S \to aaAB \qquad \text{(say } P_1\text{)}$$
$$A \to aaA/\in \qquad \text{(say } P_2\text{)}$$
$$B \to bB/\in \qquad \text{(say } P_3\text{)}$$

Let us derive the string *aaaabbb*

$$S \Rightarrow aaAB \qquad \text{(by } P_1\text{)}$$
$$\Rightarrow aaaaAB \qquad \text{(by } P_2\text{)}$$
$$\Rightarrow aaaa\in B \qquad \text{(by } P_2\text{)}$$
$$\Rightarrow aaaabB \qquad \text{(by } P_3\text{)}$$
$$\Rightarrow aaaabbB \qquad \text{(by } P_3\text{)}$$
$$\Rightarrow aaaabb\in \qquad \text{(by } P_3\text{)}$$
$$\Rightarrow aaaabb$$

which is required string.

Example 5.22. Write CFG for

$$L = \{(a^n\, b^n\, c^m\, d^m/n >= 1, m >= 1) \cup (a^n\, b^m\, c^m\, d^n/n >= 1, n >= 1\}.$$

Solution. Let suppose $L = L_1 \cup L_2$

where $L_1 = \{a^n\, b^n\, c^m\, d^m/n >= 1, m >= 1\}$

Let CFG for L_1 is G_1

$$G_1 = (V_n, V_t, P_1, S_1)$$
$$V_n = \{S_1, A, B\}$$
$$V_t = \{a, b, c, d\}$$

and P_1 is defined as follows

$$S_1 \to AB$$
$$A \to aAb/ab$$
$$B \to cBd/cd$$
$$L_2 = \{a^n\, b^m\, c^m\, d^n/n >= 1, m >= 1\}$$

Let us assume that CFG for L_2 is G_2

and $G_2 = (V_n, V_t, P_2, S_2)$

$$V_n = \{S_2, D, E\}$$
$$V_t = (a, b, c, d\}$$

P_2 is defined as follows

$$S_2 \to aDd$$
$$D \to aDd/E$$
$$E \to bEc/bc$$

By the help of the G_1 and G_2, we can define CFG for the language L, since $L = L_1 \cup L_2$

$$\text{so } G = G_1 \cup G_2$$

Let suppose

$$G = (V_n, V_t, P, S)$$
$$V_n = \{S, S_1, S_2, A, B, D, E\}$$
$$V_t = \{a, b, c, d\}$$

and P is defined as

$$S \to S_1/S_2 \qquad \text{(for } G = G_1 \cup G_2\text{)}$$
$$S_1 \to AB$$
$$A \to aAb/ab$$
$$B \to cBd/cd$$
$$S_2 \to DE$$
$$D \to aDd/E$$
$$E \to bEc/bc$$

Example 5.23. Write a CFG for language $L = \{0^i1^j2^k/k \leq i \text{ or } k \leq j\}$.

Solution. In the same fashion as we solve Example 5.21, we approach this problem.

Let us assume $L = L_1 \cup L_2$

where $L_1 = \{0^i1^j2^k/k \leq i\}$

and $L_2 = \{0^i1^j2^k/k \leq j\}$

Let as assume that CFG for L_1 is G_1

$$G_1 = (V_n, V_t, P_1, S_1)$$
$$V_n = \{S_1, X, Y\}$$
$$V_t = \{0, 1\}$$

and P_1 is defined as

$$S_1 \to 0S_12/0Y2$$
$$Y \to C/0Y/\in$$
$$C \to 1C/\in$$

Let us assume that CFG for L_2 is G_2

$$G_2 = (V_n, V_t, P_2, S_2)$$
$$V_n = \{S_2, A, B\}$$
$$V_t = \{0, 1\}$$

P_2 is defined as follows

$$S_2 \to AB$$
$$A \to 0A/\in$$
$$B \to 1B2/1B/\in$$

We can define CFG for L (since $L = L_1 \cup L_2$)

Let us assume that CFG for L is G

$$G = (V_n, V_t, P, S)$$
$$V_n = \{S_1, S_2, X, Y, Z, B\}$$
$$V_t = \{0, 1\}$$

P is defined as

$$\begin{aligned} S &\to S_1/S_2 \\ S_1 &\to 0S_12/0Y2 \\ Y &\to C/0Y/\in \\ C &\to 1C/\in \\ S_2 &\to AB \\ A &\to 0A/\in \\ B &\to 1B2/1B/\in . \end{aligned}$$

Example 5.24. Write the CFG for the langauge

$$L = \{0^i\, 1^j\, 2^k / i = j \text{ or } j = k\}.$$

Solution. Let us assume $L = L_1 \cup L_2$

where $L_1 = \{0^i\, 1^j\, 2^k / i = j\}$

and $L_2 = \{0^i\, 1^j\, 2^k / j = k\}$

Let us consider L_1 first, let CFG for L_1 be G_1

$$\begin{aligned} G_1 &= (V_n, V_t, P_1, S_1) \\ V_n &= \{S_1, A, B\} \\ V_t &= \{0, 1, 2\} \end{aligned}$$

P_1 is defined as follows

$$\begin{aligned} S_1 &\to AB \\ A &\to 0A1/\in \\ B &\to 2B/\in \end{aligned}$$

Now let us assume that CFG for language L_2 is G_2

$$\begin{aligned} G_2 &= (V_n, V_t, P_2, S_2) \\ V_n &= (S_2, C, D) \\ V_t &= (0, 1, 2) \end{aligned}$$

Productions are defined as follows :

$$\begin{aligned} S_2 &\to CD \\ C &\to 0C/\in \\ D &\to 1D2/\in \end{aligned}$$

By the help G_1 and G_2 we can define CFG for the language L, Let it is G.

$$\begin{aligned} G &= (V_n, V_t, P, S) \\ V_n &= \{S, S_1, S_2, A, B, C, D\} \\ V_t &= \{0, 1, 2\} \end{aligned}$$

Productions are defined as follows

$$\begin{aligned} S &\to S_1/S_2 \\ S_1 &\to AB \\ A &\to 0A1/\in \\ B &\to 2B/\in \end{aligned}$$

$$S_2 \rightarrow CD$$
$$C \rightarrow 0C/\in$$
$$D \rightarrow 1D2/\in$$

which is required CFG.

Example 5.25. Find the CFG for the following language

$$L = \{a^n b^{2n} c^m / n, m >= 0\}.$$

Solution. Let us analyse the language first, L is language, which contain a set of strings such that every string may start from a or c but not by b.

If string starts with 'a' then number of a's must follow b's and the, number of b's are twice then the number of a's. If string does start with a then it starts with c, followed by any number of c's. There should be no a after any b or any c, and no b after any c. Similarly there should be no b or c before 'a' and no c before any b.

On the basic of above discussion, let us assume that CFG for the languageis G.

$$G = (V_n, V_t, P, S)$$
$$V_n = \{S, A, B\}$$
$$V_t = \{a, b, c\}$$

P is defined as follows

$$S \rightarrow AB$$
$$A \rightarrow aAbb/\in$$
$$B \rightarrow cB/\in$$

which is required CFG.

Example 5.26. Write the CFG for the language

$$L = \{a^n b^m c^{2m} / n, m >= 0\}.$$

Solution. This problem is very similar to Example 5.25.

Let us assume CFG be G

$$G = (V_n, V_t, P, S)$$
$$V_n = (S, A, B)$$
$$V_t = (a, b, c)$$

P is defined as follows

$$S \rightarrow AB$$
$$A \rightarrow aA/\in$$
$$B \rightarrow bBcc/\in$$

which is required grammar.

5.2.5. Left most and Right most Derivations

In order to restrict the number of choices of replacement of variables, if at each step we replace the "left most" variable by one of its production bodies such a derivation is called a "left most derivation", and we indicate that a derivation is left most by using relation $\underset{lm}{\Rightarrow}$ and $\overset{*}{\underset{lm}{\Rightarrow}}$ for one or many steps, respectively.

Similarly, it is possible that at each step the "right most" variable is replaced by one of its bodies. If so, we call the derivation "right most" and use symbols $\underset{rm}{\Rightarrow}$ and $\overset{*}{\underset{rm}{\Rightarrow}}$ to indicate one or many right most derivation steps, respectively.

Example 5.27. ***(i)* Write a CFG for solving simple expression, such + and *.**

***(ii)* Also write CFG for regular expression**

$$r = (a + b)\ (a + b + 0 + 1)^*.$$

***(iii)* Derive the string (which is defined in part *(ii)*) *a** (*a* + b00) by applying left most derivation and right most derivation.**

Solution. *(i)* We need two variables in this grammar. One, which we call E, represents the expression, it is the start symbol and represents the language of expressions we are defining. The other variable, I, reprsents the identifiers.

Let CFG be $G = (V_n, V_t, P, S)$

$V_n = \{E, I\}$

$V_t = \{+, *, (,)\ \}$

P is defined as

$$E \to I$$
$$E \to E + E$$
$$E \to E * E$$
$$E \to (E)$$

Here I can derive any terminal symbol as

$$I \to \text{terminal.}$$

(ii) Given regular epression is

$(a + b)\ (a + b + 0 + 1)^*$

Let CFG be $G = (V_n, V_t, P, S)$

$V_n = \{E, I\}$

$V_t = \{+, *, (,), a, b, 0, 1\}$

P is defined as follows :

$$E \to I$$
$$E \to E * E$$
$$E \to E + E$$
$$E \to (E)$$
$$I \to a$$
$$I \to b$$
$$I \to Ia$$
$$I \to Ib$$
$$I \to I0$$
$$I \to I1$$
$$I \to 0$$
$$I \to 1.$$

(iii) String is $a * (a +$ b00), and we want to derive it by using left most and right most derivations, from CFG defined in part *(ii)*

By using leftmost derivation :

$$E \underset{lm}{\Rightarrow} \underline{E} * E \underset{lm}{\Rightarrow} \underline{I} * E \underset{lm}{\Rightarrow} a * \underline{E} \underset{lm}{\Rightarrow} a * (\underline{E}) \underset{lm}{\Rightarrow} a * (\underline{E} * E) \underset{lm}{\Rightarrow} a * (\underline{I} + E) \underset{lm}{\Rightarrow}$$

$$a * (a + \underline{E}) \underset{lm}{\Rightarrow} a * (a + \underline{I}\,0) \underset{lm}{\Rightarrow} a * (a + \underline{I}\,00) \underset{lm}{\Rightarrow}$$

$$\underset{lm}{\Rightarrow} a * (a + b00)$$

which is required string.

By using right most derivation :

$$E \underset{rm}{\Rightarrow} E * \underline{E} \underset{rm}{\Rightarrow} E * (\underline{E}) \underset{rm}{\Rightarrow} E * (E + \underline{E}) \underset{rm}{\Rightarrow} E * (E + \underline{I}) \underset{rm}{\Rightarrow} E * (E + \underline{I}0) \underset{rm}{\Rightarrow} E * (E + \underline{I}00) \underset{rm}{\Rightarrow}$$

$$E * (\underline{E} + b00) \underset{rm}{\Rightarrow} E * (\underline{I} + b00) \underset{rm}{\Rightarrow} \underline{E} * (a + b00) \underset{rm}{\Rightarrow}$$

$$\underset{rm}{\Rightarrow} \underline{I} * (a + b00) \underset{rm}{\Rightarrow} b * (a + b00)$$

which is required string.

5.3. PARSE TREE

There is a tree representation for derivation that has proved extremely useful. It is the second way of showing derivations, independent of the order in which productions are used, is also called "derivation tree". "A parse tree is an odered tree in which nodes are labeled with the left sides of productions and in which the children of a node represent its corresponding right sides".

5.3.1. Definition

Let $G = (V_n, V_t, P, S)$ be a context-free grammar. An ordered tree for this CFG, G is a derivation tree if and only if it has the following properties.

(*a*) The root is labeled by the starting non-terminal of the CFG that is S.

(*b*) Every leaf of the orderd ordered tree has a label from $V_t \cup \{\in\}$.

(*c*) Every Interior node of ordered tree has a label from V_n.

(*d*) Let us assume that a vertex has label $X \in V_n$, and its children are labeled (from left to right) $y_1, y_1, y_3, ..., y_n$, then production must contain a production of the form

$$X \rightarrow y_1, y_2, ..., y_n.$$

(*e*) A leaf labeled $\in$ has no siblings, that is, vertex with a child labeled $\in$ can have no other children. Clearly if the leaf is labeled $\in$, then it must be the only child of its parent.

5.3.2. The Yield of Parse Tree

If we look at the leaves of any parse tree and can concate them from the left, we get a string, called the yield of the tree, which is always a string that is derived from the root will be proved shortly of special importance are those parse trees such that :

(*a*) The yield is a terminal string. That is all leaves are labeled either with a terminal or with $\in$.

(*b*) The root is labeled by the start symbol.

Now let us see some examples based on the parse tree.

Example 5.28. Consider the CFG

$$S \rightarrow XX$$
$$X \rightarrow XXX/bX/Xb/a$$

Find the parse tree for the string *bbaaaab.*

Solution. Given CFG is

$$S \rightarrow XX$$
$$S \rightarrow XXX/bX/Xb/a$$

and string is $w = bbaaaab$

We begin with S and apply the production $S \rightarrow XX$.

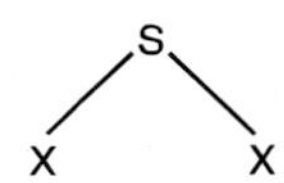

Fig. 5.1.

To the left hand X, let us apply the production $X \rightarrow bX$. To the right hand X, let us apply $X \rightarrow XXX$.

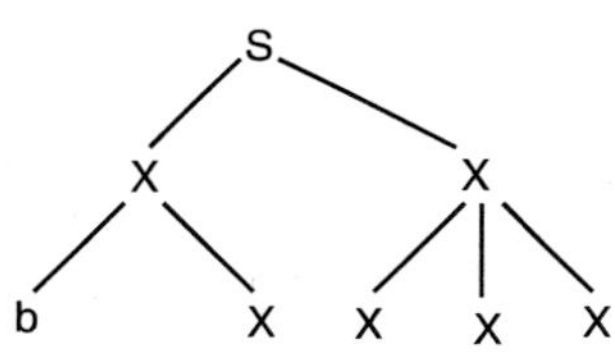

Fig. 5.2.

The b that we have on the bottom line is a terminal, so it does not descend further. In the terminology of trees, it is called a "terminal node". Let the four A's left to right, undergo the production $A \rightarrow bA$, $A \rightarrow a$, $A \rightarrow a$, and $A \rightarrow Ab$, respectively. We now have

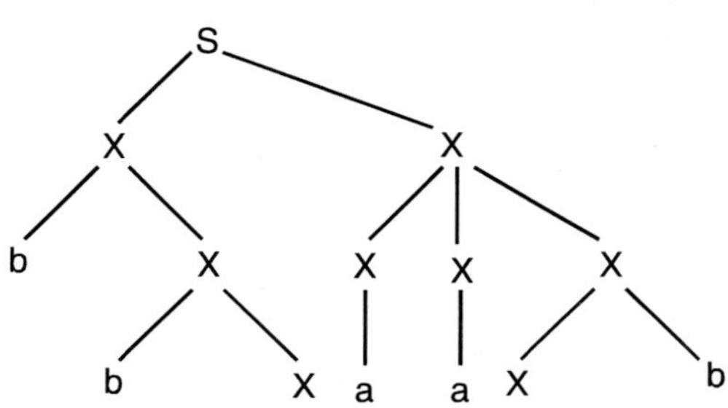

Fig. 5.3.

Let us finish off the genration of a word with the production $X \rightarrow a$ and $X \rightarrow a$:

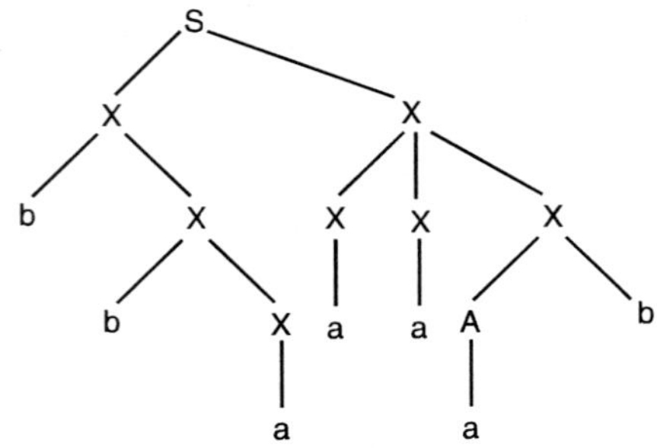

Fig. 5.4.

Reading from left to right, we see the word we have produced is *bbaaaab.*

These tree diagrams are called "syntax trees", "parse trees", "generation" trees, "production trees" or derivation trees.

Example 5.29. Consider the grammar *G*, with production

$$S \rightarrow aXY$$
$$X \rightarrow bYb$$
$$Y \rightarrow X/\in.$$

Solution. Let us see a partial derivation tree for *G*, this tree gives *abbbb* as yield.

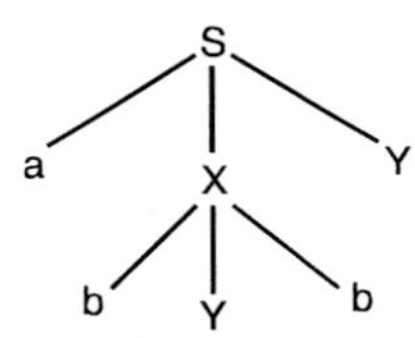

Fig. 5.5.

Now extend this partial derivative tree for the string *abbbb.*

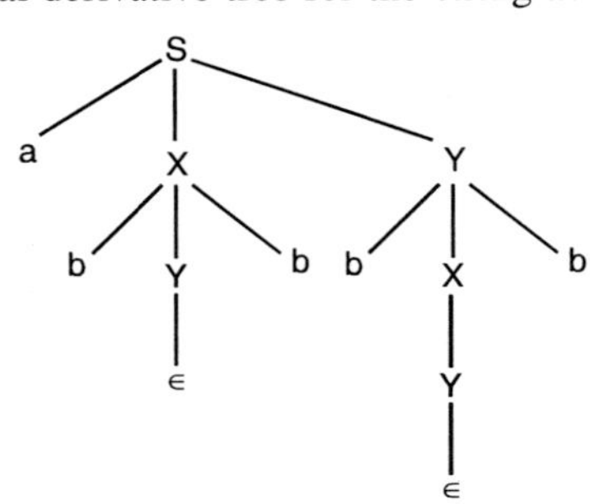

Fig. 5.6.

the yield is $ab\in bb\in b$ that is *abbbb*.

Example 5.30. Write the CFG for the language $L = \{x\ 0^n y 1^n z / n > = 0\}$, and give the parse tree for the string *x*000 *y*11*z*.

Solution. Let the CFG for the language *L* is

$$G = (V_n, V_t, P, S)$$

where $V_n = \{S, B\}$

$V_t = \{x, y, z, 0, 1\}$

and production P is defined as

$$S \to xBz$$
$$B \to y/0B1.$$

Now construct an arbitrary derivation for

$$S \overset{*}{\underset{G}{\Rightarrow}} x00\,y11z.$$

One possible derivation using the above grammar is

$$S \Rightarrow x\underline{B}z \Rightarrow x0B1z \Rightarrow x00\underline{B}11z \Rightarrow x00y11z$$

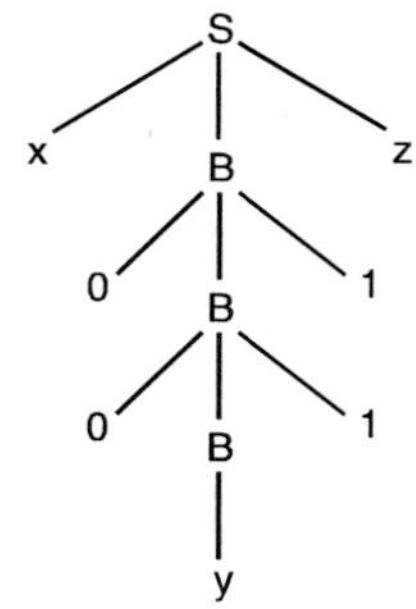

Fig. 5.7.

Now we are able to design the parse tree for the string $x00y11z$ from the above CFG. Parse tree is shown in Fig. 5.7.

Example 5.31. Consider the CFG 'G' whose productions are

$$S \to aAS/a$$
$$A \to SbA/SS/ba.$$

show that $S \overset{*}{\Rightarrow}$ ***aabbaa*** **and construct a derivation tree whose yield is** ***aabbaa*****.**

Solution. $S \Rightarrow a\underline{A}S \Rightarrow a\underline{S}bAS \Rightarrow aab\underline{A}S \Rightarrow aabbaS \Rightarrow a^2b^2a\underline{S} \Rightarrow a^2b^2a^2$

The derivation tree is given in Fig. 5.8.

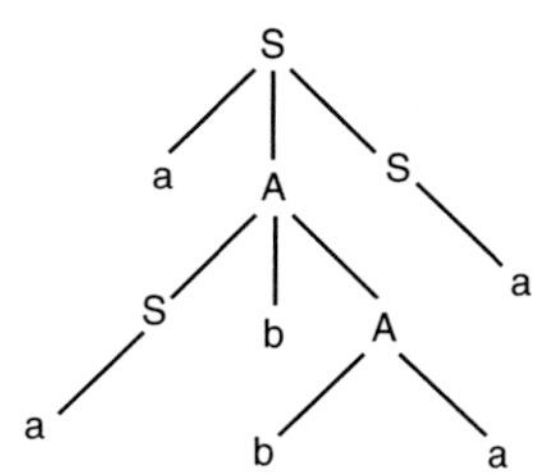

Fig. 5.8.

Derivation tree with yield *aabbaa.*

Example 5.32. Let *G* be CFG

$$S \to bB/aA,$$
$$A \to b/bS/aAA$$
$$B \to a/aS/bBB.$$

For the string *bbaababa* find

(*i*) left most derivation

(*ii*) rightmost derivation and

(*iii*) parse tree.

Solution. (*i*) Left most derivation for string $w = bbaababa$ is

$$S \Rightarrow b\underline{B} \Rightarrow bb\underline{B}B \Rightarrow bba\underline{B} \Rightarrow bbaa\underline{S} \Rightarrow b^2a^2b\underline{B} \Rightarrow b^2a^2ba\underline{S} \Rightarrow b^2a^2bab\underline{B} \Rightarrow \underline{b}^2a^2baba$$

(*ii*) The right most derivation is

$S \Rightarrow b\underline{B} \Rightarrow bbB\underline{B} \Rightarrow bbBa\underline{S} \Rightarrow bbBab\underline{B} \Rightarrow b^2Baba\underline{S} \Rightarrow b^2Babab\underline{B} \Rightarrow b^2\underline{B}ababa \Rightarrow b^2a^2baba$

The derivation tree is following in Fig. 5.9.

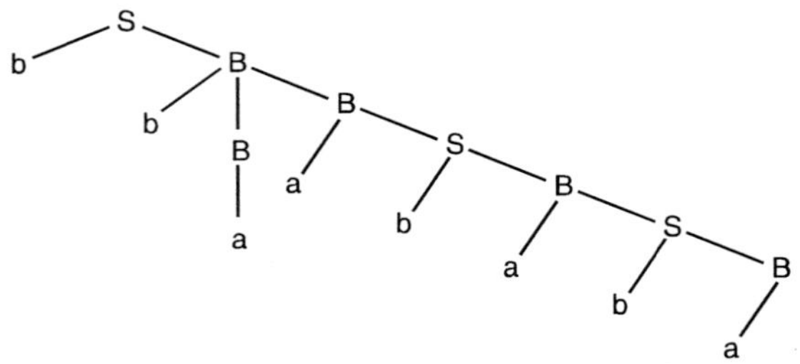

Fig. 5.9.

Yield is *bbaababa*.

5.4. AMBIGUITY IN GRAMMARS AND LANGUAGES

Grammars can be used to put structure on programs and documents. The assumption was that a grammar uniquely determines a structure for each string in its language. However not every grammar does provide unique structure.

When a grammar fails to provide unique structures, it is sometimes possible to redesign the grammar to make the structure unique for each string in the language. Unfortunately some times we cannot do so. That is, there are some CFL's that are "inherently ambigious"; every grammar for the language puts more than one structure for some strings in the language.

Definition A CFG is called ambigious if for at least one word in the lanauge that it generates, there are two possible derivations of the word that correspond to different syntax trees. If a CFG is not ambiguous, it is called unambiguous.

Example 5.33. The language of all non null strings of *a*'s can be defined by a CFG as follows :

$$S \rightarrow aS/Sa/a$$

Solution. In this case, the word a^3 can be generated by four different tree's :

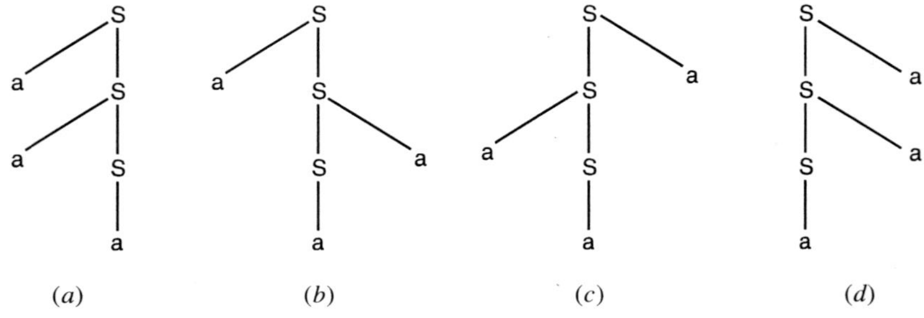

Fig. 5.10.

The CFG is therefor ambigious.

However, the same language can also be defined by the CFG

$$S \rightarrow aS/a$$

for which a^3 has only one production :

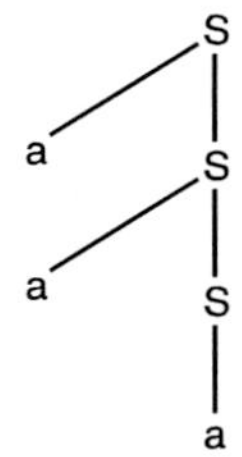

Fig. 5.11.

Example 5.34. The CFG $S \rightarrow aSb/SS/\in$ is ambigous. The sentence *aabb* has the two derivation trees shown in Fig. 5.12 (*a*) and 5.12 (*b*).

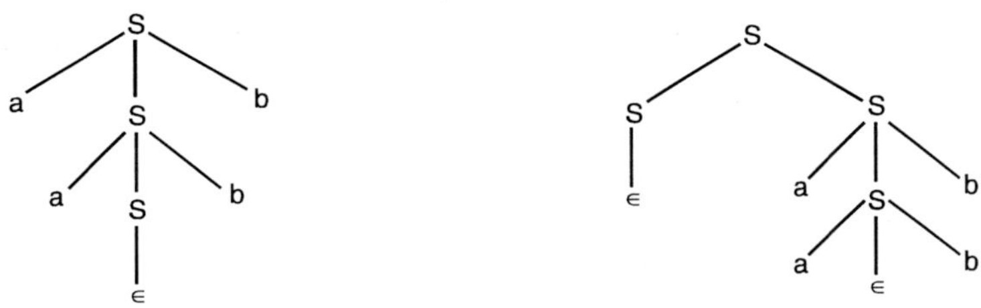

(*a*) (*b*)

Fig. 5.12.

Example 5.35. Consider the grammar

$$E \rightarrow I$$
$$E \rightarrow E + E$$
$$E \rightarrow E * E$$
$$E \rightarrow (E)$$
$$I \rightarrow a$$
$$I \rightarrow b$$
$$I \rightarrow Ia$$
$$I \rightarrow Ib$$
$$I \rightarrow I0$$
$$I \rightarrow I1$$

Solution. For instance, consider the sentential form $E + E * E$. It has two derivation from E :

(1) $E \Rightarrow E + E \Rightarrow E + E * E$

(2) $E \Rightarrow E * E \Rightarrow E + E * E$

Figure 5.13 shows the two parse trees, which we should note are distinct trees.

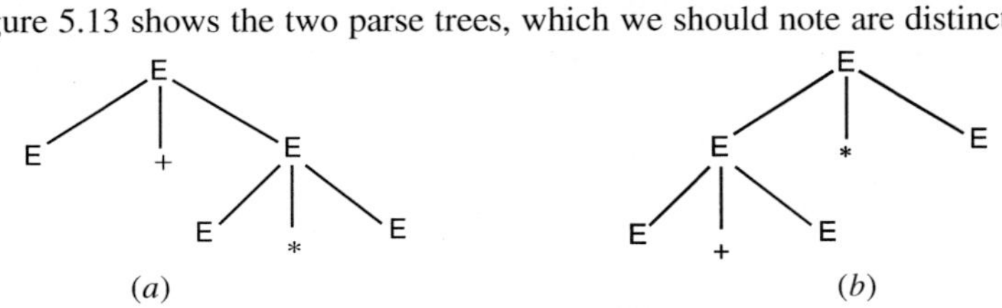

(*a*) (*b*)

Fig. 5.13. Two parse trees with the same yield.

The difference between two derivations is significant. As the structure of the expression is concerned (1) says that second and third expression are multiplied, and the result is added to the first expression, while derivation (2) adds the first two expressions and multiply the result by the third one. For example, according to first derivation 2 + 5 * 3 should be grouped 2 + (5 * 3) = 17, while the second derivation suggest the same expression should be grouped (2 + 5) * 3 = 21. Obviously, the first of these, and not the second, matches our notation of correct grooping of arithmatic expressions.

Example 5.36. Using the same grammar as in Example 5.35, we find that strng $a + b \times b$ has many different derivation trees, as follows.

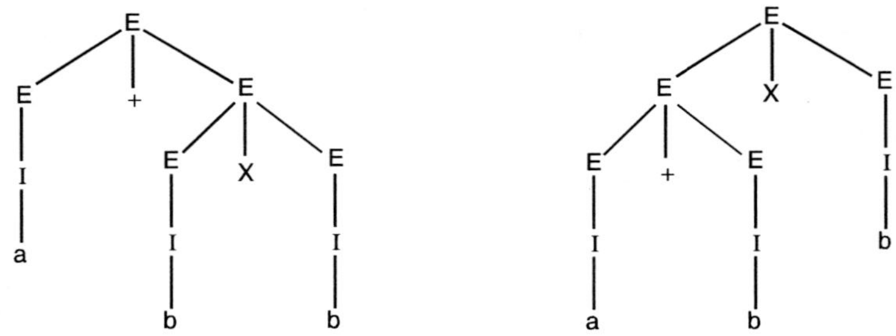

(*a*) (*b*)

Fig. 5.14.

Example 5.37. If CFG a is $S \to SbS/a$, show that G is ambigous.

Solution. To prove that G is ambigous, we have to find a string, $w \in L(G)$ which is ambigous.

Consider the w = *abababa* $\in L(G)$. Then we get two derivation tree for w. Thus grammar G is ambiguous.

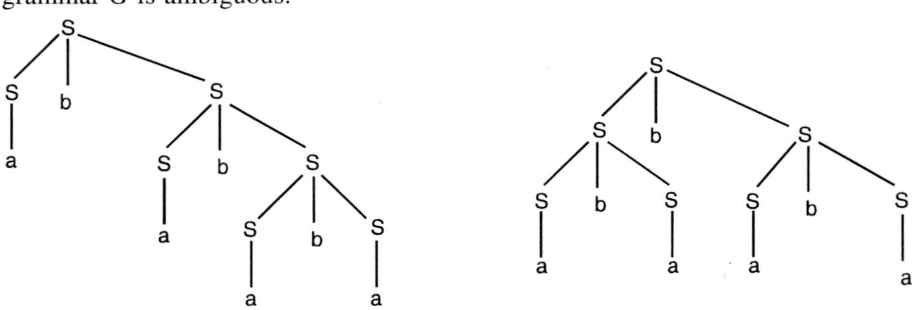

(*a*) First derivation tree for *abababa*. (*b*) Second derivation tree for *abababa*.

Fig. 5.15.

Example 5.38. Consider the productions for the *if then else* statement which are the part of programming language

<statement> → if <expression> then <st> | if <expression>
then <st> else <st> | <other statement>

Expression →exp1 | exp2 | ... exp.

<other statement> →st1 | st2 | ... | st*n*.

Prove that grammar is ambigious.

Solution. Consider the statement

if exp1 then if exp2 then st1 else st2

There are two different parse trees as shown in Fig. 5.16 and Fig. 5.17.

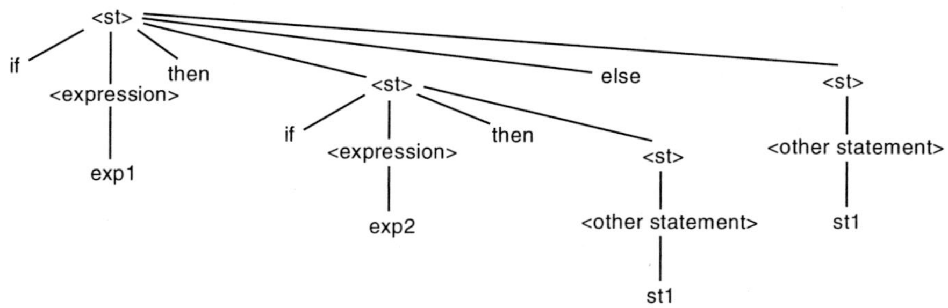

Fig. 5.16.

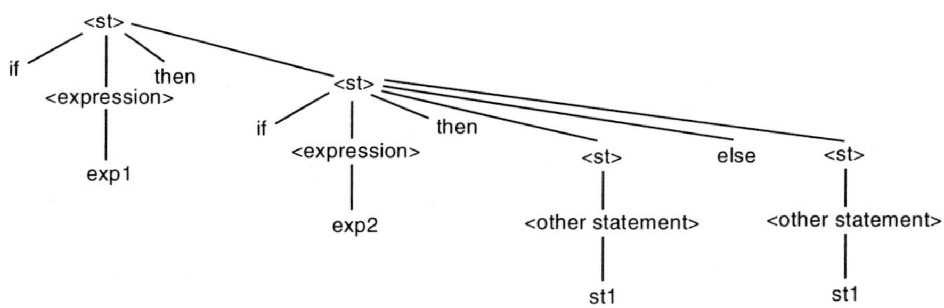

Fig. 5.17.

Hence grammar is ambiguous.

5.4.1. Removing Ambiguity from Grammars

If a context free grammar is ambiguous, it is often possible and usually desirable to find an equivalent unambiguous CFG. Although some CFGs are "**inherently ambiguous"** in the sense that they cannot be produced except by ambiguous grammars. Ambiguity is usually the property of the grammar rather than the language. Let us consider the grammar for algebraic expressions and for the string $a + a*a$ the two parse tree are possible.

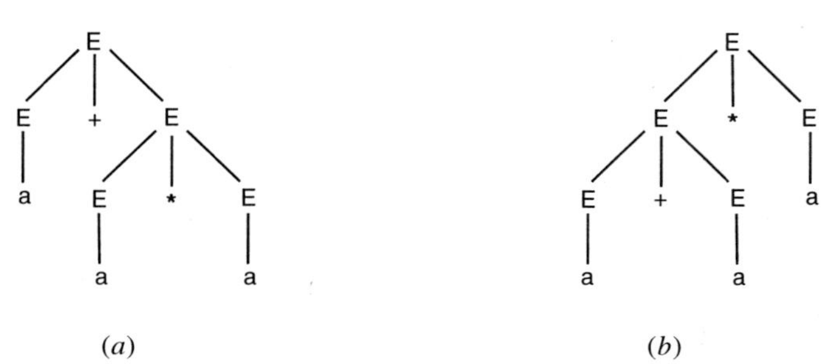

Fig. 5.18.

The first cause of ambiguity in the grammar is that the precedence of the operator + and * is not respected. Both parsing trees in Fig. 5.18 are valid. We need to force only the structure of Fig. 5.18(*a*) to be legal in an ambiguous grammar in which the * operator has higher precedence then + because $a + a*a$ is equavalent to $a + (a*a)$ and $a*a + a$ is equivalent to $(a*a) + a$.

Again, the parse trees for $a + a + a$ are

(*a*) (*b*)

Fig. 5.19.

The cause of the ambiguity in string $a + a + a$ is that the associativity it of + operator is not respected. Since, we assume the + operator is left associative the string $a + a + a$ is equivalent to $(a + a) + a$. Thus, again we need to force only the structure of Fig. 5.19(*a*) to be legal in an ambiguous grammar.

Because we do not want $E + E$ (as we have seen, $E \to E + E$, by itself is enough to produce ambiguity) we will not think of expressions involving + as being sums of other expressions. They are obviously sums of something: Let we use the word **terms** to stand for this things that are added *i.e.* combined using + to create expressions. The corresponding variable in the grammar will be T. Expressions can also be products; because the two expressions $a + b * c$ and $a * b + c$ are both sums, it is more appropriate to say that terms can be products. Let us say that "factors" are the things that are multiplied (combined using *) to produce terms. The corresponding variable in the grammar will be F.

Since the multiplication have higher precedence over addition. We can say that Expressions are sums of one or more terms, and terms are products of one or more factors.

Now, we must deal with parentheses. We might say that (E) could be an expression, a term or a factor. However, evaluation of a parenthetical expression takes precedence over any operators outside the parentheses. Factors are evaluated first in hierarchy and therefore the appropriate choice is to say that (E) is a factor and that E itself can be an arbitrary expression.

Now, expressions are sums of one or more terms, terms are products of one or more factors, and factors are either parenthetical expressions or single identifiers. Sum of terms might suggest something such as $E \to T + T \mid T$. (keep in mind that an expression can consist of single term)

However, to obtain the sum of three terms with this approach, we would be forced to try $T \to T + T$ or something comparable, and again we would have ambiguity. What we say instead is that an expression is either a single term or the sum of a term and another experssion. The only question is whether we want $E \to E$

$+ T$ or $E \to T + E$. Since + operator is left associative, we would probably choose $E \to E + T$ as more appropriate. Similarly, we choose the production $T \to T * F$ rather than $T \to F * T$.

The resulting unambiguous grammar is

$$E \to E + T \mid T$$
$$T \to T * F \mid F$$
$$F \to (E) \mid a$$

In the similar fashion we can find the unambiguous grammar for algebraic expressions having other operators.

Example 5.39. Eliminate the ambiguity from the following grammar

<statement> →if <expression> then <st>
| if <expression> then <st>
else <st> | <other statement>
<other statement> →$S_1 \mid S_2 \mid S_3 \mid \ldots S_n$
<expression> →$E_1 \mid E_2 \mid E_3 \ldots \mid E_n$

Solution. The Grammar is ambiguous since the string "if E_1 then if E_2 then S_1 else S_2" has two parse trees.

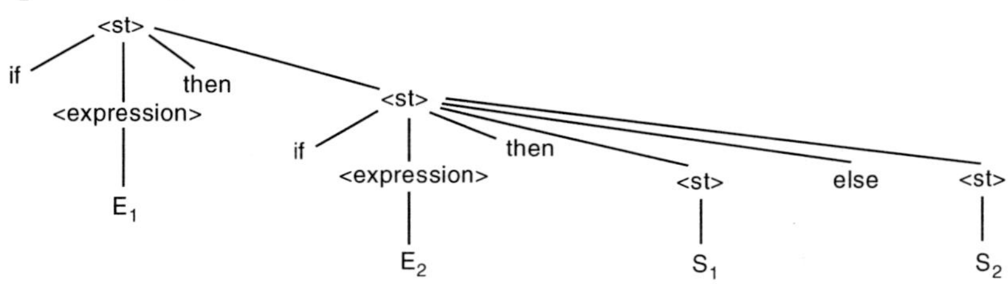

Fig. 5.20.

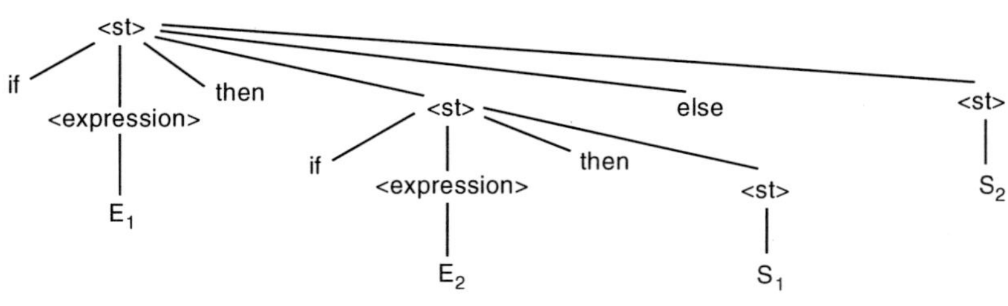

Fig. 5.21.

In all programming languages with conditional statements of this form, the first parse tree is preferred. The general rule is "Match each else with the closest previous unmatched then". Then unambiguous grammar is as follows :

<st> → <matched-statement> | <unmatched-statement>

<matched-statement> → if <expression> then <matched-statement>
else <matched statement> | <other statement>

<unmatched-statement> → if <expression> then <st>
| if <expressions> then <matched-statement>
else <unmatched-statement>

<expression> $\rightarrow E_1 \mid E_2 \mid \ldots E_n$

<unmatched statement> $\rightarrow S_1 \mid S_2 \ldots S_n$

<matched statement> $\rightarrow S_1 \mid S_2 \ldots \mid S_n$

The idea in this grammar is that a statement appearing between a *then* and an *else* must be "matched" *i.e.* it must not end with an unmatched *then* followed by *any statement*, for the *else* would then be forced to match this unmatched *then.* A matched statement is either an if then else statement containing no unmatched statements or it is any other kind of unconditional statement.

This grammar generates the same set of strings as previous one, but it allows only one parsing.

5.4.2. Inherent Ambiguity

TIPS

"If L is a context-free language for which there exists an unambiguous grammar, then L is said to be unambiguous. If every grammar that generates L is ambiguous, then the language is said to be "inherently ambiguous".

Inherently Ambiguous Context-Free Grammar

If even one grammar for L is unambiguous, then L is an unambiguous language. Let us see an example of an inherently ambiguous CFL.

$$L = \{a^n b^n c^m d^m \mid n > 1, m >= 1\} \cup \{a^n b^m c^m d^n \mid n >= 1, m >= 1\}$$

Thus, there are as many a's as b's and as many c's and d's, or there are as many a's as d's and as many b's as c's.

A grammar for L

$$\begin{aligned} S &\rightarrow AB \mid C \\ A &\rightarrow aAb \mid ab \\ B &\rightarrow cBd \mid cd \\ C &\rightarrow aCd \mid aDd \\ E &\rightarrow bDc \mid bc \end{aligned}$$

Parse Trees for *aabbccdd* :

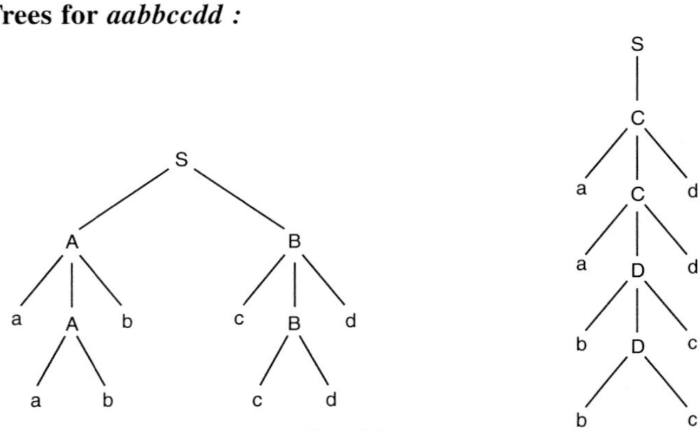

Fig. 5.22.

5.4.2.1. Proving Inherent Ambiguity

The previous parse trees and leftmost derivations prove that language L has at least some ambiguous grammars. This doesn't prove that all of L's grammars are ambiguous, so we don't know if L is ambiguous.

Proof Idea :

We argue that all but a finite number of the strings, whose counts of the four symbols a, b, c, d are all equal, must be generated in two different ways :

(*i*) a's and b's are equal and c's and d's are equal.

(*ii*) a's and d's are equal and b's and c's are equal.

(*iii*) The only way to generate strings where a's and b's are equal is with a variable like A.

(*iv*) Variations can occur, but we must follow the mechanism for generating a's in a way that matches the b's.

(*v*) This idea carries for the variables B, C, and D.

(*vi*) This argument will show that no matter what modifications we make to the basic grammar, it will generate at least some of the strings of the form $a^n b^n c^n d^n$.

(*vii*) Following this proof idea will show that L is inherently ambiguous, because no matter how the grammar is changed, it will remain ambiguous.

Let us consider an example for it :

Example 5.40. Show that

$$L = \{a^n b^n c^m\} \cup \{a^n b^m c^m\}$$

with n and m non-negative, is an inherently ambigous context-free language.

Solution. Let us say $\quad L = L_1 \cup L_2$

where $\quad L_1 = \{a^n b^n c^m\}$

and $\quad L_2 = \{a^n b^m c^m\}$

Let us write CFG for L_1, let it is G_1, as follows :

$$S_1 \rightarrow S_1 c/A$$
$$A \rightarrow aAb/\in$$

Similarly we can write CFG for L_2, Let it is G_2, as follows :

$$S_2 \rightarrow aS_2/B$$
$$B \rightarrow bBc/\in$$

Now with the help of G_1 and G_2 we can write CFG for the language L, as follows

$$S \rightarrow S_1/S_2$$

where S is starting non-terminal for CFG of language L.

The grammar is ambiguous since the string $a^n b^n c^m$ has two distinct derivation, one starting with $S \Rightarrow S_1$, and another with $S \Rightarrow S_2$. It does of course not follows that L is in herently ambiguous as there might exist some other non-ambiguous grammar for it. But in some way L_1 and L_2 have some conflicting requirements, the first putting a restriction on the number of a's and b's, while the second does the same for b's and c's. A few tries will quickly convince us of the impossiblity of combining these requirements in a single set of rules that cover the case $n = m$ uniquely.

Now let us discuss some more example about these concepts.

Example 5.41. Design a CFG for the language $L = \{a^n b^m : n \le m + 3\}$.

Solution. Let us solve this problem for $n = m + 3$. Then add more b's.

Let us assume CFG for above concept is $G = (V_n, V_t, P, S)$

where

$$V_n = \{S, A, B\}$$
$$V_t = \{a, b, \in\}$$

and productions are defined as :

$$S \to aaaA$$
$$A \to aAb/B$$
$$B \to Bb/\in$$

But when we analyse this solution, then it is found incomplete since it creates at least three a's. To take care of these cases $n = 1, 2$ we add

$$S \to \in/aA/aaA$$

So finally solution is

$$S \to \in/aA/aaA$$
$$S \to aaaA$$
$$A \to aAb/B$$
$$B \to Bb/\in$$

Example 5.42. Find the context-free grammar for the set of all regular expression on the alphabet $\{a, b\}$.

Solution. Let us analyse the regular expression on the alphabet $\{a, b\}$ regular expression by any combination of a's and b's is $r = (a + b)^*$.

Let CFG for regular expression is $G = (V_n, V_t, P, S)$

$$V_n = \{S, A, B\}$$
$$V_t = \{a, b, \in\}$$

and P is defined as

$$S \to AB/BA \qquad (P_1 \text{ say})$$
$$A \to aA/\in \qquad (P_2 \text{ say})$$
$$B \to bB/\in \qquad (P_3 \text{ say})$$

By the help of P_1, P_2 and P_3 we can derive any combination of a's and b's.

Example 5.43. Let $L = \{a^n b^n : n \ge 0\}$
Show that L^2 is context-free.

Solution. $L = \{a^n b^n : n \ge 0\}$

Let CFG for L is $G = (V_n, V_t, P, S)$

$$V_n = \{s\}$$
$$V_t = \{a, b, \in\}$$

P is defined as $S \rightarrow aSb$

$S \rightarrow \in$

Clearly L is context-free language, since there exist a CFG for given language.

Now consider a grammar, which is CFG and production of this CFG are defined as follows.

$$S_1 \rightarrow SS$$
$$S \rightarrow aSb$$
$$S \rightarrow \in$$

Now, when carefully analyse these production then we find that, it is the CFG for the language L^2. So L^2 is context free language, provided L is context free.

Example 5.44. Show that the following grammar is ambiguous.

$$\boldsymbol{S \rightarrow AB/aaB}$$
$$\boldsymbol{A \rightarrow a/Aa}$$
$$\boldsymbol{B \rightarrow b.}$$

Solution. Given CFG is

$$S \rightarrow AB/aaB$$
$$A \rightarrow a/Aa$$
$$B \rightarrow b$$

The string $w = aab$ has two distinct leftmost derivations, as follows :

(1) $S \Rightarrow aa\underline{B} \Rightarrow aab$

(2) $S \Rightarrow \underline{A}B \Rightarrow \underline{A}aB \Rightarrow aa\underline{B} \Rightarrow aab$

Now let us see parse trees for the same

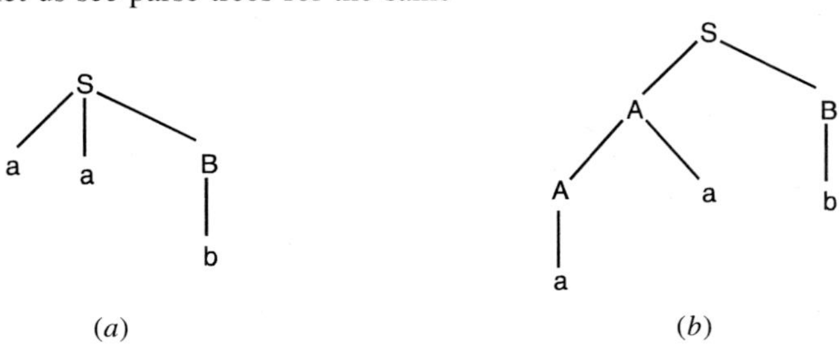

Fig. 5.23.

Example 5.45. Find a derivation free of $\boldsymbol{a * b + a * b}$ given that $\boldsymbol{a * b + a * b}$ is in $\boldsymbol{L(G)}$, where $\boldsymbol{G}$ is given by

$$S \rightarrow S + S/S * S$$
$$S \rightarrow a/b.$$

Solution. Given grammar is

$$S \rightarrow S + S/S * S$$
$$S \rightarrow a/b$$

and string is $w = a * b + a * b$

$S \Rightarrow \underline{S} + S \Rightarrow \underline{S} * S + S \Rightarrow a * \underline{S} + S \Rightarrow a + b + \underline{S} \Rightarrow a * b + \underline{S} * S \Rightarrow a * b + a * \underline{S} \Rightarrow a * b + a * b$

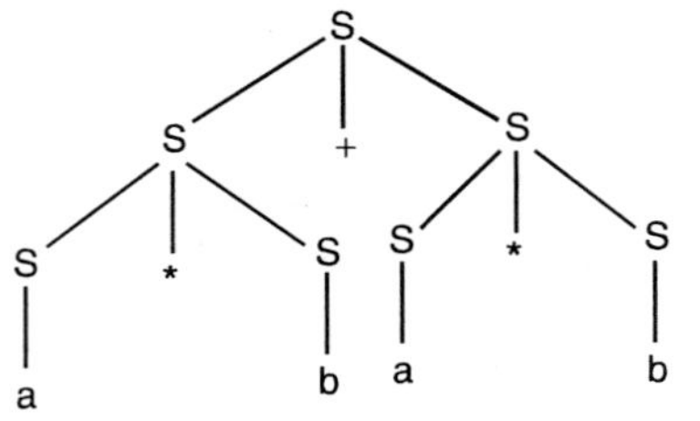

Fig. 5.24. Derivation tree for **Example** 5.44.

Example 5.46. Consider the grammar

$$S \rightarrow aB/bA$$
$$A \rightarrow aS/bAA/a$$
$$B \rightarrow bS/aBB/b$$

For the string *aaabbabbba* find

(*i*) The left most derivation and left most derivation tree.

(*ii*) The right most derivation and right most derivation tree.

Solution. Given string is *aaabbabbba.*

(*i*) Left most derivation

$S \Rightarrow a\underline{B} \Rightarrow aa\underline{B}B \Rightarrow aaa\underline{B}BB \Rightarrow aaab\underline{B}B \Rightarrow aaabb\underline{B} \Rightarrow aaabba\underline{B}B \Rightarrow aaabbab\underline{B} \Rightarrow aaabbabbS \Rightarrow aaabbabbb\underline{A} \Rightarrow aaabbabbba.$

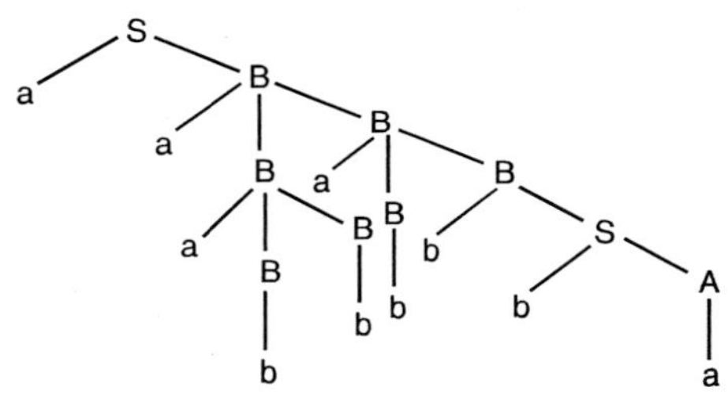

Fig. 5.25. Left most derivation tree for given string.

(*ii*) Right most derivation :

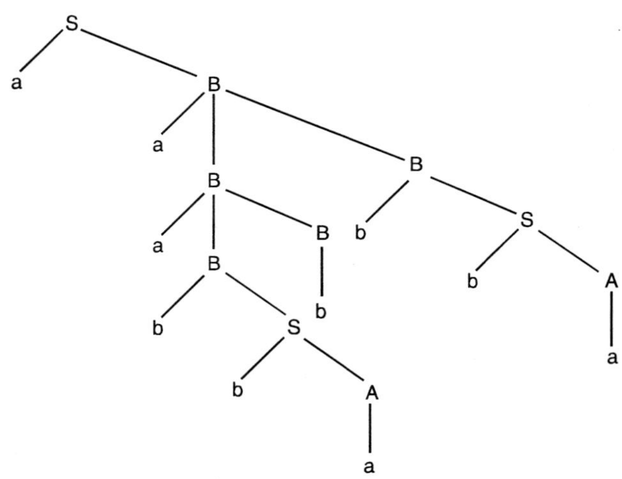

Fig. 5.26. Right most derivation tree for given string.

$S \Rightarrow a\underline{B} \Rightarrow aaB\underline{B} \Rightarrow aaBb\underline{S} \Rightarrow aaBbb\underline{A} \Rightarrow aa\underline{B}bba \Rightarrow aaaB\underline{B}bba \Rightarrow aaa\underline{B}bbba$
$\Rightarrow aaab\underline{S}bbba \Rightarrow aaabb\underline{A}bbba \Rightarrow aaabbabbba$

Example 5.47. Show that the grammar

$$S \rightarrow a/abSb/aAb$$
$$A \rightarrow bS/aAAb$$

is ambiguous.

Solution. Given grammar

$$S \rightarrow a \qquad (P_1 \text{ say})$$
$$S \rightarrow abSb \qquad (P_2 \text{ say})$$
$$S \rightarrow aAb \qquad (P_3 \text{ say})$$
$$A \rightarrow bS \qquad (P_4 \text{ say})$$
$$A \rightarrow aAAb \qquad (P_5 \text{ say})$$

Let us consider a string $w = abab$, it has two different derivations, as follows :

(*i*) $S \Rightarrow abSb \Rightarrow abab$ (by using P_2 and P_1)

(*ii*) $S \Rightarrow aAb \Rightarrow abSb \Rightarrow abab$ (by using P_3, P_4 and P_1)

Example 5.48. Show that grammar

$$S \rightarrow aB/ab$$
$$A \rightarrow aAB/a$$
$$B \rightarrow ABb/b$$

is ambiguous.

Solution. Let us consider a string

$$w = ab$$

which can be easily generated by given grammar.

Now $w = ab$ as two different derivation tree as follows :

(*a*)

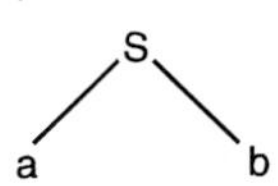

(*b*)

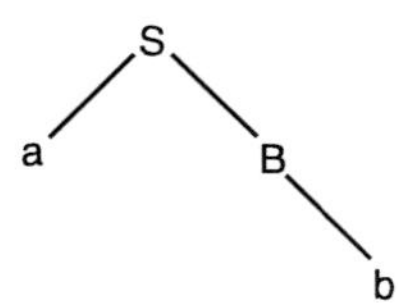

So clearly given grammar is ambiguous.

Example 5.49. Construct CFG to generate the following :

(*i*) $\{0^m 1^n / m \neq n, m, n \geq 1\}$

(*ii*) $\{a^l b^m c^n$/**one of the** l, m, n **equals** *1* **and the remaining two are equal}.**

(*iii*) $\{0^m 1^n / 1 \leq m \leq n\}$.

(*iv*) $\{a^l b^m c^n / l + m = n\}$.

Solution. (*i*) $\quad L = \{0^m 1^n / m \neq n, m, n \geq 1\}$.

Let CFG is $\quad G = (V_n, V_t, P, S)$

$$S \rightarrow 0S1/0A/1B/0A$$
$$A \rightarrow 0A/0$$
$$B \rightarrow 1B/1$$

(*ii*) Let $G = \{V_n, V_t, P, S)$

$$V_n = \{S, A, B, C\}$$
$$V_t = \{a, b, c\},$$

S is starting non-terminal production rules are given below :

$$S \rightarrow aA/aBc/Cc$$
$$A \rightarrow bAc/bc$$
$$B \rightarrow aBc/b$$
$$C \rightarrow aCb/ab$$

(*iii*) Let CFG for $L = \{0^m 1^n / 1 \leq m \leq n\}$ is

$$G = (V_n, V_t, P, S)$$
$$V_n = \{S, A\}$$
$$V_t = \{0, 1\}$$

Productions are as follows :

$$S \rightarrow 0S1/01/0A1$$
$$A \rightarrow 1A/1$$

By applying first production. $S \rightarrow 0S1/01$, we can get $0^m 1^n$ where $m = n$.

When $m = n = 1$, we have $S \rightarrow 01$. Thus $01 \in L(G)$.

By applying $S \rightarrow 0S1$, $S \rightarrow 0A1$, $A \rightarrow 1A$ and $A \rightarrow 1$, we can generate $0^m 1^n$ where $m < n$. Thus $L(G) = L$.

(*iv*) Let CFG for $L = \{a^l b^m c^n \,/\, l + m = n\}$ is

$$G = (V_n, V_t, P, S)$$
$$V_n = \{S, A\}$$
$$V_t = \{a, b, c\}$$

Production 'P' are given below :

$$S \rightarrow aSc/ac/bc/bAc$$
$$A \rightarrow bAc/bc$$

A typical string in the given language can be written in the form $a^l b^m c^m c^l$ where $l, m \geq 0$.

$S \rightarrow aSc$, $S \rightarrow ac$ general $a^l c^l$ for $l \geq 1$

$S \rightarrow bAc$, $A \rightarrow bAc$, $A \rightarrow bc$ generate $b^m c^m$ for $m \geq 1$. For getting $a^l b^m$, $c^m c^l$, we have to apply $S \rightarrow aSc$, $S \rightarrow bc$, $S \rightarrow bAc$, $A \rightarrow bAc$ and $A \rightarrow bc$ are to be applied. For $m = 1$, $S \rightarrow$ bc has to be applied. For $m > 1$, we have to apply $S \rightarrow bAc$, $A \rightarrow bAc$ and $A \rightarrow bc$ repeatedly. The terminal c is added whenever the terminal a or b is added in the course derivation. This take a care of the condition $l + m = n$. Thus $L = L(G)$.

EXERCISES

1. Find CFG's that generate these regular language over the alphabet $\Sigma = \{a, b\}$.
   (*i*) The language defined by $(aaa + b)^*$.
   (*ii*) The language defined by $(a + b)^* (bbb + aaa) (a + b)^*$.
   (*iii*) All the strings without the substring *aaa*.
   (*iv*) All strings that end in *b* and have an even number of *b*'s in total.
   (*v*) The set of all strings of odd length.
   (*vi*) All strings with exactly one a or exactly one *b*.
   (*vii*) All strings with an odd number of *a*'s or even number of *b*'s.
2. Write CFG for the following languages :
   (*i*) $\{a^m b^n / m >= n\}$
   (*ii*) $\{a^m b^n c^p d^q / m + n = p + q\}$
   (*iii*) $\{u\ a\ w\ b / \mu, w \in \{a, b\}^*, |\mu| = |w|\}$
3. Consider the grammar

$$G = (V_n, V_t, P, S)$$

where
$$V_n = \{E, I\}$$
$$V_t = \{a, b, c, +, *, (,)\},$$
$$S = E, \text{ and } P \text{ is given by}$$
$$E \to I$$
$$E \to E + E$$
$$E \to E * E$$
$$E \to (E)$$
$$I \to a/b/c$$

show that this grammar is ambiguous by deriving two parse tree for string $a + b * c$.
4. Consider the grammar with production

$$S \to aaB$$
$$A \to bBb/\in$$
$$B \to Aa$$

Show that the string *aabbabba* is not accepted by this grammar.
5. The following grammar generates prefix expression with operands x and y and binary operators +, –, and *

$$E \to +EE/*EE/-EE/x/y.$$

Find the leftmost and rightmost derivations for the string $+ * - xyxy$.
6. (*i*) Find the left most derivation for the word *abba* in the grammar

$$S \to AA$$
$$A \to aB$$
$$B \to bB/\in$$

(*ii*) Find the left most derivation for the word *abbabaabbbabbab* in the CFG

$$S \to SSS/aXb$$
$$X \to ba/bba/abb$$

7. Find the regular expression and defined the language of following CFG's.

(i) $S \rightarrow aX/bS/a/b$

$X \rightarrow aX/a$

(ii) $S \rightarrow bS/aX/b$

$X \rightarrow bX/aS/a$

(iii) $S \rightarrow aaS/abS/baS/bbS/\in$

(iv) $S \rightarrow aB/bA/\in$

$A \rightarrow aS$

$B \rightarrow bS$

(v) $S \rightarrow aB/bA$

$A \rightarrow aB/a$

$B \rightarrow bA/b$

8. Starting with the alphabet

$$\Sigma = \{a, b, (,), +, *\}$$

Find a CFG that generates all regular expressions.

9. Give the derivation tree for $(((a + b)(c)) + a + b$, using the following grammar

$E \rightarrow T$

$T \rightarrow F$

$F \rightarrow I$

$E \rightarrow E + T$

$T \rightarrow T * F$

$F \rightarrow (E)$

$I \rightarrow a/b/c$

10. Show that the language $L = \{wwR : w \in \{a, b\}^*\}$ is not inherently ambiguous.

11. Write the CFG for the following language (with $n \geq 0$, $m \geq 0$)

(i) $L = \{a^n b^m : n \neq m - 1\}$

(ii) $L = \{w \in \{a, b\}^* : n_a(w) = 2n_b(w) + 1\}$.

12. Find the CFG for the following language (with $n \geq 0$, $m \geq 0$).

(i) $L = \{a^n b^m c^k, k \neq n + m\}$

(ii) $L = \{a^n b^n c^k : k \geq 3\}$

(iii) $L = \{w \in \{a, b, c\}^* : n_a(w) + n_b(w) \neq n_c(w)\}$

13. Define what one right mean by properly rested parenthesis structures involving two kind of paranthesis, say ( ) and [ ]. Intuitively, properly rested strings in this situation are ([ ]), ([[ ]]) [( )], but not ([ )] or (( ]]. Using your definition, give a CFG for generating all properly nested parentheses.

14. Find a CFG that generate all the production rules for CFG with $T = \{a, b\}$ and $V = \{A, B, C\}$.

*****

CHAPTER 6

Simplified Context-Free Grammar and Its Normal Form

INSIDE THIS CHAPTER

6.1. INTRODUCTION

We have already seen, that definition of a context grammar imposes no restriction whatsoever on the right side of a production. Eliminating rules of the form $X \to \in$ and $X \to Y$ made the arguments simpler. In many instances, it is desirable to place even more restriction on the grammar. Due to this, we need to look at method for transforming an arbitrary context free grammar into an equivalent one that satisfies certain restriction on its form. In this chapter, we are going to study several transformation and substitutions that will be useful in subsequent discussion.

We will also see normal forms for context free grammar. These are the form of context free gramars, which are broad enough so that any grammar has an equivalent normal form version. We will study two of the most useful of these, the Chomsky Normal Form (CNF) and the Greibach Normal Form (GNF).

6.2. REDUCTION OF CONTEXT FREE GRAMMAR

There are several ways in which one can restrict the format of context free grammar without reducing the language generation power of context free grammar. Let L be a non-empty context free language, then it can be generated by a context-free grammar G with the following properties :

(*a*) We must eliminate useless symbols, those variables or terminals that do not appears in any derivation of a terminal string from the start symbol.

(*b*) We must eliminate $\in$-productions, those of the form $X \rightarrow \in$ for some variable X.

(*c*) We must eliminate unit production, those of the form $X \rightarrow Y$ for variables X and Y.

6.2.1. Eliminating Use Less Symbols

Here we are going to identified those symbols, which do not play any role in the derivation of any string w in $L(G)$ (these symbols are called use less symbols) and then eliminate the identified production, which contains useless symbols, from the context free grammar.

A symbols Y in a context-free grammar is useful it and only if :

(*a*) $Y \overset{*}{\Rightarrow} w$, where $w \in L(G)$ and w in V_t^*, that is Y leads to a string of terminals. Here Y is said to be "generating".

(*b*) If there is a derivation $S \overset{*}{\Rightarrow} \alpha Y \beta \overset{*}{\Rightarrow} w$, $w \in L(G)$, for same α and β, then Y is said to be reachable.

Surely a symbol that is useful will be both generating and reachable. If we eliminate the symbols that are not generating first, and then eliminate from the remaining grammar those symbols that are not reachable. Then, after this process, context free grammar will have only useful symbols.

There for reduction of a given grammar G, involves following steps :

(*a*) Identified non-generating symbols in given CFG and eliminate those productions which contains non-generating symbols.

(*b*) Identified non-reachable symbols in grammar and eliminate those productions which contains non-reachable symbols.

Let us consider examples :

Example 6.1. Consider a CFG

$$S \rightarrow AB/a$$
$$A \rightarrow b$$

Identified and eliminate useless symbols.

Solution. Given CFG is

$$S \rightarrow AB/a$$
$$A \rightarrow b$$

Here by observing the given CFG, it becomes very clear that B is non-generating symbol. Since A derives b, S derives a but B is not deriving any string w in the V_t^*.

So we can eliminate $S \rightarrow AB$ from the context free grammar, now CFG becomes

$$S \rightarrow a$$
$$A \rightarrow b$$

Here A is non-reachable symbol, since it can not be reached by starting non-terminal S.

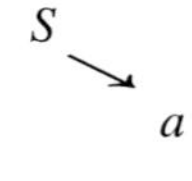

So we can eliminate $A \to b$, now grammar is

$$S \to a$$

which is reduced grammar.

Example 6.2. Remove the useless symbol from the given context free grammar :

$$S \to aB/bX$$
$$A \to BAd/bSX/a$$
$$B \to aSB/bBX$$
$$X \to SBD/aBx/ad.$$

Solution. First we choose those non-terminals which are deriving to the strings of terminals. The non-terminals A and X directly deriving to the strings of terminal a and ad respectively. Hence they are useful symbols. Since $S \to bX$ and X is useful symbol, hence S is also useful symbol. But B does not derive any string w in V_t^* so clearly B is a non-generating symbol, so eliminate those productions which contains B. Now grammar becomes :

$$S \to bX$$
$$A \to bSX/a$$
$$X \to ad$$

Again here A is non-reachable, since A can not be reached from S (starting non-terminal), it can be understand by following parse tree in Fig. 6.1.

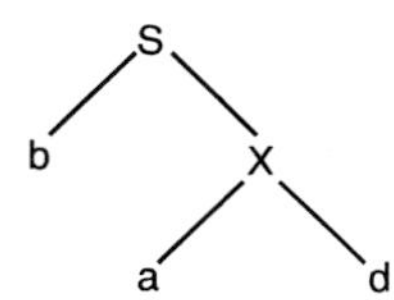

Fig. 6.1.

So eliminate $A \to bSX/a$, then reduced CFG is

$$S \to bX$$
$$X \to ad.$$

Example 6.3. Consider the following grammar and obtain an equivalent grammar containing no useless grammar symbol.

$$A \to xyz/Xyzz$$
$$X \to Xz/xYz$$
$$Y \to yYy/Xz$$
$$Z \to Zy/z.$$

Solution. Since $A \to xyz$ and $Z \to z$, hence A and Z are directly deriving to the string of terminals. Hence they are useful symbols.

Hence X and Y do not lead to a string of terminals, that is, X and Y are useless symbols. Therefore by eliminating these productions we get :

$$A \to xyz$$
$$Z \to Zy/z$$

Now since A is the starting non-terminal and right side of A does not contain Z, it means Z is not reachable. Hence after eliminating $Z \to Zy/z$, grammar becomes

$$A \to xyz.$$

which is a grammar containing no useless symbols.

Example 6.4. Find the reduced grammar that is equivalent to the CFG given below :

$$S \to aC/SB$$
$$A \to bSCa$$
$$B \to aSB/bBC$$
$$C \to aBC/ad.$$

Solution. Since $C \to ad$, therefore, C is generating symbol. Since $S \to aC$, therefore S is also a useful symbol according to production $A \to bSCa$, A is also generating. Right side of $B \to aSB$ and $B \to bBC$ contains B, and B is not terminating so B is not a generating symbol so we can eliminate those productions and grammar becomes

$$S \to aC$$
$$A \to bSCa$$
$$C \to ad$$

Now since S is the start symbol and right side of S does not contain A, hence A is not reachable as :

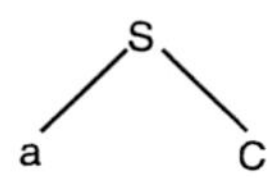

Fig. 6.2.

So by eliminating A we get

$$S \to aC$$
$$C \to ad$$

which is reduced grammar equivalent to the given grammar, containing no useless symbol.

6.2.2. Removal of Unit Production

Let us first define unit production :

A production of the form

Non-terminal $\to$ One non-terminal

that is a production of the form $A \to B$ (where A and B, both are non-terminals) is called unit production. Unit production increase the cost of derivation in a grammar. Following algorthim can be used to eliminate the unit production.

Algorithm : Removal of unit productions $\Rightarrow$

While (there exist a unit production, $A \to B$)

{

select a unit production $A \to B$, such that there exist a production $B \to \alpha$, where α is a terminal.

For (every non-unit production, $B \to \alpha$)

Add production $A \to \alpha$ to the grammar

Eliminate $A \to B$ from the grammar.

}.

Now let us apply this algorthim on the given CFG's.

Example 6.5. Consider the context free grammar G.

$$S \to AB$$
$$A \to a$$
$$B \to C/b$$
$$C \to D$$
$$D \to E$$
$$E \to a$$

remove the unit production.

Solution. Given CFG

$$S \to AB$$
$$A \to a$$
$$B \to C/b$$
$$C \to D$$
$$D \to E$$
$$E \to a$$

Contain three unit productions

$$B \to C$$
$$C \to D$$
$$D \to E$$

Now to remove unit production $B \to C$, we see if there exists a production whose left side has C and right side contains a terminal (*i.e.* $C \to a$), but there is no such productions in G. Similar things holds for production $C \to D$. Now we try to remove unit production $D \to E$, because there is a production $E \to a$. Therefore, eliminate $D \to E$ and introduce $D \to a$, grammar becomes

$$S \to AB$$
$$A \to a$$
$$B \to C/b$$
$$C \to D$$
$$D \to a$$
$$E \to a$$

Now we can remove $C \to D$ by using $D \to a$, we get

$$S \to AB$$
$$A \to a$$
$$B \to C/b$$
$$C \to a$$
$$D \to a$$
$$E \to a$$

Similarly, we can remove $B \to C$ by using $C \to a$, we obtain

$$S \to AB$$
$$A \to a$$
$$B \to a/b$$

$$C \to a$$
$$D \to a$$
$$E \to a$$

Now it can be easily seen that productions $C \to a$, $D \to a$ $E \to a$ are useless because if we start deriving from S, these productions will never be used. Hence eliminating them gives,

$$S \to AB$$
$$A \to a$$
$$B \to a/b$$

which is completely reduced grammar.

Example 6.6. Consider the following unambiguous expression grammar

$$I \to a/b/Ia/Ib/I0/I1$$
$$F \to I/(E)$$
$$T \to F/T*F$$
$$E \to T/E + T$$

Identified unit production and remove them.

Solution. In the given grammar

$$F \to I$$
$$T \to F$$
$$E \to T$$

are three unit production.

Unit production $F \to I$ can be removed by the help of

$$I \to a/b/Ia/Ib/I0/I1$$

and grammar becomes

$$I \to a/b/Ia/Ib/I0/I1$$
$$F \to (E)/a/b/Ia/Ib/I0/I1$$
$$T \to F/T*F$$
$$E \to T/E + T$$

Now we can eliminate $T \to F$ by the help of

$$F \to (E)/a/b/Ia/Ib/I0/I1$$

and now grammar becomes

$$I \to a/b/Ia/Ib/I0/I1$$
$$F \to (E)/a/b/Ia/Ib/I0/I1$$
$$T \to T*F/(E)/a/b/Ia/Ib/I0/I1$$
$$E \to T/E + T$$

Now let us remove $E \to T$ by the help

$$T \to T*F/(E)/a/b/Ia/Ib/I0/I1$$

and finally, unit production free grammar is

$$I \rightarrow a/b/Ia/Ib/I0/I1$$
$$F \rightarrow (E)/a/b/Ia/Ib/I0/I1$$
$$T \rightarrow T*F/(E)/a/b/Ia/Ib/I0/I1$$
$$E \rightarrow E + T/T * F/(E)/a/b/Ia/Ib/I0/I1$$

Example 6.7. Identified and remove the unit productions from following grammar

$$S \rightarrow A/bb$$
$$A \rightarrow B/b$$
$$B \rightarrow S/a$$

Solution. Here unit productions are

$$S \rightarrow A$$
$$A \rightarrow B$$
$$B \rightarrow S$$

We list all unit productions and sequences of unit productions.Then we will handle one unit production or sequence at a time and try to prove generating and reachable for each non-terminal involve in the unit productions, as follows :

$S \rightarrow A$	generate $S \rightarrow b$
$S \rightarrow A \rightarrow B$	generate $S \rightarrow a$
$A \rightarrow B$	generate $B \rightarrow a$
$A \rightarrow B \rightarrow S$	generate $A \rightarrow bb$
$B \rightarrow S$	generate $B \rightarrow bb$
$B \rightarrow S \rightarrow A$	generate $B \rightarrow b$

The new CFG becomes

$$S \rightarrow bb/b/a$$
$$A \rightarrow b/a/bb$$
$$B \rightarrow a/bb/b$$

which hold no unit productions.

6.2.3. Removal of $\in$-productions and Nullable Non-terminal

Any context-free language in which $\in$ is a word, must have some $\in$-productions in its grammar since otherwise we could never derive the word $\in$ from S. The statement is obvious, but it should be given some justification, mathematically, this is easy; we observe that $\in$-productions are only productions that shorten the working string. If we begin with the string S and apply only non-$\in$-productions, we never develop a word of length zero.

We now form our attention to the eliminate the productions of the form $A \rightarrow \in$, which are called $\in$-productions. Surely if $\in$ is in $L(G)$, we can not eliminate all $\in$-productions from G, but if $\in$ is not in $L(G)$, we can eliminate all $\in$-productions from G.

"In a given CFG, we call a non-terminal N Nullable if

There is a production $N \rightarrow \in$ or

There is a derivation that starts at N and leads to $\in$:

$$N \Rightarrow \ldots \Rightarrow \in"$$

To eliminate $\in$-productions from a grammar G we use the following technique :

"If $A \rightarrow \in$ is a prodoction to be eliminated then we look for all productions, whose right side contains A, and replace each occurrence of A in each of these productions to obtain the non $\in$-productions. Now these resultant non $\in$-production must be added to the grammar to keep the language generated, the same."

Let us apply this concept :

Example 6.8. Consider the following grammar

$$S \rightarrow aA$$
$$A \rightarrow b/\in$$

remove the Nullable non-terminal.

Solution. Given CFG is $S \rightarrow aA$

$$A \rightarrow b/\in$$

Here $A \rightarrow \in$ is $\in$-production, so let us apply above described concept, put $\in$ in place of A, at the right side of productions and add the resulted productions to the grammar.

Here is only one productions $S \rightarrow aA$ whose right side contains A. So, replacing A by $\in$ in this production, we get $S \rightarrow a$; now add this new productions **to keep the language generated by this grammar same.** Therefore, the $\in$-free grammar, equivalent to above grammar is

$$S \rightarrow aA$$
$$S \rightarrow a$$
$$A \rightarrow b$$

or

$$S \rightarrow aA/a$$
$$A \rightarrow b$$

there is no $\in$-production in the above context-free grammar.

Example 6.9. Consider the following grammar G

$$S \rightarrow ABAC$$
$$A \rightarrow aA/\in$$
$$B \rightarrow bB/\in$$
$$C \rightarrow c$$

remove the $\in$-production from the above grammar.

Solution. Here we have two $\in$-production to remove $A \rightarrow \in$ and $B \rightarrow \in$. To eliminate $A \rightarrow \in$ from the grammar, the non $\in$-productions to be added are obtained as follows :

List of productions whose right side contain A are :

$$S \rightarrow ABAC$$
$$A \rightarrow aA$$

So on replacing each occurrence of A by $\in$ (one by one), we get four new productions to be added to the grammar :

$$S \rightarrow ABC/BAC/BC$$
$$A \rightarrow a$$

add these productions to the grammar and eliminate $A \rightarrow \in$, to obtain the following grammar

$$S \rightarrow ABAC/ABC/BAC/BC$$
$$A \rightarrow aA/a$$
$$B \rightarrow bB/\in$$
$$C \rightarrow c$$

Now to eliminate $B \rightarrow \in$, list all productions whose right side contains B, these are :

$$S \rightarrow ABAC/BAC/ABC/BC$$
$$B \rightarrow bB$$

Replace occurrence of B in each of these productions to obtain non $\in$-productions to be added to the grammar, we get following new productions.

$$S \rightarrow AAC/AC/C$$
$$B \rightarrow b$$

Add these productions to the grammar and eliminate $B \rightarrow \in$ from the grammar, to obtain

$$S \rightarrow ABAC/BAC/ABC/BC/AAC/AC/C$$
$$A \rightarrow aA/a$$
$$B \rightarrow bB/b$$
$$C \rightarrow c$$

which is the grammar without any $\in$-production.

Example 6.10. Consider the following grammar and remove $\in$-productions

$$S \rightarrow aSa$$
$$S \rightarrow bSb/\in$$

Solution. Here $S \rightarrow \in$ is the $\in$-production, so let us replace occurrence of S by $\in$ in the following productions

$$S \rightarrow aSa$$
$$S \rightarrow bSb$$

We will get following new productions

$$S \rightarrow aa/bb$$

add these productions in the grammar and remove $S \rightarrow \in$ from the grammar

$$S \rightarrow aSa/aa/bb/bSb$$

It is $\in$-production free grammar.

Example 6.11. Consider the following grammar

$$S \rightarrow a/Xb/aYa$$
$$X \rightarrow Y/\in$$
$$Y \rightarrow b/X$$

eliminate $\in$-productions.

Solution. Here $X \rightarrow \in$ is $\in$-productions. So this problem can also solved in similar fashion and grammar without $\in$-productions will be

$$S \rightarrow a/Xb/aYa/b/aa$$
$$X \rightarrow Y$$
$$Y \rightarrow b/X.$$

Example 6.12. Design a CFG for regular expression
$r = (a + b)^* \, bb \, (a + b)^*$, which is free from $\in$-productions.

Solution. Let CFG G be required context free grammar with $\in$-productions

$$S \rightarrow XY$$
$$X \rightarrow Zb$$
$$Y \rightarrow bW$$
$$Z \rightarrow AB$$
$$W \rightarrow Z$$
$$A \rightarrow aA/bA/\in$$
$$B \rightarrow Ba/Bb/\in$$

Finally here $A \rightarrow \in$ and $B \rightarrow \in$ are $\in$-productions and A, B are nullable non-terminals.

The CFG without $\in$-production can be achieved from G in similar fashion, as above discussed. Let it be G', then G' is

$$S \rightarrow XY$$
$$X \rightarrow Zb/b$$
$$Y \rightarrow bW/b$$
$$Z \rightarrow AB/A/B$$
$$W \rightarrow Z$$
$$A \rightarrow aA/bA/a/b$$
$$B \rightarrow Ba/Bb/a/b$$

Because $\in$ was not generated by G, the new CFG G' generates exactly the same language.

6.3. CHOMSKY NORMAL FORM

Definition

If a CFG has only production of the form
Non-terminals $\rightarrow$ string of exactly two non-terminals or of the form
Non-terminals $\rightarrow$ one terminal
is said to be chomsky normal form or CNF.

6.3.1. Conversion to Chomsky Normal Form

"Conversion of the grammar to CNF can be understood by the following example"

Let us conside a grammar

$G = (\{S, A, B\}, \{a, b\}, P, S)$, where S is the start symbol and P is given by

$$S \to bA/aB$$
$$A \to bAA/aS/a$$
$$B \to aBB/bS/a$$

Now let us find an equivalent chomsky normal form of it.

As we know that right side in CNF either contains two non-terminals or one terminal. So it is clear that in the first production above, we have to replace terminal 'b' by a non-terminal say C_b and 'a' by C_a, hence grammar becomes

$$S \to C_bA/C_aB$$
$$A \to C_bAA/C_aS/a$$
$$B \to C_aBB/C_bS/b$$
$$C_a \to a$$
$$C_b \to b$$

Now first rule is in CNF. In second rule C_bAA is not in CNF. So we can replace AA by a non-terminal say D and similarly BB by E, we obtain

$$S \to C_bA/C_aB$$
$$A \to C_bD/C_aS/a$$
$$B \to C_aE/C_bS/b$$
$$C_a \to a$$
$$C_b \to b$$
$$D \to AA$$
$$E \to BB$$

The above grammar is equivalent Chomsky normal form of the given grammar.

Example 6.13. Change the following grammar in to CNF

$$\mathbf{S \to 1A/0B}$$
$$\mathbf{A \to 1AA/0S/0}$$
$$\mathbf{B \to 0BB/1}$$

Solution. We know that in CNF, either two non-terminal or one terminal is on the right side.

Therefore replace AA by C_a and BB by C_b, 1 by C_1, 0 by C_0, we get

$$S \to C_1A/C_0B$$
$$A \to C_1C_a/C_0S/0$$

$$B \rightarrow C_0C_b/1$$
$$C_1 \rightarrow 1$$
$$C_0 \rightarrow 0$$
$$C_a \rightarrow AA$$
$$C_b \rightarrow BB$$

while the CNF form.

Example 6.14. Change the following grammar in to CNF

$$G = (\{S\}, \{a, b, c\}, \{S \rightarrow a/b/CSS\}, S\}.$$

Solution. Here productions are

$$S \rightarrow a$$
$$S \rightarrow b$$
$$S \rightarrow CSS$$

Replace SS by D, we get

$$S \rightarrow a$$
$$S \rightarrow b$$
$$S \rightarrow CD$$
$$D \rightarrow SS$$

which is in CNF form.

Example 6.15. Change the following grammar in to CNF

$$S \rightarrow abSb/a/aAb$$
$$A \rightarrow bS/aAAb.$$

Solution. Replace aA by B_a and Ab by B_b, we get

$$S \rightarrow abSb/a/aB_b$$
$$A \rightarrow bS/B_aB_b$$
$$B_a \rightarrow aA$$
$$B_b \rightarrow Ab$$

Now replace ab by C and Sb by D, we get

$$S \rightarrow CD/a/aB_b$$
$$A \rightarrow bS/B_aB_b$$
$$B_a \rightarrow aA$$
$$B_b \rightarrow aA$$
$$B_b \rightarrow Ab$$
$$C \rightarrow ab$$
$$D \rightarrow Sb$$

Now replace a by X_a and b by X_b, we get

$$S \rightarrow CD/a/X_aB_b$$
$$A \rightarrow X_bS/B_aB_b$$

$$B_a \rightarrow X_aA$$
$$B_b \rightarrow X_aB$$
$$C \rightarrow AX_b$$
$$D \rightarrow SX_b$$
$$X_a \rightarrow a$$
$$X_b \rightarrow b.$$

It is required CNF form.

Example 6.16. Convert CFG which is given below in to CNF form.

$$S \rightarrow bA/aB$$
$$A \rightarrow bAA/aS/a$$
$$B \rightarrow aBB/bS/b.$$

Solution. Let us replace b by C_b and a by C_a, then CFG becomes

$$S \rightarrow C_bA/C_aB$$
$$A \rightarrow C_bAA/C_aS/a$$
$$B \rightarrow C_aBB/C_bS/b$$
$$C_b \rightarrow b$$
$$C_a \rightarrow a$$

Now let us replace C_bA by D and C_aB by E then grammar becomes as follows :

$$S \rightarrow C_bA/C_aB$$
$$A \rightarrow DA/C_aS/a$$
$$B \rightarrow EB/C_bS/b$$
$$C_b \rightarrow b$$
$$C_a \rightarrow a$$
$$D \rightarrow C_bA$$
$$E \rightarrow C_aB$$

Now every production of the grammar is in the CNF form.

Example 6.17. Design a CFG for the language $L = \{a^{4n} \mid n \geq 1\}$ and convert that CFG into CNF form.

Solution. Let CFG for the language $L = \{a^{4n}/n \geq 1\}$ is G

$$G = (V_n, V_t, P, S)$$

where

$$V_n = \{S\}$$
$$V_t = \{a\}$$

and P is defined as

$$S \rightarrow aaaaS/aaaa$$

Now let us convert this CFG in to CNF form, replace each a by A then CFG becomes

$$S \rightarrow AAAAS/AAAA$$
$$A \rightarrow a$$

or we can also write

$$S \rightarrow AAAAS$$
$$S \rightarrow AAAA$$
$$A \rightarrow a$$

then CNF form will be

$$S \rightarrow AR_1$$
$$R_1 \rightarrow AR_2$$
$$R_2 \rightarrow AR_3$$
$$R_3 \rightarrow AS$$
$$S \rightarrow AR_4$$
$$R_4 \rightarrow AR_5$$
$$R_5 \rightarrow AA$$
$$A \rightarrow a.$$

6.4. GREIBACH NORMAL FORM

Definition For every context free language L without $\in$, there exist a grammar in which every production is of the form $A \rightarrow aV$, where 'A' is a variable, 'a' is exactly one terminal and 'V' is the string of none or more variables, clearly $V \in V_n^*$.

"In other words if every production of the context free grammar is of the form $A \rightarrow aV/a$, then it is in Greibach Normal Form."

Greibach normal form will be used to construct a push down atuomata that recognize the language generated by a context free grammar.

To convert a grammar to GNF we start with a production in which the left side has a higher numbered variable than first variable in the right side and make replacements in right side.

6.4.1. Left Recursion and its Removal

A CFG containing rules of the form

$$A \rightarrow A\alpha/\beta$$

is called left recursive in A. The language geneated by such rules is of the form $A \overset{*}{\Rightarrow} \beta\alpha^n$. If we replace the rules $A \rightarrow A\alpha/\beta$ by

$$A \rightarrow \beta A'/\beta$$
$$A' \rightarrow \alpha A'/\alpha$$

where A' is a new variable, then language generated by A is the same while no left-recursive A-rules are used in the derivation.

Left-recursion can be eliminuated by the following scheme :

Let production under left-recursion are

$$A \to A\alpha_1/A\alpha_2 \;..../A\alpha_m$$
$$A \to \beta_1/\beta_2/..../\beta_n$$

Choose a new non-terminal, say A′;

Add the new A'-rule as follows :

$$A' \to \alpha_i/\alpha_i A', \quad 1 \le i \le m$$

Replace the A-rule as follows :

$$A \to \beta_i/\beta_i A', \quad 1 \le i \le n$$

or, we can write as follows :

$$A \to \beta_1 A'/\beta_2 A'/..../\beta_n A'/\beta_1/\beta_2/...../\beta_n$$
$$A' \to \alpha_1 A'/\alpha_2 A'/..../\alpha_m A'/\alpha_1/\alpha_2/..../\alpha_m$$

or

$$A \to \beta_1 A'/\beta_2 A'/..../\beta_n A'$$
$$A \to \beta_1/\beta_2/..../\beta_n$$
$$A' \to \alpha_1 A'/\alpha_1 A'/..../\alpha_m A'$$
$$A' \to \alpha_1/\alpha_2/..../\alpha_m$$

6.4.2. Conversion of CFG into GNF

We know that every CFL 'L' without $\in$ can be generated by a grammar for which every productions are in Greibach Normal form.

We can convert CFG into GNF by following the number of steps discussed below :

Step 1 : Every CFG must be in CNF form, if not then convert it into CNF. Now rename all variables (V_n) by $A_1, A_2, ..., A_n$ where A_1 is start symbol.

Step 2 : Now modify the grammar so that every production are in following form :

$$A_i \to a\gamma$$

or

$$A_i \to A_j\gamma \text{ where } j > i \text{ and } \gamma \in V_n^*$$

To get CNF in above discussed format, we should follow the method :

Start with A_1 and proceed to A_n. Suppose upto A_{m-1} the above condition $1 \le i \le m-1$, $A_i \to A_j\gamma$ then $j > i$ is satisfied and for A_m we have production $A_m \to A_j\gamma$ such that $j < m$. To convert A_m production so that it satisfy the above condition we apply the substitute rule.

(that is if P contains productions like $A \to \alpha_1\beta\alpha_2$ and

$$B \to \beta_1 \mid \beta_2 \mid \beta_3 \; ... \; \beta_n$$

then we can write $\quad A \to \alpha_1\beta_1\alpha_2 \mid \alpha_2\beta_2\alpha_2 \mid ... \; \alpha_1\beta_n\alpha_2$)

Here we can generate new productions by substituting A_j, the right hand side of each A_j proproduction *i.e.* if $A_2 \to A_3A_1 \mid a$ and $A_3 \to A_2A_1 \mid b$ then use A_2 production in R.H.S. of A_3. Now it becomes $A_3 \rightarrow A_3A_1 \mid aA_1 \mid b$. By repeating the process at most $m - 1$ times, we obtain productions of the form

$$A_m \rightarrow A_p\gamma \text{ where } p \geq m.$$

If $m = p$, we have $A_m \rightarrow A_m\gamma$, in left recursive form, which are then replaced by method of elimination of left recursion introducing a new variable B_m.

By repeating the step 2 for each variable in G, we have only productions of the form

$$A_i \rightarrow A_j\gamma \quad \text{where } j > i$$

$$A_i \rightarrow a\gamma \quad a \in V_t$$

$$B_i \rightarrow \gamma \quad \text{where } \gamma \in (V_n \cup \{B_1B_2 \ldots B_{i-1}\})^*$$

Step 3 : Since A_n is the highest numbered variable then productions are of the form $A_n \rightarrow a\gamma$. Now the leftmost symbol of right side of any production for A_{m-1} must be either terminal or A_m (since all productions have been converted $A_i \rightarrow A_j$, $j > i$ or $A_i \rightarrow a\gamma$ in step 2) replace A_m on right side of production of A_{m-1} by replacement rule.

Now, the leftmost symbol of the right hand side of productions for A_{m-1}, is terminal. Repeat this process for $A_{m-2}, A_{m-3} \ldots A_1$. Now right hand side of each production for an A_i starts with a terminal symbol.

Step 4 : In step 2; when we have productions of the form $A_i \rightarrow A_i\gamma$, we eliminate left recursion and each time when we eliminate left recursion, we produced new variables B_i. Since in first step we have already converted the grammar into CNF and on elimination method of left recursion (if $A \rightarrow A\alpha \mid \beta$ we convert it into $A \rightarrow \beta B \mid \beta$, $B \rightarrow \alpha B \mid \alpha$), we cannot find new variable B_i in right hand side as leftmost symbol of right hand side. Thus, after step 3 every production of $A_i (1 \leq i \leq n)$ is in the form $A_i \rightarrow a\gamma$ or $A_i \rightarrow A_jA_k$. Again, in production $B_i(B_i \rightarrow \alpha, B_i \rightarrow \alpha B_i)$, all B_i productions have right hand side beginning with terminals or variable A_i's. So, no B_i production can start with another B_j. Finally, we apply the substitution rule and get the grammar which is in GNF.

In the above steps, it is clear to see that we start with grammar in CNF and then we apply substitution rule and elimination of left recession.

Example 6.18. Convert the grammar.

$$S \rightarrow AB \mid BC$$

$$A \rightarrow aB \mid bA \mid a$$

$$B \rightarrow bB \mid cC \mid b$$

$$C \rightarrow c$$

into GNF.

Solution. Here the production $S \rightarrow AB \mid BC$ is not in GNF. On applying the substitution rule we immediately get equivalent grammar.

$$S \rightarrow aBB \mid bAB \mid aB \mid bBC \mid cCC \mid bC$$

which is in GNF.

Example 6.19. Convert the grammar.

$$S \rightarrow abaSa \mid aba$$

into GNF.

Solution. If we introduce new variables A and B and productions as $A \rightarrow a$, $B \rightarrow b$ and substitute into given grammar as

$$S \rightarrow aBASA \mid aBA$$
$$A \rightarrow a,\ B \rightarrow b$$

which is in GNF.

Example 6.20. Convert the grammar

$$S \rightarrow AB$$
$$A \rightarrow BS \mid a$$
$$B \rightarrow SA \mid b$$

into GNF.

Solution. Step 1 : The given grammar is in CNF. Let us change the name of variables as A_1, A_2, A_3 for S, A, B respectively. Now the given productions are in the form (with A_1 as start symbol)

$$A_1 \rightarrow A_2A_3$$
$$A_2 \rightarrow A_3A_1 \mid a$$
$$A_3 \rightarrow A_1A_2 \mid b$$

Step 2 : In this step we need productions must be in the form that the R.H.S. of productions must starts with terminal or with higher numbered variable. The productions.

$$A_1 \rightarrow A_2A_3$$
$$A_2 \rightarrow A_3A_1 \mid a$$
$$A_3 \rightarrow b$$

are in required form. But $A_3 \rightarrow A_1A_2$ is not. Now we want to convert $A_3 \rightarrow A_1A_2$ into required form.

By applying substitution in R.H.S. for A_1, we get $A_3 \rightarrow A_2A_3A_2$

Again applying substitution rule for A_2, we get

$$A_3 \rightarrow A_3A_1A_3A_2 \mid aA_3A_2$$

Thus, we get :

$$A_1 \rightarrow A_2A_3$$
$$A_2 \rightarrow A_3A_1 \mid a$$
$$A_3 \rightarrow A_3A_1A_3A_2 \mid aA_3A_2 \mid b.$$

Eliminating the left recursion from A_3 production, we get

$$A_3 \rightarrow aA_3A_2B_3 \mid bB_3 \mid aA_3A_2 \mid b$$

and

$$B_3 \rightarrow A_1A_3A_2B_3 \mid A_1A_3A_2$$

where B_3 is a new variable introduced.

Thus, we have set of productions as :

$$A_1 \to A_2A_3$$
$$A_2 \to A_3A_1 \mid a$$
$$A_3 \to aA_3A_2B_3 \mid bB_3 \mid aA_3A_2 \mid b$$
$$B_3 \to A_1A_3A_2B_3 \mid A_1A_3A_2$$

Step 3 : Now all the A_3 productions start with terminals. With the help of substitution rule we can replace A_3 in R.H.S. of A_2 production. We get

$$A_2 \to aA_3A_2B_3A_1 \mid bB_3A_1 \mid aA_3A_2A_1 \mid bA_1 \mid a.$$

Now, all A_2 productions start with terminal. Again we replace A_2 from $A_1 \to A_2A_3$, we have

$$A_1 \to aA_3A_2B_3A_1A_3 \mid bB_3A_1A_3 \mid aA_3A_2A_1A_3 \mid bA_1A_3 \mid aA_3$$

The resulting set of productions are the following :

$$A_1 \to aA_3A_2B_3A_1A_3 \mid bB_3A_1A_3 \mid aA_3A_2A_1A_3 \mid bA_1A_3 \mid aA_3$$
$$A_2 \to aA_3A_2B_3A_1 \mid bB_3A_1 \mid aA_3A_2A_1 \mid bA_1 \mid a$$
$$A_3 \to aA_3A_2B_3 \mid bB_3 \mid aA_3A_2 \mid b$$
$$B_3 \to A_1A_3A_2B_3 \mid A_1A_3A_2.$$

Step 4 : In step 3 only B productions are not in GNF. Applying substitution rule in production

$$B_3 \to A_1A_3A_2B_3$$

We get

$$B_3 \to aA_3A_2B_3A_1A_3A_3A_2B_3 \mid bB_3A_1A_3A_3A_2B_3 \mid aA_3A_2A_1A_3A_3A_2B_3$$
$$\mid bA_1A_3A_3A_2B_3 \mid aA_3A_3A_2B_3$$

And from $B_3 \to A_1A_2A_3$, we get

$$B_3 \to aA_3A_2B_3A_1A_3A_3A_2B_3 \mid bB_3A_1A_3A_3A_2B_3 \mid aA_3A_2A_1A_3A_3A_2A_3$$
$$\mid bA_1A_3A_3A_2B_3 \mid aA_3A_3A_2$$

The final sets of productions are

$$A_1 \to aA_3A_2B_3A_1A_3 \mid bB_3A_1A_3 \mid aA_3A_2A_1A_3 \mid bA_1A_3 \mid aA_3$$
$$A_2 \to aA_3A_2B_3A_1 \mid aB_3A_1 \mid aA_3A_2A_1 \mid bA_1 \mid a$$
$$A_3 \to aA_3A_2B_3 \mid bB_3 \mid aA_3A_2 \mid \mathrm{b}$$
$$B_3 \to aA_3A_2B_3A_1A_3A_2A_3B_3 \mid bB_3A_1A_3A_2A_3B_3 \mid aA_3A_2A_1A_3A_2A_3B_3$$
$$\mid bA_1A_3A_2A_3B_3 \mid aA_3A_2A_3B_3 \mid aA_3A_2B_3A_1A_3A_2A_3$$
$$\mid bB_3A_1A_3A_2A_3 \mid aA_2A_3A_1A_3A_2A_3 \mid bA_1A_3A_2A_3 \mid aA_3A_2A_3$$

In which all productions are in GNF.

Example 6.21. Consider the following grammar and write its equivalent GNF

$$S \rightarrow AB$$
$$A \rightarrow aA/bB/b$$
$$B \rightarrow b.$$

Solution. We know that in the GNF each production should be of the $A \rightarrow aV/a$ form where $V \in V_n^*$.

We can replace A in $S \rightarrow AB$ by aA, or bB or b.

$$S \rightarrow aAB/bBB/bB$$
$$A \rightarrow aA/bB/b$$
$$B \rightarrow b$$

which is in Greibach normal form.

Example 6.22. Convert the following grammar

$$S \rightarrow abSb/aa$$

in to GNF.

Solution. Here we can use a method similar to the one introduced in the construction of Chomsky normal form. We introduce new variables A and B that are essentially synonyms for a and b, respectively, substituting for the terminals with their associated variables leads to the equivalent grammar.

$$S \rightarrow aBSB/aA$$
$$A \rightarrow a$$
$$B \rightarrow b$$

which is in GNF.

Example 6.23. Convert the following grammar into GNF

$$S \rightarrow aAS$$
$$S \rightarrow a$$
$$A \rightarrow SbA$$
$$A \rightarrow SS$$
$$A \rightarrow ba.$$

Solution. Let us introduced new non-terminals as B, C and D, then GNF will be

$$A \rightarrow aBC$$
$$A \rightarrow aBS$$
$$A \rightarrow aC$$
$$A \rightarrow aS$$
$$A \rightarrow bD$$
$$B \rightarrow aBCS$$
$$B \rightarrow aBSS$$
$$B \rightarrow aCS$$
$$B \rightarrow aSS$$

$$B \rightarrow bDS$$
$$C \rightarrow bA$$
$$S \rightarrow aB$$
$$S \rightarrow a$$
$$D \rightarrow a.$$

6.5. REGULAR GRAMMARS

Definition

A regular grammar may be left linear or right linear.

"If all production of a CFG are of the form $A \rightarrow wB$ or $A \rightarrow w$, where A and B are variables and $w \in V_t^*$, then we say that grammar is right liner".

"If all the production of a CFG are of the form $A \rightarrow Bw$ or $A \rightarrow w$, we call it left linear."

A right or left linear grammar is called a regular grammar.

Example 6.24. Write the left linear and right linear grammar for the regular expression

$$r = 0(10)^*.$$

Solution. Given regular expression $r = 0(10)^*$, it means any string which starts with 0 followed by any number of 10's is in regular expression.

Left Linear. Let grammar be $G = (V, V_t, P, S)$

here $V_n = \{S\}$

$V_t = \{0, 1\}$

and P is defined as

$$S \rightarrow S10/0$$

Right Linear. Let grammar be $G' = (V_n, V_t, P, S)$

where $V_n = \{S, A\}$

$V_t = (0, 1\}$

and P is defined as

$$S \rightarrow 0A$$
$$A \rightarrow 10\,A/\in.$$

Example 6.25. Write the right linear and left linear regular grammar for the regular expression $r = (ab)^*\,a$.

Solution. Let us analyse the language of this regular expression, clearly it contains set of strings in which every string start with any number of ab's and end with single a.

Right Linear. Let grammar be $G_1 = (\{S\}, \{a, b\}, S, P)$

where P_1 is defined as follows :

$$S \rightarrow abS/a$$

Left Linear. Let grammar be $G_2 = (\{S, S_1, S_2\}, \{a, b\}, S, P_2)$

with production

$$S \to S_1 a$$
$$S_1 \to S_1 ab/\in$$

is left-linear. Both G_1 and G_2 are regular grammars.

6.5.1. Equivalence of Regular Grammars and Finite Automata

The regular grammars characterize the regular sets, in the sense that a language is regular if and only if it has a left-linear grammar or if and only if it has a right linear grammar. These results are proved in the next two theorems.

Theorem 6.1. If L has a regular grammar, then L is a regular set ?

Proof. First, suppose $L = L(G)$ for some right-linear grammar

$$G = (V_n, V_t, P, S).$$

Let us construct an NFA with $\in$-moves, $M = (Q, V_t, \delta, \{S\}, \{\in\})$ that simulates derivates derivation in G.

Consists of the symbols $[\alpha]$ such that α is S or a suffix of some right hand side of a production in P, we define δ by :

(a) If A is a variable, then $\delta([A], \in) = \{[\alpha]/A \to \alpha \text{ is a production}\}$.

(b) If a is in V_t and α in $V_t^* \cup V_t^* V$, then

$$\delta([a\alpha], a) = \{[\alpha]\}.$$

Then it is easy to show that $\delta([S], w)$ contains $[\alpha]$ if and only if $S \overset{*}{\Rightarrow} xA \Rightarrow xy\alpha$, where $A \to y\alpha$ is a production and $xy = w$, or if $\alpha = S$ and $w = \in$. As $[\in]$ is the unique final state, M accepts w if and only if $S \overset{*}{\Rightarrow} xA \Rightarrow w$. But since every derivation of a terminal string has atleast on step, we see that on accepts w if and only if G generates w. Hence every right-linear grammar generates a regular set.

Theorem 6.2. If L is a regular set, then L is generated by some left-linear grammar and by some right-linear grammar.

Proof. Let $L = L(M)$ for DFA $M = (Q, \Sigma, \delta, q_0, F)$. First suppose that q_0 is not a fixed state. Then $L = L(a)$ for right linear grammar $G = (Q, \Sigma, P, q_0)$, where P consists of production $P \to aq$ whenever $\delta(P, a) = q$ and also $P \to a$ whenever $\delta(P, a)$ is a final state. Then clearly $\delta(P, w) = q$ if and only if $P \overset{*}{\Rightarrow} wq$. If wa is accepted by M, let $\delta(q_0, w) = P$, implying $q_0 \overset{*}{\Rightarrow} wP$. Also $\delta(P, a)$ is final, so $P \to a$ is a production. Thus $q_0 \overset{*}{\Rightarrow} wa$. Conversely, let $q_0 \overset{*}{\Rightarrow} x$. Then $x = wa$ and $q_0 \overset{*}{\Rightarrow} wP \Rightarrow wa$ for some state (variable) P. Then $\delta(q_0, w) = P$, and $\delta(P, a)$ is final. Then x is in $L(M)$. Hence $L(M) = L(G) = L$.

Now let q_0 be in F, so $\in$ is in L. We note that the grammar G defined above generates $L - \{\in\}$. We may modify G by adding a new start symbol S with productions $S \to q_0/\in$. The resulting grammar is still right-linear and generates L.

To produce a left-linear grammar for L, start with an NFA for L^R and then reverse the right sides of all productions of the resulting right-linear grammar.

6.5.2. Conversion of Regular Grammar into Finite Automata

Given a regular grammar G, a finite automata accepting $L(G)$ can be obtained as follows :

(a) The number of states in the automata will be equal to the number of non-terminals plus one. Each state in automata represents each non-terminal in the regulr grammar. The additional state will be the final state of the automata. The state corresponding to the start symbol of the grammar will be the initial state of automata. If $L(G)$ contains $\in$ that is start symbol in grammar derives to $\in$, then make start state also as final state.

(b) The transitions for automata are obtained as follows

(i) For every production $A \to aB$ make $\delta(A, a) = B$ that is make an are labelled 'a' from A to B.

(ii) For every production $A \to a$ make $\delta(A, a)$ = final state.

(iii) For every production $A \to \in$, make $\delta(A, \in) = A$ and A will be final state.

Example 6.26. Consider the following grammar

$$S \to 0A/1B/0/1$$
$$A \to 0S/1B/1$$
$$B \to 0A/1S$$

Find the automation for this grammar.

Solution. The automation for this grammar will have four states : S, A, B and one additional say C, which will be the final state.

Step 1. For production $S \to 0A$ and $S \to 1B$

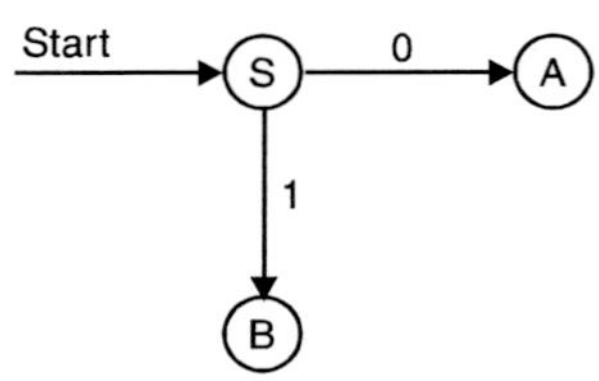

Fig. 6.3.

Step 2. For production $S \to 0/1$

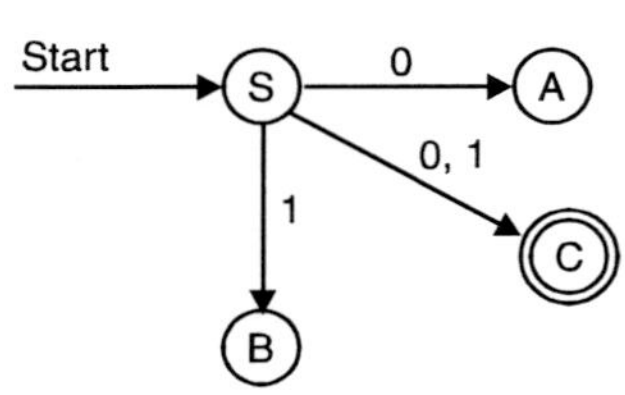

Fig. 6.4.

Step 3. For production $A \to 0S$

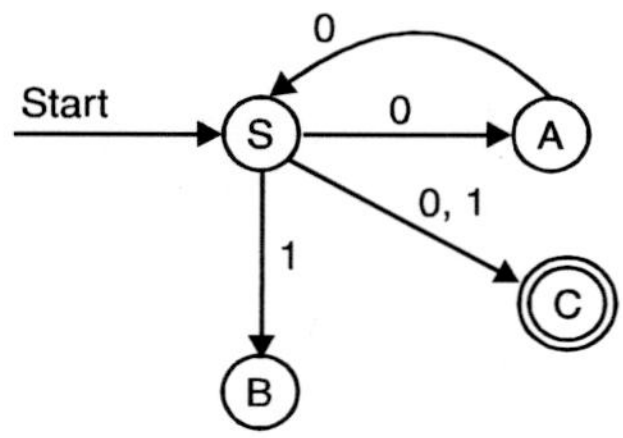

Fig. 6.5.

Step 4. For the production $A \to 1B$

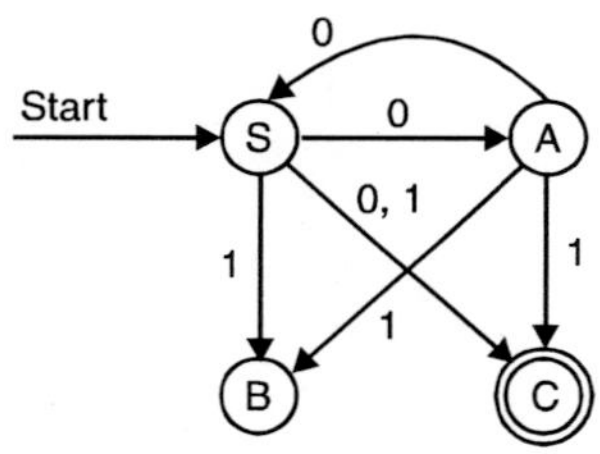

Fig. 6.6.

Step 5. For the production $B \to 0A$ and $B \to 1S$

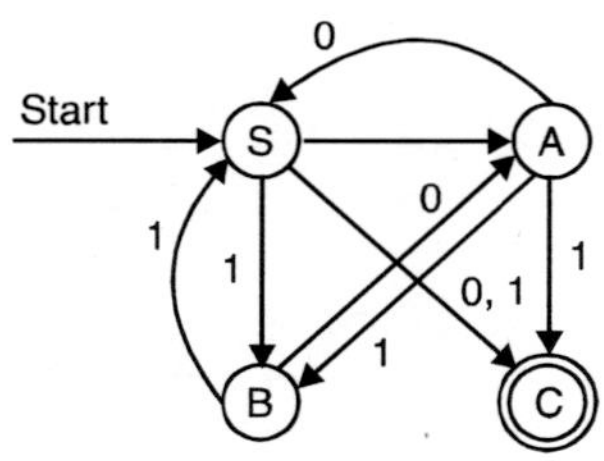

Fig. 6.7.

There is no need to show all the steps every time and FA can be generated directly.

Example 6.27. Give the automation for the following grammar :

$$S \to 0S/1A/1$$

$$A \to 0A/1A/0/1.$$

Solution. The automation for the above grammar will be

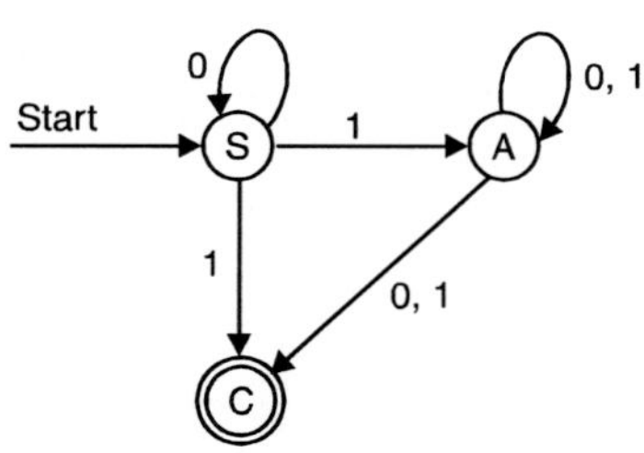

Fig. 6.8.

Example 6.28. Consider the regular grammar $G = (V_n, V_\in, P, S)$, where $V_n = \{A, B, S\}$, $V_t = \{a, b\}$ and P is defined as

$$S \to abA$$
$$S \to B$$
$$S \to baB$$
$$S \to \in$$
$$A \to bS$$
$$B \to aS$$
$$A \to b$$

Construct a NFA M such that $L(M) = L(G)$. Trace the transitions of M to accept the string *abba*.

Solution. The NFA M is

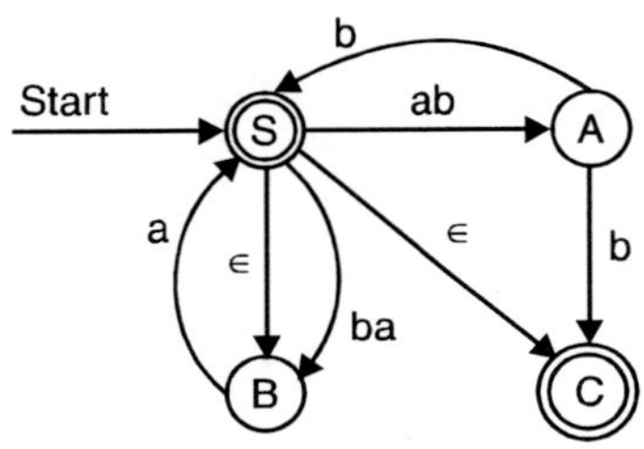

Fig. 6.9.

Here C is final state.

Given string $BW = abba$

$$S \xrightarrow{ab} A \xrightarrow{b} S \xrightarrow{\in} B \xrightarrow{a} S \xrightarrow{\in} C$$

Example 6.29. Find whether the following grammars denote the same language.

Grammar G_1 is

$$A \to 0B/1E$$
$$B \to 0A/1F/\in$$
$$C \to 0C/1A$$
$$D \to 0A/1D/\in$$
$$E \to 0C/1A$$
$$F \to 0A/1B/\in$$

where A is the start symbol for G_1.

Grammar G_2 is :

$$X \to 0Y/0/1Z$$
$$Y \to 0X/1Y/1$$
$$Z \to 0Z/1X$$

where X is the start symbol of G_2.

Solution. In order to prove that G_1 and G_2 denote the same language, *i.e.*, $L(G_1) = L(G_2)$, we only need to prove that minimum state automata accepting $L(G_1)$ and the minimal state automata accepting $L(G_2)$ are identical.

The finite state automata for grammar G_1 is given below :

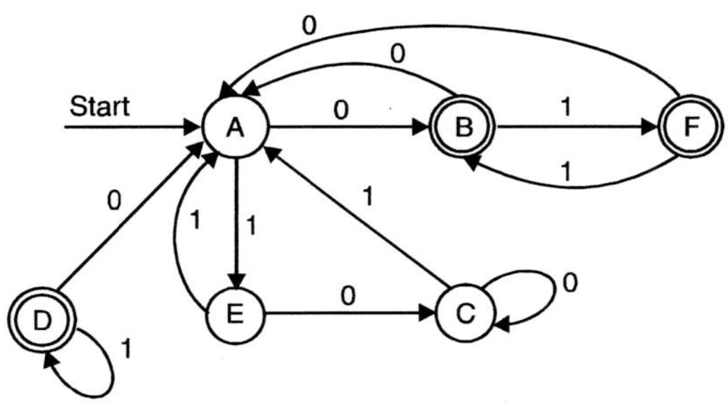

Fig. 6.10.

This automation shown in Fig. 6.10 is determinstic. To minimise this, we write it in tabular forms as follows :

State	*0*	*1*
$\to A$	B	E
$*B$	A	F
C	C	A
$*D$	A	D
E	C	A
$*F$	A	B

Here, two sets of states (final and non-final) are $\{A, C, E\}$ and $\{B, D, F\}$.

Since the output C and E are same, hence replacing every occurrence of E by C, we get

State	*0*	*1*
$\to A$	B	C
$*B$	A	F
C	C	A
$*D$	A	D
$*F$	A	B

and now the two sets become $\{A, C\}$ and $\{B, D, F\}$.

Now since in set $\{B, D, F\}$ each state enters in a final state 'A' at input 0 and enters in a final state at input 1. Hence they can be reduced as follows :

State	*0*	*1*
$\to A$	B	C
$*B$	A	B
C	C	A

Similarly automation for grammar G_2 is

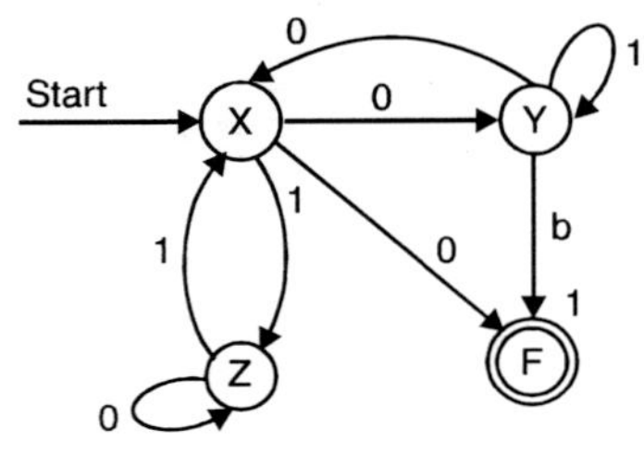

Fig. 6.11.

Above automata is not DFA, so convert it to DFA as follows :

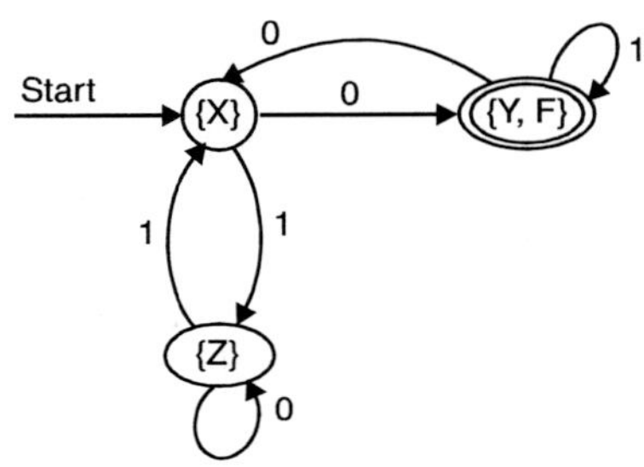

Fig. 6.12.

To minimize it, write it in tabular form

State	*0*	*1*
$\rightarrow \{X\}$	$\{Y, F\}$	$\{Z\}$
$*\{Y, F\}$	$\{X\}$	$\{Y, F\}$
$\{Z\}$	$\{Z\}$	$\{X\}$

It is already minimised.

It is clear that these two automation for G_1 and G_2 are similar, hence G_1 and G_2 represent the same language.

Example 6.30. Convert the following grammar G into GNF

$$S \rightarrow XA \mid BB$$

$$B \rightarrow b \mid SB$$

$$X \rightarrow b$$

$$A \rightarrow a$$

Solution.

1. Rewrite G in Chomsky Normal Form (CNF)

   It is already in CNF.

2. Re-label the variables

   S with A_1

X with A_2

A with A_3

B with A_4

$$A \to A_2A_3 \mid A_4A_4$$
$$A_4 \to b \mid A_1A_4$$
$$A_2 \to b$$
$$A_3 \to a$$

3. Identify all productions in which left side has a lower numbered variable than first variable in the right side.

4. $A_4 \to A_1A_4$ ...identified

5. $A_4 \to A_1A_4/b$

To eliminate A_1 we will use the substitution rule

Substituting for $A_1 \to A_2A_3 \mid A_4A_4$

$$A_4 \to A_2A_3A_4 \mid A_4A_4A_4 \mid b$$

The above two productions still do not conform to any of the types in step 3.

Substituing for $A_2 \to b$

$$A_4 \to bA_3A_4 \mid A_4A_4A_4 \mid \text{b}$$

Now we have to remove left recursive production

$$A_4 \to A_4A_4A_4$$
$$A_4 \to bA_3A_4 \mid b \mid bA_3A_4Z \mid bZ$$
$$Z \to A_4A_4 \mid A_4A_4Z$$

6. At this stage our grammar now looks like

$$A_1 \to A_2A_3 \mid A_4A_4$$
$$A_4 \to bA_3A_4 \mid b \mid bA_3A_4Z \mid bZ$$
$$Z \to A_4A_4 \mid A_4A_4Z$$
$$A_2 \to b$$
$$A_3 \to a$$

All rules now conform to one of the types in step 3.

But the grammer is still not in Greibach Normal Form!

7. All productions for A_2, A_3 and A_4 are in GNF

for $A_1 \to A_2A_3 \mid A_4A_4$

Substitute for A_2 and A_4 to convert it to GNF

$$A_1 \to bA_3 \mid bA_3A_4A_4/bA_4/bA_3A_4ZA_4/bZA_4$$

for $Z \to A_4A_4 \mid A_4A_4Z$

Substitute for A_4 to convert it to GNF

$Z \to bA_3A_4A_4 \mid bA_4 \mid bA_3A_4ZA_4 \mid bZA_4 \mid bA_3A_4A_4Z \mid bA_4Z \mid bA_3A_4ZA_4Z \mid bZA_4Z$

8. Finally the Grammar in GNF is

$$A_1 \to bA_3 \mid bA_3A_4A_4 \mid bA_4 \mid bA_3A_4ZA_4 \mid bZA_4$$
$$A_4 \to bA_3A_4 \mid b \mid bA_3A_4Z \mid bZ$$
$$Z \to bA_3A_4A_4 \mid bA_4 \mid bA_3A_4ZA_4 \mid bZA_4 \mid bA_3A_4A_4Z \mid bA_4Z \mid bA_3A_4ZA_4Z \mid bZA_4Z$$
$$A_2 \to b$$
$$A_3 \to a$$

Example 6.31. Construct GNF for following CFG

$$S \to aSb$$
$$S \to ab$$

Solution.

Step 1 : Let us find CNF of above GNF as follows :

$$S \to ASB$$
$$S \to AB$$
$$A \to a$$
$$B \to b$$

Step 2 : Rename the non-terminals as A_1, A_2, A_3, by putting $S = A_1$, $A = A_2$, $B = A_3$, then CNF can b written as follows :

$$A_1 \to A_2A_1A_3$$
$$A_1 \to A_2A_3$$
$$A_2 \to a$$
$$A_3 \to b$$

Step 3 : $A_1 \to A_2A_1A_3$ and $A_1 \to A_2A_3$ are not in GNF, so let us replace them as follows :

$$A_1 \to aA_1A_3$$
$$A_1 \to aA_3$$

So GNF is

$$A \to aA_1A_3 \mid aA_3$$
$$A_2 \to 9$$
$$A_3 \to b$$

Example 6.32. Reduce the following grammar to GNF.

$$S \to SS, \ S \to 0S1101.$$

Solution.

Step 1: Rename S as A_1. It is enough to replace terminals by new variables to get an equivalent grammar G_1 as follows :

$$A_1 \to A_1A_1 \mid A_2A_1A_3 \mid A_2A_3$$
$$A_2 \to 0$$
$$A_3 \to 1$$

Step 2 : All production of G_1 except $A_1 \to A_1A_1$ are in proper form, since it suffered with left recursion.

$$A_1 \to A_1A_1 \mid A_2A_1A_3 \mid A_2A_3$$

Compare it with

$$A \to A\alpha/\beta_1/\beta_2$$

Let us chose a new non terminal A'

Add the new A'-rule as follows :

$$A_1 \to A_2A_1A_3A' \mid A_2A_3A' \mid A_2A_1A_3 \mid A_2A_3$$
$$A' \to A_1A_1 \mid A_1$$

Now grammar is

$$A_1 \to A_2A_1A_3 \mid A_2A_3 \mid A_2A_1A_3A' \mid A_2A_3A'$$
$$A' \to A_1A' \mid A_1$$
$$A_2 \to 0$$
$$A_3 \to 1$$

Step 3 : $A' \to A_1$ and $A' \to A_1A'$ are modified by replacing A_1 so

$$A' \to A_2A_1A_3A' \mid A_2A_3A' \mid A_2A_1A_3A'A' \mid A_2A_3A'A'$$
$$\mid A_2A_1A_3 \mid A_2A_3 \mid A_2A_1A_3A' \mid A_2A_3A'$$

Finally GNF is

$$A_1 \to 0A_1A_3 \mid 0A_3 \mid 0A_1A_3A' \mid 0A_3A'$$
$$A_2 \to 0,\ A_3 \to 1$$
$$A' \to 0A_1A_3/0A_3/0A_1A_3A' \mid 0A_3A'$$
$$A' \to 0A_1A_3A' \mid 0A_3A' \mid 0A_1A_3A'A' \mid 0A_3A'A'$$

Example 6.33. Obtain the language generated by the CFG's with the given production rules

(i) $\{A \to a, A \to aB, A \to \in \}$

(ii) $\{S \to aS, S \to \in\}$

(iii) $\{A \to aS, S \to bS, S \to \in\}$

(iv) $\{S \to aS, S \to bS, S \to a\}$

(v) $\{S \to ab, S \to bS, S \to a, S \to b\}$.

Solution.

(i) $L(G) = \{ww^k \mid w \in (a, b)^*\}$

(ii) $L(G) = \{a^n \mid n \geq 0\}$

(iii) $L(G) = \{a, b\}^*$

(iv) $L(G) = \{$set of all strings over (a, b) ended with $a\}$

(v) $L(G) = \{$set of all strings over $(a, b)\}$.

Example 6.34. Construct CFG for $L = \{w \in (0, 1)^* \mid n_a(w) \neq n_b(w)\}$

Solution. Let CFG
$$G = (V_n, V_t, P, S)$$
$$V_n = \{S, A, B\}$$
$$V_t = \{a, b\}$$

S is starting non-terminal and P is detired as below :

$$S \to A \mid B$$
$$A \to a \mid aA \mid bAA \mid AAb \mid AbA$$
$$B \to b \mid bB \mid aBB \mid BBa \mid BaB$$

Strings derived from A has more a's than b's. And strings derivable from B has more b's than a's.

Example 6.35. Construct the CFG for the $L = (011 + 1)^* (0)^*$.

Solution. Let $G = (V_n, V_t, P, S)$ is CFG for given language L, where

$$V_n = \{S, A, B, C, D\}$$
$$V_t = \{0, 1\}$$
$$S = \text{Starting non-terminal.}$$

and productions 'P' are defined as below :

$$S \to AB \mid \in$$
$$A \to CA \mid \in$$
$$B \to DB \mid \in$$
$$C \to 011 \mid 1$$
$$D \to 01$$

Example 6.36. Convert the following in GNF $S \to ab/aS/aaS$.

Solution. Modify the grammar in to CNF

$$S \to B_aB_b/B_aS/B_aB_aS$$
$$B_a \to a$$
$$B_b \to b$$

Rename the variables S, B_a, B_b as A_1, A_2 and A_3 respectively.

$$A_1 \to A_2A_3 \mid A_2A_1 \mid A_2A_2A_1$$
$$A_2 \to a$$
$$A_3 \to b$$

All the productions are in the required form

$$A_i \rightarrow A_j, \ i < j$$

Now let us modify the productions to obtain each productions of the form $A \rightarrow a\alpha$ where $a \in V_t$, $\alpha \in V^*_n$.

In $A_1 \rightarrow A_2A_3 \mid A_2A_1 \mid A_2A_2A_1$ substitute for the left most variable

$$A_1 \rightarrow aA_3 \mid aA_1 \mid aA_2A_1$$
$$A_2 \rightarrow a$$
$$A_3 \rightarrow b$$

which is required GNF form.

EXERCISES

1. Each of the following CFG's has a production using the symbol $\in$ and yet $\in$ is not a word in the language. Using the method, disccussed in this chapter, show that there are other CFGs for these languages that do not use $\in$-productions.

(*i*) $S \rightarrow aX/bX$
$X \rightarrow a/b/\in$

(*ii*) $S \rightarrow aX/bS/a/b$
$X \rightarrow aX/a/\in$

(*iii*) $S \rightarrow aS/bX$
$X \rightarrow aX/\in$

(*iv*) $S \rightarrow XaX/bX$
$X \rightarrow XaX/XbX/\in$

2. Each of the following CFG's has unit productions. Using the algorthim presented in this chapter, find CFG's for these same languages that do not have unit productions.

(*i*) $S \rightarrow aX/Yb$
$X \rightarrow S$
$Y \rightarrow bY/b$

(*ii*) $S \rightarrow AA$
$A \rightarrow B/BB$
$B \rightarrow abB/b/bb$

(*iii*) $S \rightarrow AB$
$A \rightarrow B$
$B \rightarrow aB/Bb/\in$

3. Convert the following CFG's to CNF :

(*i*) $S \rightarrow SS/a$

(*ii*) $S \rightarrow aSa/SSa/a$

(*iii*) $S \rightarrow aXX$
$X \rightarrow aS/bS/a$

(*iv*) $E \rightarrow E + E$

$$E \rightarrow E * E$$

$$E \rightarrow (E)$$

$$E \rightarrow 7$$

The terminal here are +, *, (,), 7.

(*v*) $S \rightarrow ABABAB$

$$A \rightarrow a/\in$$

$$B \rightarrow b/\in$$

Note that $\in$ is the word in this language, but when converted in to CNF, the grammar will no longer generate it.

(*vi*) $S \rightarrow SaS/SaSbS/SbSaS/\in$

(*vii*) $S \rightarrow AS/SB$

$$A \rightarrow BS/SA$$

$$B \rightarrow SS$$

4. Convert the following CFG's with unit production in to CNF :

(*i*) $S \rightarrow X$

$$X \rightarrow Y$$

$$Y \rightarrow Z$$

$$Z \rightarrow aa$$

(*ii*) $S \rightarrow SS/A$

$$A \rightarrow SS/AS/a$$

5. Construct a DFA that accepts the language generated by the grammar

$$S \rightarrow abA$$

$$A \rightarrow baB$$

$$B \rightarrow aA/bb.$$

6. Find the regular grammar that generates the language $L(aa^* (ab + a)^*)$.

7. Construct the left linear grammar and right-linear grammar for the language

$$L = \{a^n b^m : n \geq 2, m \geq 3\}.$$

8. Find a regular grammar that generates the language on $\Sigma = \{a, b\}$ consisting of all strings with no more than a's.

9. Find a regular grammar for the language

$$L = \{a^n \ b^m : n + m \text{ is ever}\}.$$

10. Find regular grammars for the following languages on $\{a, b\}$.

(*a*) $L = \{w : n_a(w) \text{ and } n_b(w) \text{ are both even}\}$

(*b*) $L = \{w : (n_a(w) - n_b(w)) \bmod 3 = 1\}.$

*****

SECTION-C

- Pushdown Automata
- Properties of Context-Free Languages
- Turing Machine

SYLLABUS

Pushdown Automata : Introduction to Pushdown Machines, Application of Pushdown Machines.

Turing Machines : Deterministic and Non-Deterministic Turing Machines, Design of T.M., Halting Problem of T.M., PCP Problem.

CHAPTER 7

Pushdown Automata

7.1. INTRODUCTION

INSIDE THIS CHAPTER

The finite automata that we have studied earlier are not capable to recognise the context free languages such as $\{W\ C\ W^R / \ W \in \Sigma^*\}$. Because by reading a string in this language from left to right, automata must "remember" the string before it encounters symbol '*C*' and then compare it to the string after '*C*'. But our finite automata is not capable to remember any thing.

From above example it becomes clear that the finite automata can be extended by addition of an auxiliary storage to accept context free languages.

A familiar example of the stack is the set of plates that we see in cafeteria. We can remove plates only from the top of this set (stack) not from the middle. Similarly we can add plates only at the top of the stack not in the middle.

TIPS

The pushdown automata is essentially a finite automata with control of both an input tape and a stack to store what it has read. A stack is a "first in last out" list, that is, symbols can be entered or removed only at the top of the list. When a new symbol is entered at the top, the symbol previously at the top becomes second and so on.

We shall now formalise the concept of a PDA. The PDA will have 3 things : an input tape, a finite control, and a stack. The device will be non-deterministic, having some finite number of moves in each situation.

We can also see pictorial representation of pushdown automata.

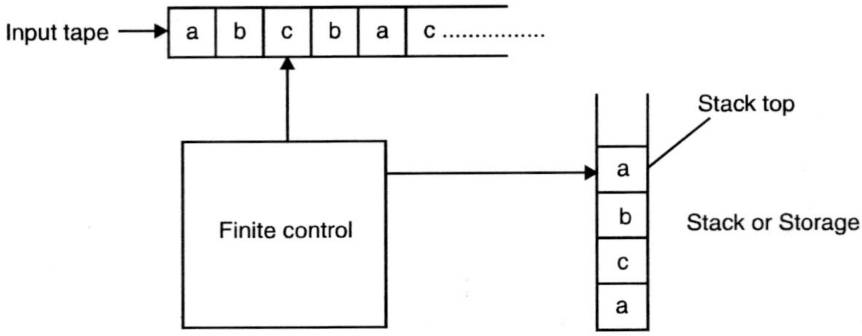

Fig. 7.1.

Here is input tape, from which finite control (which works like finite control of finite automata), reads the input, and same time it reads the symbol from stack top. Now it depends on finite control, that what is the next state and what will happen with stack top.

7.2. DEFINITION OF PUSHDOWN AUTOMATA

Definition A pushdown automata is a system, which is mathematically defined as follows :

$$P = (Q, \Sigma, \Gamma, \delta, s, F)$$

where Q : non-empty finite set of states

Σ : non-empty finite set of input symbols

Γ : is finite set of pushdown symbols

s : is the initial state, $s \in Q$

F : is the set of final states and $F \subseteq Q$

δ : it is a transition function or transition relation which maps.

$$(Q \times \Sigma^* \times \Gamma^*) \rightarrow (Q \times \Gamma^*)$$

As we have already seen in case of automation, that it may be nondeterministic or deterministic, pushdown automata also behaves in similar fashion.

"If it follows all the properties of non-determinision then it is said to be non-deterministic pushdown automata".

Clearly for the "NPDA" transition relation must behave as follows :

$$(Q \times (\Sigma \cup \{\in\} \times \Gamma^*) \rightarrow \text{finite subset of } (Q \times \Gamma^*).$$

TIPS

From the above relation it becomes very clear that $\in$-transitions are allowed in case of non-deterministic pushdown automata. Similar to the non-deterministic finite automata, non-deterministic pushdown automata also have the power of guess.

Now let us analyse the case of deterministic pushdown automata.

"A pushdown automata $P = (Q, \Sigma, \Gamma, \delta, S, F)$ is said to be deterministic if it is an automation as defined in definition of PDA, subject to the restrictions that for every $q \in Q$, $a \in (\Sigma \cup \{\in\})$ and $b \in \Gamma$.

(*a*) $\delta(q, a, b)$ contains at most one element,

(*b*) if $\delta(q, \in, b)$ is not empty, then $\delta(q, c, b)$ must be empty for every $c \in \Sigma$.

The first of these conditions simply requires that for any given input symbol and any stack top, at most one move can be made. The second condition is that when a ∈-transition is possible for some configuration, no input-consuming alternative is available.

TIPS

"It is interesting to note the difference between the definitions of deterministic and non-deterministic pushdown automata. The domain of the transition function is still same rather than $(Q \times \Sigma \times \Gamma)$ because we want to retain ∈-transitions. Since the top of the stack plays a role in determining the next move, the presence of ∈-transition does not automatically simply non-determinism. **Also some transition of DPDA may be to the empty set, that is undefined, so there may be dead configurations.** This is not going to affect the definition; the only criterian for determinism is that all the times at most one possible move exists.

7.2.1. Explanation for the Moves of Pushdown Automata

Let us see some moves of PDA in the form of mapping, that

$$(Q \times \Sigma^* \times \Gamma^*) \rightarrow (Q \times \Gamma^*).$$

(*a*) The interpretation of $((q, a, z), (p, y)) \in \delta$ (where $p, q, \in Q$, a is an alphabet, z and y in Γ^*) is that PDA whenever is in state q, with z on the top of the stack may read 'a' from the input tape, replace z by y on the top of the stack and enter state P.

(*b*) To push a symbol on the stack, just add it on the top of the stack. This can be achieved by the transition $((q, a, \in), (p, a))$.

(*c*) Similarly to pop a symbol is to remove it from the top of the stack. The transition $((q, a, z), (p, \in))$ pops z from the stack.

(*d*) PDA can also behave as do nothing machine, just read the input from the input and don't make any change in the state and symbol at the stack.

Let PDA in state P and it has any symbol say x on the stack top, now it read a input 'a' from the input tape then write a move for PDA which behaves as do nothing machine.

$$((P, a, z), (p, z))$$

Clearly PDA just read input 'a', and without making any change, the state or at stack top, waits for another input from input tape.

7.3. ACCEPTANCE BY PDA

There are two kind of possible acceptance by PDA for any string. A set of words is accepted by final state by some PDA, it can be accepted by empty store by some other PDA and vice versa. So there are following two kind of acceptance by PDA.

(*i*) Acceptance by final state.

(*ii*) Acceptance by empty store (or stack).

7.3.1. Acceptance by Final State

A string is accepted by a PDA if starting with the starting state in the finite control, the start push down symbol on the top of the pushdown store, by a sequence of moves, the word w is read from left to right letter by letter and PDA finally enters a final state after reading the right most symbol. This kind of acceptance is called acceptance by final state.

Let PDA be $P = \{Q, \Sigma, \Gamma, \delta, q_0, F)$.

Let $w \in \Sigma^*$ is a string, which is accepted by PDA '*P*' by final state. It will be represented by as follows :

$$(q_0, w, \alpha) \overset{*}{\vdash} (q_i, \in, \alpha)$$

where $q_i \in F$

and $\alpha \in \Gamma^*$

7.3.2. Acceptance by Empty Stack

A string w may be accepted by emptying the pushdown store also. Here the action of the PDA is similar but instead of entering a final state when the last input letter is read, the PDA accepts the word w if the pushdown stack is empty after reading the rightmost input letter.

Let $P = (Q, \Sigma, \Gamma, \delta, q_0, F)$ be a PDA. A given string $w \in \Sigma^*$ is accepted by empty stack is defined as follows :

$$(q_0, w, \alpha) \overset{*}{\vdash} (q, \in, \in)$$

↓

Empty stack

where $q \in Q$

and $\alpha \in \Gamma^*$

Now I think we are able to handle some practical problems :

Example 7.1. Design a PDA which accepts the language

$L = \{w \in \{a, b\}^*/ w$ has the equal number of *a*'s and *b*'s}.

Solution. To solve any problem, it becomes very important that we should be able to analyse the language very clearly.

Here Language $L = \{w \in \{a, b\}^*/w$ has equal number of *a*'s and *b*'s}

is having a set of string in which every string is having equal number of *a*'s and *b*'s, there is no concern of the position of *a*'s and *b*'s, the machine can read input '*a*' first or input b's first, a after b, b after a, a followed by a or b followed by b and of course null string is also accepted.

Now let us assume that PDA be P.

$$P = (Q, \Sigma, \Gamma, \delta, S, F)$$

where

$$Q = \{s, q, f\}$$
$$\Sigma = \{a, b\}$$
$$\Gamma = \{a, b, c\}$$
$$F = \{f\} \ \delta \text{ is follows.}$$

(Note that we are introducing another symbol other than Σ that is 'c' to make the solution logical.)

1. $((s, \in, \in), (q, c))$
2. $((q, a, c), (q, ac))$
3. $((q, a, a), (q, aa))$
4. $((q, a, b), (q, \in))$
5. $((q, b, c), (q, bc))$
6. $((q, b, b), (q, bb))$
7. $((q, b, a), (q, \in))$
8. $((q, \in, c), (f, \in))$

Here we are using symbol c to mark the bottom of the stack. PDA intially pushes 'c' in to the stack (using transition 1). Now if PDA reads symbol 'a' from the input tape and if stack is empty or has 'a' on the stack top then push 'a' in to the stack (by using transition 2 and 3). Now if it reads 'a' from input and 'b' is on the top of the stack then it simply pops the symbol 'b' and pushes nothing. (This is achieved by transition 4) same reason holds true for symbol 'b'. When it has read all the input and finds empty symbol 'c' on the top of the stack, it enters in state f, and pop the symbol c. **So final state is defined as, PDA is in state *f* and stack is empty.**

Take a string *abab* for example. Now if we want to check whether it contains equal number of a's and b's, we simply store a's or b's whatever we read. Now when we read next a, we delete a corresponding 'b' from our store. The same process follows for b's. If at the end we have nothing to read and nothing in store then we can say that the string contains equal number of a's and b's. Following table shows the operation of PDA on the string *abbbaaba*.

S. No.	*State*	*Unread input*	*Stack*	*Transition used*
1	s	*abbbaaba*	$\in$	.
2	q	*abbbaaba*	c	1
3	q	*bbbaaba*	ac	2
4	q	*bbaaba*	c	7
5	q	*baaba*	bc	5
6	q	*aaba*	bbc	6
7	q	*aba*	bc	4
8	q	*ba*	c	4
9	q	*a*	bc	5
10	q	$\in$	c	4
11	f	$\in$	$\in$	8

This problem can also be solved without introducing the symbol 'c' at the stack bottom, let us see another solution.

Let PDA be P_1

where
$$\begin{aligned} P_1 &= (Q, \Sigma, \Gamma, \delta, S, F) \\ Q &= \{s, q, f\} \\ \Sigma &= \{a, b\} \\ \Gamma &= \{a, b\} \\ F &= \{f\} \end{aligned}$$

and transition relation δ is defined as follows :

1. $((s, a, \in), (q, a))$
2. $((s, b, \in), (q, b))$
3. $((q, a, a), (q, aa))$
4. $((q, b, b), (q, bb))$
5. $((q, a, b), (q, \in))$
6. $((q, b, a), (q, \in))$
7. $((q, \in, \in), (f, \in))$

It will also work.

Example 7.2. Design PDA for the language $L = \{wcw^r / w \in \{a, b\}^*\}$.

Solution. By the analysis of the language it becomes clear that $ababcbaba \in L$, but $abc\ ab \notin L$, and $cbc \notin L$.

Let pushdown automata be P

where
$$\begin{aligned} P &= (Q, \Sigma, \Gamma, \delta, S, F) \\ Q &= \{s, f\} \\ \Sigma &= \{a, b, c\} \\ \Gamma &= \{a, b\} \end{aligned}$$

$F = \{f\}$ and transition relation δ is defined as follows :

(1) $((s, a, \in), (s, a))$

(2) $((s, b, \in), (s, b))$

(3) $((s, c, \in), (f, \in))$

(4) $((f, a, a), (f, \in))$

(5) $((f, b, b), (f, \in))$.

This automation operates in the following way. As it reads the first half of its input, if remains its in intial state and uses transition 1 and 2 to transfer symbols from the input string on to the pushdown store. **Note that these transitions are applicable regardless of the current content of the stack**. Since the "string" to be matched on the top of the pushdown store is the empty string, when machine sees *a* '*c*' in the input string, it switches from state *s* to state *f* without operating on the stack. Thereafter only transition 4 and 5 are operative; these permit the removal of the top symbol on the stack, provided that it is the same as the next input symbol. If the input symbol does not match the top symbol on the stack, no further operation is possible. If the automation reaches in this way the configuration $(f, \in, \in)$. Final state, end of input, empty stack then the input was indeed of the form $wc\ w^R$, and the automation accepts on the other hand if automation or if the input is exhausted before the stack is emptied, then it does not accept.

The operation of P can be seen by following table, we have taken the string *abacaba* for example

S.No.	State	Unread input	Stack	Transition used
1	s	*abacaba*	$\in$	–
2	s	*bababa*	a	1
3	s	*acaba*	ba	2
4	s	*caba*	aba	1
5	f	*aba*	aba	3
6	f	*ba*	ba	4
7	f	*a*	a	5
8	f	$\in$	$\in$	4

Example 7.3. Construct a push down automata to accept the following language

$$L = \{ww^R : w \in \{a, b\}^*\}.$$

Solution. By the analysis of given language it becomes very clear that the strings accepted by the machine are the same as those accepted by the machine of the previous example, except that the symbol c that marked the center of the string is missing. **Therefore, the machine must "gess" when it has reached the middle of the input string and change from state 's' to state 'f' in a non-deterministic fashion.**

Let PDA be $\quad P = \{Q, \Sigma, \Gamma, \delta, S, F)$

where $\quad Q = \{s, f, f'\}$

$$\Sigma = \{a, b\}$$

$$F = \{f'\}$$

and transition relation δ is defined as follows :

(1) $((s, a, \in), (S, a))$

(2) $((s, b, \in), (S, b))$

(3) $((s, \in, \in), (f, \in)) \longrightarrow$ (middle of the input string)

(4) $((f, a, a), (f, \in))$

(5) $((f, b, b), (f, \in))$

(6) $((f, \in, \in), (f', \in))$

Clearly whenever the machine is in state 's' it can non-deterministically choose either to push the next input symbol on the stack or to switch to state f without consuming any input. Move 6 is for accepting the string.

Example 7.4. Design a PDA for the following language

$$L = \{a^n b^n : n > 0\}.$$

Solution. By the analysis of the language it is clear that PDA for the given language will accept the set of strings in which every string starts with 'a' followed by any number of a's and followed by same number of b's, clearly 'a' should not be encountered whenever b is encountered strings like *ab, aabb, aaabbb* ... should be accepted and $\in$, *aa, bb*, *aabbb* and *bbaa* should be rejected.

Let push down automata be P

$$P = (Q, \Sigma, \Gamma, \delta, S, F)$$

where $Q = \{s, f, q\}$

$$F = \{f\}$$

$$\Sigma = \{a, b\}$$

$$\Gamma = \{a\}$$

and transition relation δ is defined as follows :

(1) $((s, a, \in), (S, a))$

(2) $((s, a, a), (S, aa))$

(3) $(s, b, a), (q, \in))$

(4) $((q, b, a), (q, \in))$

(5) $((q, \in, \in), (f, \in))$

Initially machine is at state s, and it can read input a then pushes all a's encountered before any 'b' to the stack (by using transition relation on 1 and 2). When first 'b' is encountered, corresponding 'a' is poped up from the stack and now state becomes q, and this process can be repeat with the help of transition relation number '4'. Finally when there is no input an stack is empty PDA charges its state from q to final state 'f'.

Example 7.5. In Fig. 7.2 a DFA is given, write the regular expression for it and design a pushdown automata

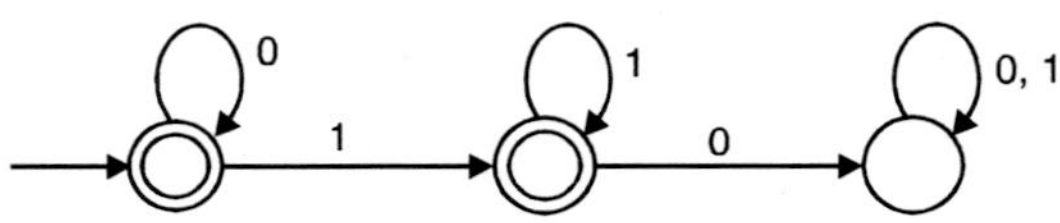

Fig. 7.2.

Solution. By the observation of DFA defined by the transition diagram in Fig. 7.2, we can write the regular expression for it, let it is r then

$$r = 0^* \, 1^*$$

Clearly all the strings which start with any number of 0's followed by any number of 1's and there is no '0' after 1 is encountered is the string of regular set.

Let pushdown automata be P

$$P = (Q, \Sigma, \Gamma, \delta, s, F)$$

$$Q = \{s, p, r\}$$

$$\Sigma = \{0, 1\}$$

$$\Gamma = \{\in\}$$

$$F = \{s, P\}$$

and the transition relations δ are defined as follows :

(1) $((s, \in, \in), (s, \in))$

(2) $(s, 0, \in), (s, \in))$

(3) $((s, 1, \in), (p, \in))$

(4) $(p, 1, \in), (p, \in))$

(5) $((p, 0, \in), (r, \in))$

Now let us see the explanation of transitions, initial state of machine is s, it is also final state so $\in$ is accepted. Now any number of 0's can be accepted (by the help of transition relation 2). As soon as 1 is encountered state s is changed to p from s (by the help of transition 3). Now any number of 1 can be accepted by the machine with the help of transition number 4. Whenever '0' is encountered after reading 1, state is change to r from P and r state behaves like trap state and here process stopped and string is not accepted, since r is not final state.

Example 7.6. Design a pushdown automata for the following language, $L = \{a^n b^{2n} : n > 0\}$.

Solution. Language $L = \{a^n b^{2n} : n > 0\}$ is the set of strings, in which every string starts with a, ends with b, no 'a' comes after b, no b comes before a and number of b's are double than number of a's. For example, strings like *aabbbb, abb, aaabbbbbb* will be accepted by PDA and strings like *a, b, ab, ba, aab, aaabbb* will be rejected.

Let PDA be $P = (Q, \Sigma, \Gamma, \delta, S, F)$

where

$$Q = (p, q, f, r)$$
$$S = \{P\}$$
$$F = \{f\}$$

and transition relations are defined as follows :

(1) $((p, a, \in), (p, a))$

(2) $((p, a, a), (p, a))$

(3) $((p, b, a), (q, b))$

(4) $((q, b, ba), (r, \in))$

(5) $((r, b, a), (q, ba))$ → Here these two transition 4 and 5 reprsent the looping.

(6) $((r, \in, \in), (f, \in))$

Now let us see explanation for above discussed transition relation. Here first we pushed all the a's by the help of moves 1 and 2. Now as first b is encountered, then it is pushed on the stack by the help of move number 3 and when next 'b' is encountered then we pop the ba from the stack and change the state from q to r. It should be clear that moves 4 and 5 represent loop that is move 4 and 5 works on the value of 'n'. Finally when input becomes empty and stack also becomes empty then state r changes to final state that is f and string is accepted.

Example 7.7. Construct a PDA for the regular expression

$$r = 0^* 1^+.$$

Solution. Regular expression is $r = 0^* 1^+$, let us write the language for this regular expression, let us say it is L,

$$L = \{0^m 1^n / m \geq 0, n \geq 1\}$$

Clearly regular expression r contains the set of string in which every string starts with any number of 0's but certainly having atleast one 1, and it must be last symbol of the string. As 1 encountered then 0's should not come.

Let PDA be P

$$P = (Q, \Sigma, \Gamma, \delta, S, F)$$

where
$$Q = \{q_0, f\}$$
$$S = \{q_0\}$$
$$F = \{f\}$$
$$\Sigma = \{0, 1\}$$

and P is defined as

(1) $((q_0, 0, \in), (q_0, \in))$

(2) $((q_0, 1, \in), (f, \in))$

(3) $((f, 1, \in), (f, \in))$

It is very important to understand that in above three moves we are not using the stack. Initially machine is at state q_0 when it reads 0's then it does nothing, but as soon as 1 is encountered then state changes to final state, with the help of move number '2'. Now after it, machine can accept any number of time 1's in the string by the help of move '3'.

Example 7.8. Construct the pushdown automata for the following language :

$$L = \{a^n b^{n+1} / n = 1, 2, 3, ...\}.$$

Solution. Language L is the set of strings in which number of b's are exactly one greater number of a's and every string starts with 'a' and ends with 'b'. No a is encountered after b.

Let PDA be P

$$P = (Q, \Sigma, \Gamma, \delta, S, F)$$

where
$$Q = \{p, q, q_1, f\}$$
$$\Sigma = \{a, b\}$$
$$S = \{p\}$$
$$F = \{f\}$$

and transitions relations are defined as follows :

(1) $((p, a, \in), (p, a))$

(2) $((p, a, a), (p, aa))$

(3) $((p, b, a), (q, \in))$

(4) $((q, b, a), (q, \in))$

(5) $((q, b, \in), (q_1, \in))$

(6) $(q_1, \in, \in), (f, \in))$.

Let us analyse the solution, first of all by the help of move number '1' and '2' we are pushing all the a's on the stack then by the help of move '3' and '4' we pop all a's from the stack when machine starts reading b's from the input tape. When machine reads last b (at this time stack is empty), the change the state to q_1, and now at state q_1, input is null string and stack is empty, then machine goes to final state, so string is accepted.

Example 7.9. Design a pushdown automata for the following language

$$L = \{a^n b^m / n > m \geq 0\}.$$

Solution. As usual, let us first analyse the language. Here language is the set of strings in which every string starts with 'a' (there is at least one a symbol in the strings) and if b is there then number of a's are always greater than number of b's and there should be no '*a*' when any *b* is encountered clearly strings *aaabb, a, aab* are in language but *ab, abb, b, aaabbb* ... are not accepted.

Let PDA for the language is P

$$P = (Q, \Sigma, \Gamma, \delta, S, F)$$

where

$$Q = (q_0, q_1, q_2, q_3, f)$$

$$S = \{q_0\}$$

$$F = \{f\}.$$

$$\Sigma = \{a, b\}$$

and here transition relations are defined as follows :

(1) $((q_0, a, \in), (q_1, a))$

(2) $((q_1, a, a), (q_1, a))$

(3) $((q_1, \in, a), (q_1, \in))$

(4) $((q_1, \in, \in), (f, \in))$

(5) $((q_1, b, a), (q_2, \in))$

(6) $((q_2, b, a), (q_2, \in))$

(7) $((q_2, \in, a), (q_3, \in))$

(8) $((q_3, \in, a), (q_3, \in))$

(9) $((q_3, \in, \in), (f, \in))$

7.4. PUSHDOWN AUTOMATA AND CONTEXT-FREE GRAMMARS

It should be clear now that PDA can recognise any language for which there exists a CFG. "That is class of language accepted by pushdown automata is exactly the class of context-free languages".

7.4.1. Construction of PDA Equivalent of a CFG

Let $G = (V_n, V_t, P, S)$ be a context-free grammar; we must construct a pushdown automata P such that $L(P) = L(G)$. The machine we construct has only two states, p and q, and remains permanently in state q after its first move. Also, p uses V_n, the set of non-terminals and V_t the set of terminals, as its stack alphabet. We let $P = (Q, \Sigma, \Gamma, \delta, S, F)$

where

$$Q = (p, q\}$$

$$\Sigma = V_t$$

$$\Gamma = V_n \cup V_t$$

that is the set of terminals and non-terminals.

$$S = p$$

and transition relation δ is defined as follows :

(1) $((p, \in, \in), (q, S))$ (as S is strating non-terminal of CFG)

(2) $((q, \in, A), (q, x))$ for each rule $A \to x$ in CFG.

(3) $((q, a, a), (q, \in))$ for each $a \in V_t$.

The pushdown automata P begins by pushing S, the starts symbol of grammar G, on its initially empty pushdown store, and entering state q (transition 1). On each subsequent step, it either replaces the top most symbol A on the stack, provided that it is a non-terminal, by the right hand side x of some rule $A \to x$ in grammar (transition of type 2), or POPS the top most symbol from the stack, provide that it is a terminal symbol that matches the next input symbol (transition of type 3).

Example 7.10. Design a PDA for the following CFG,

$$G = (V_n, V_t, P, S) \text{ with}$$

$V_n = \{S\}$, $V_t = \{(,)\}$ **and P is define δ as follows :**

$$S \to \in$$
$$S \to SS$$
$$S \to (S).$$

Solution. Let us assume corresponding PDA will be

$$P = (Q, \Sigma, \Gamma, \delta, S, F)$$

where $Q = \{p, q\}$

$$\Sigma = \{(,)\}$$

$\Gamma = (S, (,)\}$ that is terminals and non-terminals in grammar

$$S = p$$

$F = \{q\}$, and δ is as follows.

(1) $((p, \in, \in), (q, S))$

(2) $((q, \in S), (q, \in))$ because $S \to \in$ is a rule of CFG

(3) $((q, \in S), (q, SS))$

(4) $((q, \in, S), (q, (S))$

(5) $((q, (, ( ), (q, \in))$

(6) $((q,),)), (q, \in))$ for each $a \in V_t$

Let us apply these transition relations on the string

$$w = ( ) ( )$$

S. No.	State	Unread input	Stack	Transition used
1	P	( ) ( )	∈	–
2	q	( ) ( )	S	(1)
3	q	( ) ( )	SS	(3)
4	q	( ) ( )	$(S)\ SS$	(4)
5	q	) ( )	$S)\ S$	(5)
6	q	) ( )	$)\ S$	(2)
7	q	( )	S	(6)
8	q	( )	(S)	(4)
9	q	)	$S)$	(5)
10	q	)	)	(2)
11	q	∈	∈	(6)

Example 7.11. Design PDA for the grammar $G = (V_n, V_t, P, S)$

where $V_n = \{S\}$

$$V_t = \{a, b, c\}$$

and P is defined as

$$S \to aSa$$
$$S \to bSb$$
$$S \to c.$$

Solution. Let PDA be

$$P = (Q, \Sigma, \Gamma, \delta, S, F)$$

where

$$Q = \{p, q\}$$
$$\Sigma = \{a, b, c\}$$
$$\Gamma = \{S, a, b, c\}$$
$$S = \{p\}$$
$$F = \{q\}$$

and δ is defined as :

(1) $((p, \in, \in), (q, S))$

(2) $((q, \in, S), (q, aSa))$

(3) $((q, \in, S), (q, bSb))$

(4) $((q, \in, S), (q, c))$

(5) $((q, a, a), (q, \in))$

(6) $((q, b, b), (q, \in))$

(7) $((q, c, c), (q, \in))$

Let us apply this on string

$$w = abbcbba$$

S.No.	*State*	*Unread input*	*Stack*	*Transform used*
1	p	*abbcbba*	$\in$	–
2	q	*abbcbba*	*S*	(1)
3	q	*abbcbba*	*aSa*	(2)
4	q	*bbcbba*	*Sa*	(5)
5	q	*bbcbba*	*bSba*	(3)
6	q	*bcbba*	*Sba*	(6)
7	q	*bcbba*	*bSbba*	(3)
8	q	*cbba*	*Sbba*	(6)
9	q	*cbba*	*cbba*	(7)
10	q	*bba*	*bba*	(6)
11	q	*ba*	*ba*	(6)
12	q	*a*	*a*	(6)
13	q	$\in$	$\in$	(5)

7.4.2. Construction of CFG Equivalent of a PDA

Let $P = (Q, \Sigma, \Gamma, \delta, q_0)$ be a PDA that accepts a language L by empty stack. Then a CFG $G = (V_n, V_t, P, S)$ that generates L can be constructed using following rules :

Assumption : Here we are taking a assumption that stack bottom is indicated by a special symbol \$.

Let S be the start symbol of grammar.

(*i*) For every $q \in Q$, add a production $S \rightarrow [q_0 \ \$ \ q]$ in P. Thus if there are n states in a PDA P, then there will be 'n' new productions added in P.

(*ii*) For every $q, r \in Q$, $a \in \{\Sigma \cup \in\}$, $X \in \Gamma$, if $\delta(q, a, X) = (r, \in)$, then add a production

$$[q \ X \ r] \rightarrow a \text{ in } P.$$

(*iii*) For every $q, r \in Q$, $a \in \{\Sigma \cup \in\}$, $X \in \Gamma$, and $x \geq 1$, if $\delta(q, a, X) = (r, X_1, X_2, \ldots X_k)$ where $X_1, X_2, \ldots, X_k \in \Gamma$, then for every choice of $q_1, q_2, \ldots, q_k \in Q$, add the production

$$[q \ X \ q_k] \rightarrow a[r \ X_1 \ q_1] \ [q_1 \ X_2 \ q_2] \ldots [q_{k-1} \ X_k \ q_k] \text{ in } P.$$

The basic idea behind this construction is to recognize that the current string in the derivation will consist of two parts, the string of input symbols read by the PDA so far and a remaining portion corresponding to the stack contents. Next we introduce variables of the proposed grammar in the form $[P \ X \ q]$ where P and q are states of the PDA. For the variable $[P \ X \ q]$ to be replaced by a symbol 'a' (a may be $\in$ also), it may be the case there is a move in PDA that reads a, pops X from the stack, and takes machine from state P to q moves. For example, it may go from state P to state P_1, that reads a and replaced X on stack by $Y_1, Y_2, Y_3, \ldots Y_k$. Then after a sequence of moves it may Pop off Y_1 and end up in state say P_2. Similarly it may pop off Y_2 and go into state P_2 and through many such steps finally entering into state P_k and popping of symbol Y_k. All these intermediate move, in their cumulative effect, are equivalent to derive a portion of strings say w and reaching a state P_k. We can represent this fact by the production

$$[P \ X \ q_k] \rightarrow a[r \ X_1 \ q_1] \ [q_1 \ X_2] \ldots [q_{k-1} \ X_k \ q_k]$$

Example 7.12. Consider the following PDA and construct equivalent CFG.

1. $((q_0, a, \$) \rightarrow (q_0, a\$))$
2. $((q_0, b, \$) \rightarrow (q_0, b\$))$
3. $((q_0, a, a) \rightarrow (q_0, aa))$
4. $((q_0, b, b) \rightarrow (q_0, bb))$
5. $((q_0, a, b) \rightarrow (q_0, ab))$
6. $((q_0, b, a) \rightarrow (q_0, ba))$
7. $((q_0, c, \$) \rightarrow (q_1, \$))$
8. $((q_0, c, a) \rightarrow (q_1, a))$
9. $((q_0, c, b) \rightarrow (q_1, b))$
10. $((q_1, a, a) \rightarrow (q_1, \in))$
11. $((q_1, b, b) \rightarrow (q_1, \in))$
12. $((q_1, \in, \$) \rightarrow (q_1, \in))$

Solution. CFG using above construction production rules is as follows :

$$S \rightarrow [q_0 \$ q_0]/[q_0 \$ q_1] \qquad \text{(from rule } (i)\text{)}$$

$$\left.\begin{aligned} [q_1\, a\, q_1] &\rightarrow a \\ [q_1\, b\, q_1] &\rightarrow b \\ [q_1 \$ q_1] &\rightarrow \in \end{aligned}\right\} \text{from rule } (ii)$$

Now we will get remaining productions using rule (*iii*).

$$[q_0 \$ q_0] \rightarrow a[q_0\, a\, q_0]\, [q_0 \$ q_0]$$
$$[q_0 \$ q_1] \rightarrow a[q_0\, a\, q_0]\, [q_0 \$ q_1]$$
$$[q_0 \$ q_1] \rightarrow a[q_0\, a\, q_1]\, [q_1 \$ q_1]$$
$$[q_0 \$ q_1] \rightarrow b[q_0\, b\, q_0]\, [q_0 \$ q_0]$$
$$[q_0 \$ q_1] \rightarrow b[q_0\, b\, q_0]\, [q_0 \$ q_1]$$
$$[q_0 \$ q_1] \rightarrow b[q_0\, b\, q_1]\, [q_1 \$ q_1]$$
$$[q_0\, a\, q_0] \rightarrow a[q_0\, a\, q_0]\, [q_0\, a\, q_0]$$
$$[q_0\, a\, q_1] \rightarrow a[q_0\, a\, q_0]\, [q_0\, a\, q_1]$$
$$[q_0\, a\, q_1] \rightarrow a[q_0\, a\, q_1]\, [q_1\, a\, q_1]$$
$$[q_0\, a\, q_0] \rightarrow b[q_0\, b\, q_0]\, [q_0\, a\, q_0]$$
$$[q_0\, a\, q_1] \rightarrow b[q_0\, b\, q_0]\, [q_0\, a\, q_1]$$
$$[q_0\, a\, q_1] \rightarrow b[q_0\, b\, q_1]\, [q_1\, a\, q_1]$$
$$[q_0\, b\, q_0] \rightarrow a[q_0\, a\, q_0]\, [q_0\, b\, q_0]$$
$$[q_0\, b\, q_1] \rightarrow a[q_0\, a\, q_0]\, [q_0\, b\, q_1]$$
$$[q_0\, b\, q_1] \rightarrow a[q_0\, a\, q_1]\, [q_1\, b\, q_1]$$
$$[q_0\, b\, q_0] \rightarrow b[q_0\, b\, q_0]\, [q_0\, b\, q_0]$$
$$[q_0\, b\, q_1] \rightarrow b[q_0\, b\, q_0]\, [q_0\, b\, q_1]$$
$$[q_0\, b\, q_1] \rightarrow b[q_0\, b\, q_1]\, [q_1\, b\, q_1]$$
$$[q_0 \$ q_1] \rightarrow c[q_1\, b\, q_1]$$
$$[q_0\, a\, q_1] \rightarrow c[q_1\, a\, q_1]$$
$$[q_0\, b\, q_1] \rightarrow c[q_1\, b\, q_1]$$

Example 7.13. Construct PDA for the language

$$L = \{a^3\, b^n\, c^n : n \geq 0\}.$$

Solution. By the observation we can say that language contains only that set of strings, in which every string contains exactly 3*a*'s in the starting of string (like *aaa* ...). If $n > 0$, then *a*'s are followed by *b*'s and *b*'s are followed by *c*'s and number of *b*'s and *c*'s are equal.

String like *aaa, aaabc, aaabbbccc* ... are accepted and strings like *a, aa, abc, aabc, aaabbbc, aaaccc, aacb* are not accepted.

Let PDA be $P = (Q, \Sigma, \Gamma, \delta, S, F)$

where

$$Q = \{q_0, q_1, q_2, q_3, q_4, q_5 f\}$$
$$\Gamma = \{b\}$$
$$\Sigma = \{a, b, c\}$$
$$F = \{f\}$$
$$S = \{q_0\}$$

and δ is defined as follows :

(1) $((q_0, a, \in), (q_1, \in))$

(2) $((q_2, a, \in), (q_2, \in))$

(3) $((q_2, a, \in), (q_3, \in))$

(4) $((q_3, \in, \in), (f, \in))$

(5) $((q_3, b, \in), (q_4, b))$

(6) $(q_4, b, b), (q_4, b))$

(7) $(q_4, c, b), (q_5, \in))$

(8) $((q_5, c, b), (q_5, \in))$

(9) $((q_5, \in, \in), (f, \in))$

Here first three moves read exactly three a's and machine moves to state q_3, now if $n = 0$ then q_3 change to final state and if $n > 0$ then we push all b's on stack with the help of moves 5 and 6. Now when c encountered then we POP all corresponding b's (by the help of moves 7 and 8). If b's and c's are equal then machine finally transit to final state.

Example 7.14. Construct a NPDA for Language $L + \{w \in (a, b)^*/n_a(w) > n_b(w)\}$.

Solution. Given language is $L = \{w \in (a, b)^*/n_a(w) > n_b(w)\}$. Here $n_a(w)$ means number of a's in w and $n_b(w)$ means number of b's in w.

Let NPDA be $P = \{Q, \Sigma, \Gamma, \delta, q_0, F)$

$$Q = \{q_0, q_1, f\}$$
$$\Sigma = \{a, b\}$$
$$\Gamma = \{a, b\}$$
$$q_0 = \text{Initial state}$$
$$F = \text{Non-empty set of final states}\{f\}.$$

δ is defined as follows :

1. $((q_0, \in, a), (q_0, a))$
2. $((q_0, a, a), (q_0, a))$
3. $((q_0, \in, b), (q_0, b))$
4. $((q_0, b, b), (q_0, b))$
5. $((q_0, a, b), (q_1, \in))$

6. $((q_1, a, b), (q_1, \in))$
7. $((q_0, b, a), (q_1, \in))$
8. $((q_1, b, a), (q_1, \in))$
9. $((q_0, \in, a), (f, \in))$
10. $((q_1, \in, a), (f, \in))$
11. $((f, \in, a), (f, \in))$

Example 7.15. Design a PDA for language $L = \{a^n b^m a^n / m, n \geq 1\}$.

Solution. Let PDA be $P = (Q, \Sigma, \Gamma, \delta, q_0, F)$.

Here δ is defined as follows :

1. $((q_0, a, \in), (q_0, a))$
2. $((q_0, a, a), (q_0, a))$
3. $((q_0, b, a), (q_1, a))$
4. $((q_1, b, a), (q_1, a))$
5. $((q_1, a, a), (q_2, \in))$
6. $((q_2, a, a), (q_2, \in))$
7. $((q_2, \in, \in), (f, \in))$

$$Q = \{q_0, q_1, q_2, f\}$$
$$\Sigma = \{a, b\}, \Gamma = \{a, b\}$$
$$F = \{f\}$$

Here we are reading all a's first and putting them on the stack. After it all b's are read without making any change on stack. Now again we are reading a's from input tape and one 'a' from stake is poped for every single 'a' of input tape. If all a's are poped and there is no 'a' on input tape then final state is achieved.

Example 7.16. Construct a PDA to accept the language L = {set of all words in a and b with an even number of a's}.

Solution. Let PDA be $P = \{Q, \Sigma, \delta, \Gamma, q_0, F)$

$$Q = \{q_0, f\}$$
$$\Sigma = \Gamma = \{a, b\}$$
$$q_0 = \text{Initial state}$$
$$F = \{f\}$$

S is defined as follows :

1. $((q_0, a, \in), (q_0, a))$
2. $((q_0, a, a), (q_0, \in))$
3. $((q_0, b, \in), (q_0, \in))$
4. $((q_0, b, a), (q_0, a))$
5. $((q_0, \in, \in), (f, a))$

Example 7.17. Construct PDA for language $L = \{(ab)^n / n \geq 1\}$.

Solution. Let PDA be $P = (Q, \Sigma, \Gamma, \delta, q_0, F)$

$$Q = \{q_0, q_1, q_2, q_3, f\}$$

$$\Sigma = \Gamma = \{a, b\}$$

q_0 = Intial state

$$F = \{f\}$$

δ is defined as follows :

1. $((q_0, a, \in), (q_1, \in))$
2. $((q_1, b, \in), (q_2, \in))$
3. $((q_2, a, \in), (q_3, \in))$
4. $((q_3, b, \in), (q_2, \in))$
5. $((q_2, \in, \in), (f, \in))$

Example 7.18. Construct PDA for language $L = \{a^n b^{3n} / n \geq 1\}$.

Solution. Let PDA be $P = \{Q, \Sigma, \Gamma, \delta\ q_0, F)$, where

$$Q = \{q_0$$

$$\Gamma = \Sigma = \{a, b\}$$

q_0 = Intial state

$$F = \{f\}$$

δ is defined as follows :

1. $((q_0, a, \in), (q_0, a))$
2. $((q_0, a, a), (q_0, a))$
3. $((q_0, b, a), (q_1, ba))$
4. $(q_1, b, b), (q_2, bba))$
5. $((q_2, b, bba), (q_3, \in))$
6. $((q_3, b, a), (q_1, ba))$
7. $((q_3, \in, \in), (f, \in))$

Now let us see explanation for above discussed transition relation. Here first we pushed all the a's by the help of moves 1 and 2. Now as first 'b' encountered and it has been pushed by the help of more 3, second b has been pushed by the help of more 4. Now third b is encountered then bba has been poped corresponding this 'b' with the help of more 5. Move 4, 5, and 6 generates a loop. Finally when input tape becomes empty and stack also empty then state q_3 changes to final state (f) and string is accepted.

Example 7.19. Let $L = \{a^m b^n / n < m\}$. Construct (*i*) A CFG generation L (*ii*) PDA for CFG.

Solution. (*i*) Given language is $L = \{a^m b^n / n < m\}$.

Let CFG be $G = (V_n, V_t, P, S)$

$$V_n = \{S\}$$
$$V_t = \{a, b\}$$
$$P = \{S \to aSb, S \to aS, S \to a\}$$

(*ii*) Let equivalent PDA is $P = \{Q, \Sigma, \Gamma, \delta, q_0, F)$

where
$$Q = (P, q)$$
$$\Sigma = \{a, b\} = V_t$$
$$\Gamma = V_n \cup V_t = \{S, a, b\}$$
$$q_0 = \{P\}$$
$$F = \{q\}$$

The transition relation 'δ' is defined as follows :

1. $((P, \in, \in), (q, S))$ (*S* is starting non-terminal of CFG)
2. $((q, \in, S), (q, aSb))$
3. $((q, \in, S), (q, aS))$
4. $((q, \in, S), (q, a))$
5. $((q, a, a), (q, \in))$
6. $((q, b, b), (q, \in))$

Example 7.20. Construct a CFG which accepts *L*(*P*) where

$$\mathbf{P = (Q, \Sigma, G, \delta, q_0, F)}$$
$$\mathbf{Q = \{q_0, q_1\}}$$
$$\mathbf{\Sigma = \{a, b\}}$$
$$\mathbf{G = \{a, \$\}}$$
$$\mathbf{q_0 = \text{Intial state}}$$
$$\mathbf{F = \{q_1\}}$$

and δ is defined as follows :

1. $((q_0, a, \$), (q_0, a\$))$
2. $((q_0, a, a), (q_0, aa))$
3. $((q_0, b, a), (q_1, a))$
4. $((q_1, b, a), (q_1, a))$
5. $((q_1, a, a), (q_1, \in))$
6. $((q_1, \in, \$), (q_1, \in))$

Solution. Let equivalent CFG is $G = (V_n, V_t, P, S)$ when

$V_n = \{(q_0, \$, q_1), (q_1, \$, q_0), (q_0, a, q_0), (q_1, a, q_0), (q_0, \$, q_1), (q_1, \$, q_1), (q_1, \$, q_1), (q_0, a, q_1), (q_1, a, q_1)$.

The production in *P* are constructed as follows :

The *S*-productions are

$$S \to [q_0, \$, q_0]/[q_0, \$, q_1]$$

$((q_0, a, \$), (q_0, a\$))$ induces

$$[q_0, \$, q_0] \rightarrow a[q_0, a, q_0]\,[q_0, \$, q_0]$$
$$[q_0, \$, q_0] \rightarrow a[q_0, a, q_1]\,[q_1, \$, q_0]$$
$$[q_0, \$, q_1] \rightarrow a[q_0, a, q_0]\,[q_0, \$, q_1]$$
$$[q_0, \$, q_1] \rightarrow a[q_0, a, q_1]\,[q_1, \$, q_1]$$

$((q_0, a, a), (q_0, aa))$ yields

$$[q_0, a, q_0] \rightarrow a[q_0, a, q_0]\,[q_0, a, q_0]$$
$$[q_0, a, q_0] \rightarrow a[q_0, a, q_1]\,[q_1, a, q_0]$$
$$[q_0, a, q_1] \rightarrow a[q_0, a, q_0]\,[q_0, a, q_1]$$
$$[q_0, a, q_1] \rightarrow a[q_0, a_1, q_1]\,[q_1, a, q_1]$$

$((q_0, b, a), (q_1, a))$ gives

$$[q_0, a, q_0] \rightarrow b[q_1, a, q_0]$$
$$[q_0, a, q_1] \rightarrow b[q_1, a, q_1]$$

$((q_1, b, a), (q_1, a))$ yields

$$[q_1, a, q_0] \rightarrow b[q_1, a, q_0]$$
$$[q_1, a, q_1] \rightarrow b[q_1, a, q_1]$$

$((q_1, a, a), (q_1, \in))$ gives

$$[q_1, a, q_1] \rightarrow a$$

$((q_1, \in, \$), (q_1, \in))$ yields

$$[q_1, \$, q_1] \rightarrow \in$$

EXERCISES

1. Let input alphabet be $\Sigma = \{a, b, c\}$ and L be the language of all words in which all the a's come before the b's and there are same number of a's as b's and arbitrarily many c's that can be in front, behind or among the a's and b's. Some words in L are *abc, caabcb, ccacaabcccbccbc*.

(*i*) Show the language is not regular.

(*ii*) Design a PDA for above given language.

2. Let L be some regular language in which all the words happen to have an even length. Let us define the new language second (L) to be set of all the words of L second, where second we mean the first and second letters have been interchanged, the third and fourth letters have been interchanged, and so on.

If $L = \{ba, abba, babb, ...\}$

Second $(L) = \{ab, baab, abbb, ...\}$

Build a PDA that accepts second (L).

3. Construct a PDA accepting the following languages :

(*i*) $\{0^n 1^m 0^n / m, n \geq 1\}$

(*ii*) $\{0^n 1^{2n} / n \geq 1\}$

(*iii*) $\{0^m 1^m 2^n / m, n \geq 1\}$

(*iv*) $\{0^m\ 1^n / m > n \geq 1\}$.

4. Let given language $L = \{a^m b^n / n < m\}$
   (i) Construct a PDA for the language
   (ii) Write the CFG for the language and
   (iii) Convert this CFG into pushdown automata.
5. Starting with the CFG for $\{a^n b^n\}$

$$S \rightarrow aSb/ab$$

   (i) put this CFG into CNF form
   (ii) Give the equivalent PDA for this CNF form.
6. For each of the CFGs, construct PDA that accepts the same language they generate.
   (i) $S \rightarrow aSbb/abb$
   (ii) $S \rightarrow SS/a/b$
   (iii) $S \rightarrow XaaX$
   $X \rightarrow aX/bX/\in$
   (iv) $S \rightarrow aS/aSbS/a$
   (v) $S \rightarrow XY$
   $X \rightarrow aX/bX/a$
   $Y \rightarrow Ya/Yb/a$
   (vi) $S \rightarrow Xa/Yb$
   $X \rightarrow Sb/b$
   $Y \rightarrow Sa/a$
7. Construct the equivalent PDA for the following CFG's
   (i) $S \rightarrow Saa/aSa/aaS$
   (ii) $S \rightarrow (S)\ (S)/a$
   (iii) $S \rightarrow XaY/YbX$
   $X \rightarrow YY/aY/b$
   $Y \rightarrow b/bb$
8. Construct NPDA's that accepts the following languages :
   (i) $L_1 = L(aaa^*b)$
   (ii) $L_2 = L(aab^*aba^*)$
9. Find the NPDA for the language :

$$L = \{ab(ab)^n b(ba)^n : n \geq 0\}$$

10. Find the PDA for the following languages :
   (i) $L = \{w : n_a(w) < n_b(w)\}$
   (ii) $L = \{w : n_a(w) = 2n_b(w)\}$
   (iii) $L = \{w : n_a(w) + n_b(w) = n_c(w)\}$.
11. Design a PDA which convert infix to prefix.
12. Design a PDA which convert prefix to postfix.

*****

CHAPTER 8

Properties of Context-Free Languages

8.1. INTRODUCTION

INSIDE THIS CHAPTER

We have seen till now that there are two ways to recognise a context-free language. One way is to find a context-free grammar for that CFL and another is to generate a PDA for them. Sometimes one representation is more easier than the other. For example, we can easily generate a PDA for the language in which all strings contains equal number of a's and b's, but it is difficult to find a CFG for this language.

In this section, we study the general properties of CFL. So that we will be more familiar with them. These properties can help us to recognise whether a language is a CFL or not. These properties also help us to generate a CFG for complex CFLS.

Theorem 8.1. The family of context free languages is closed under union, concatenation and Kleene star-closure.

Proof. Let L_1 and L_2 be two context-free languages generated by the following context free grammars respectively.

$$G_1 = (V_{n_1}, V_{t_1}, P_1, S_1)$$

and

$$G_2 = (V_{n_2}, V_{t_2}, P_2, S_2)$$

Union. Consider the language $L(G)$, generated by the following grammar

$$G = (V_n, V_t, P, S)$$

where

$$V_n = V_{n_1} \cup V_{n_2} \cup \{S\}$$

$$V_t = V_{t_1} \cup V_{t_2}$$

S is the start symbol, and productions P is defined as follows :

$$P = \{P_1 \cup P_2 \cup \{S \rightarrow S_1/S_2\}\}$$

Now let us choose a string $W \in (V_{t_1} \cup V_{t_2})^*$

It $S_1 \overset{*}{\Rightarrow} W$ or $S_2 \overset{*}{\Rightarrow} W$, and in our grammar

$S \rightarrow S_1/S_2$, hence S will lead to W.

Hence G is a context-free grammar, so that $L(G)$ is a context free languages that is

$$L(G) = L_1 \cup L_2 \text{ is a CFL.}$$

Concatenation. Now again consider a language $L(G)$ generated by the CFG.

$$G = (V_n, V_t, P, S)$$

where

$$V_n = V_{n_1} \cup V_{n_2} \cup \{S\}$$

$$V_t = V_{t_1} \cup V_{t_2}$$

S is the starting symbol of this grammar and production P is defined as follows :

$$P = P_1 \cup P_2 \cup \{S \rightarrow S_1\ S_2\}$$

Let W_1 is a string, $W_1 \in L_1$ and W_2 is a string, $W_2 \in L_2$ we know that $S_1 \overset{*}{\Rightarrow} W_1$ and $S_2 \overset{*}{\Rightarrow} W_2$ but in above grammar G, $S \Rightarrow S_1S_2$ so S will lead to W_1W_2 in broader way the language of grammar G will be $L_1\ L_2$, since L_1 is CFL and L_2 is CFL and G is context-free grammar so $L_1\ L_2$ is also context-free language.

Kleene-Star. Let L be context free language, generated by the following grammar

$$G = (V_n, V_t, P, S)$$

Now let Kleene star of language L that is L^* is generated by a grammar G'.

$$G' = (V'_n, V'_t, P', S')$$

where

$$V'_n = V_n$$

$$V'_t = V_t$$

S' is the starting non terminal.

and set of production is given by

$$P' = P \cup \{S' \rightarrow S, S' \rightarrow \in, S' \rightarrow SS\}$$

When we carefully analyse P' then it is very clear that if follow all the properties of CFG since P is production of given CFG and $S' \rightarrow S$, $S' \rightarrow \in$, $S' \rightarrow SS$ also fulfill the requirement so we can say that G' is a context-free grammar, then L^* will be context free.

Theorem 8.2. The family of context free language is not closed under intersection and complementation.

Proof. Intersection :

We know that languages

$$L_1 = \{a^m\ b^m\ c^n/n, m \geq 0\}$$

and $$L_2 = \{a^n b^m c^m / n, m \geq 0\}$$

are context free.

Now let see $L_1 \cap L_2 = \{a^n b^n c^n / n \geq 0\}$ is not context-free (we will prove it in the next section by pumping lemma that $a^n b^n c^n / n \geq 0$ is not CFL). So it is proved that family of context-free languages is not closed under intersection.

Complementation. Let L_1 and L_2 be two context free languages, their union $L_1 \cup L_2$ also is context-free (by the theorem).

Now let assume that family of context-free language is closed under complementation, that is complement of a context-free language is also context free.

So according to our assumption $\overline{L_1}$ and $\overline{L_2}$ are context free and their union $\overline{L_1} \cup \overline{L_2}$ is also CFL (by theorem). Again $\overline{\overline{L_1} \cup \overline{L_2}}$ will be CFL (by the assumption).

We know that Demorgan's law is

$$L_1 \cap L_2 = \overline{\overline{L_1} \cup \overline{L_2}}$$

We have right now prove that $\overline{\overline{L_1} \cup \overline{L_2}}$ is context-free. So $L_1 \cap L_2$ should also be context free but it is not true, we have already seen it so our assumption that complement of context-free language is also context-free is wrong.

So we can say that, family of context-free languages is not closed under complementation.

Theorem 8.3. The intersection of a context-free language and a regular language is a context-free language.

Proof. Here we give a proof based on finite and push down automata; which would be simpler than one based on grammars. Let L be context-free language, and R is a regular set. So three exists a PDA $A_1 = (Q_1, \Sigma_1, \Gamma_1, \delta_1, S_1, F_1)$ and a finite automata $A_2 = (Q_2, \Sigma_2, \delta_2, S_2, F_2)$ for these languages respectively. Now we will combine these machines into a single PDA P that carries out computations by A_1 and A_2 in parallel and accepts a string if both would have accepted. Let combined PDA be

$$A = (Q, \Sigma, \Gamma, \delta, S, F)$$

where $Q = Q_1 \times Q_2$ (Cartesian product of the states of A_1 and A_2)

$\Sigma = \Sigma_1 \cup \Sigma_2$

$\Gamma = \Gamma_1$

$F = F_1 \times F_2$ and δ is defined as follows.

$(((P_1, P_2), W, \alpha), ((q_1, q_2), \beta))) \in \delta$, it and only if, $((P_1, W, \alpha), (q_1, \beta)) \in \delta_1$ and $(P_2, W) \vdash^*_{A_2} (q_2, \in)$

It means that combined PDA 'P' goes from state (P_1, P_2) to (q_1, q_2) in the same way that A_1 passes from state p_1 and q_1 and also keeps track that A_2 changes its state on the same input. It is easy to see that indeed $W \in L(A)$ if and only if $W \in L(A_1) \cap L(A_2)$.

8.2. PERIODICITY PROPERTIES

CFL's also exhibit periodicity that is paths (substring) may be repeated periodically is the derivation. It can be seen is parse trees also :

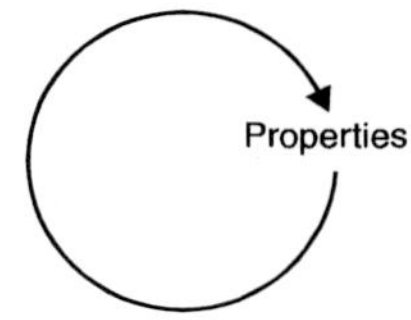

Fig. 8.1

Let G is a grammar

$$S \rightarrow aB / bA / \in$$
$$A \rightarrow aS / bAA$$
$$B \rightarrow bS / aBB$$

Derivations :

$$S \Rightarrow a\underline{B} \Rightarrow aa\underline{B}B$$
$$\Rightarrow aabS\underline{B}$$
$$\Rightarrow aabSb\underline{S}$$
$$\Rightarrow aab\underline{S}b$$
$$\Rightarrow aaba\underline{B}b$$
$$\Rightarrow aabab\underline{S}b$$
$$\Rightarrow aababb$$

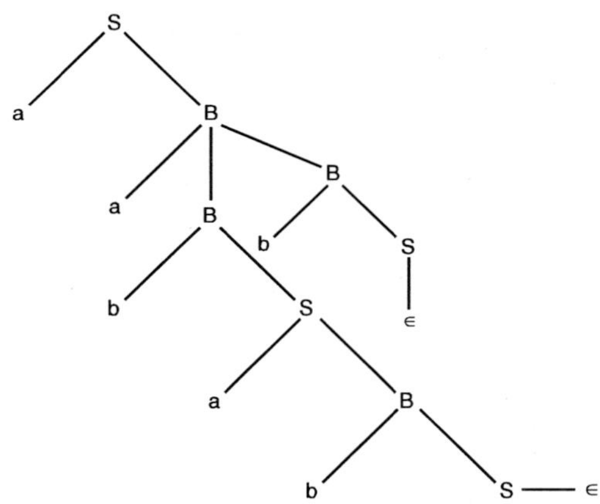

Fig. 8.2

So repeations can be seen in both parse tree as well as derivations.

8.3. THE PUMPING LEMMA FOR CFL's

Let G be any CFG, then there is a constant, depending only on G, such that if W is in $L(G)$ and $|W| \geq n$ then we may write $W = \mu vxyz$ such that

(*a*) $|vy| \geq 1$ that is either v or y is non-empty

(*b*) $|vxy| \leq n$ then for all $i \geq 0$

$\mu v^i x y^i z$ is in $L(G)$.

Proof. Let our grammar be $G = (V_n, V_t, P, S)$. Now in order to prove the theorem it is sufficient to show that there is a constant n such that any $W \in L(G)$ of length greater than or equal to n has a derivation of the form.

$$S \overset{*}{\Rightarrow} \mu X z$$

$$X \overset{*}{\Rightarrow} vXy$$

$$X \Rightarrow x$$

Now if we use above productions as :

$$S \overset{*}{\Rightarrow} \mu Xz \overset{*}{\Rightarrow} \mu vXyz \overset{i-1}{\Rightarrow} \mu\, v^i Xy^i z \Rightarrow \mu v^i Xy^i z$$

where $\mu, v, x, y, z \in \Sigma^*$ and $X, S \in V_n$ and $|\,vy\,| \geq 1$.

Then derivation $X \overset{*}{\Rightarrow} vXy$ can be repeated any number of times to obtain $\mu v^i xy^i z$.

So Now we get the line to prove the theorem, Now we will concentrated to prove above discussed terminology.

Let P be the largest number of symbols on the right hand side of any Rule $X \to \alpha$ in G. Now we know it by the theorem that a parse tree of height m can have at most P^m leaves where $m \geq 1$, and can therefore have yield of length at most P^m.

Now let m = number of non-terminals and $n = P^m$ and suppose $|\,W\,| \geq n$. Let T be a parse tree with root S and which derives the string W. Then T has at least one path with more than $(m + 1)$ nodes, and therefore, at least one path including two nodes labeled with the same number of X of V_n. Now let us see this concept picturally.

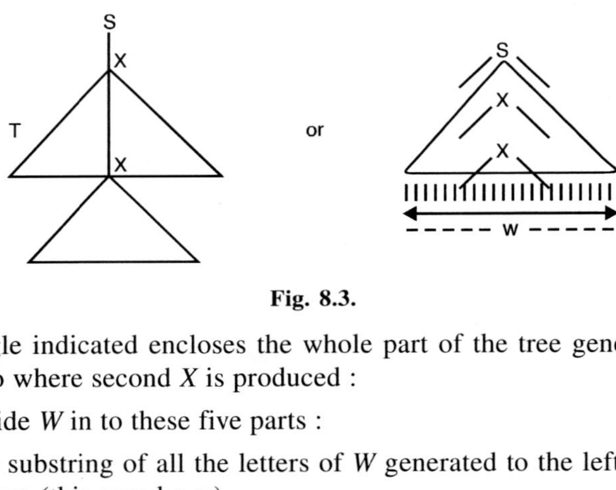

Fig. 8.3.

The triangle indicated encloses the whole part of the tree generated from the first X down to where second X is produced :

Let us divide W in to these five parts :

μ = the substring of all the letters of W generated to the left of the triangle above (this may be $\in$).

v = the substring of all the letters of W descended from the first X but to the left of the letters generated by the second X (this may be $\in$).

x = the substring of W descended from the lower X (this may not be $\in$ because this non-terminal must turn in to some terminal).

y = the substring of W of all letters generated by the first X but the right of the letters descending from the second X (this may be $\in$, but as we shall see, not if $v = \in$).

z = the substring of all the letters of w generated to the right of the triangle (this may be $\in$).

Picturally,

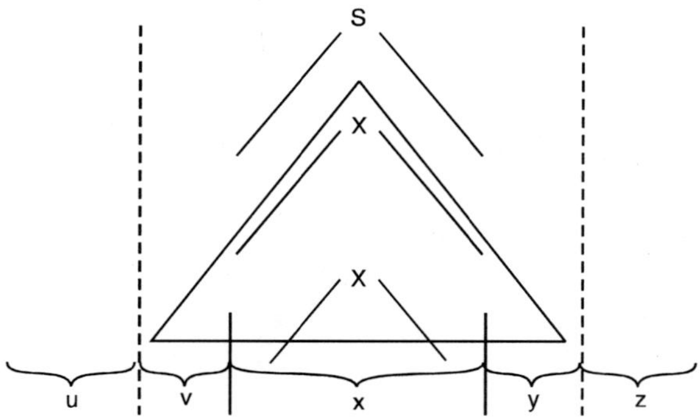

Fig. 8.4

For example, the following is a complete tree in an unspecified grammar :

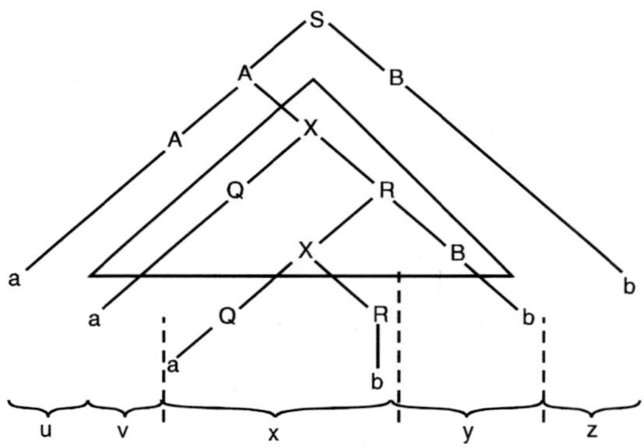

Fig. 8.5.

It is possible that either μ or z or both might be ∈, as in the following example

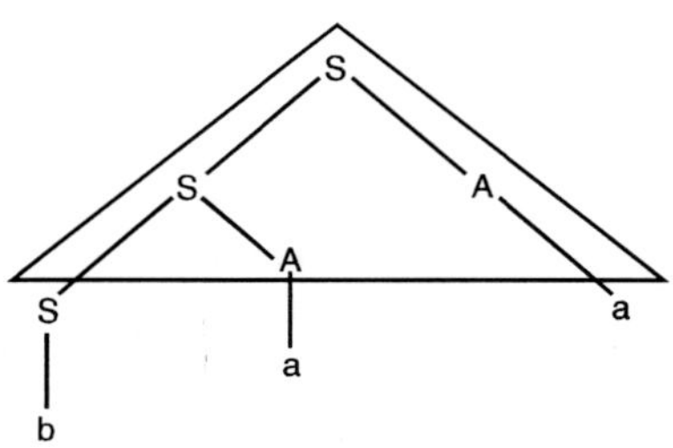

$\mu = \in$ $\qquad v = \in$ $\qquad x = ba$ $\qquad y = a$ $\qquad z = \in$

Fig. 8.6.

Now let us see the situation picturally it we change the derivation tree by reiterating productions inside the triangle :

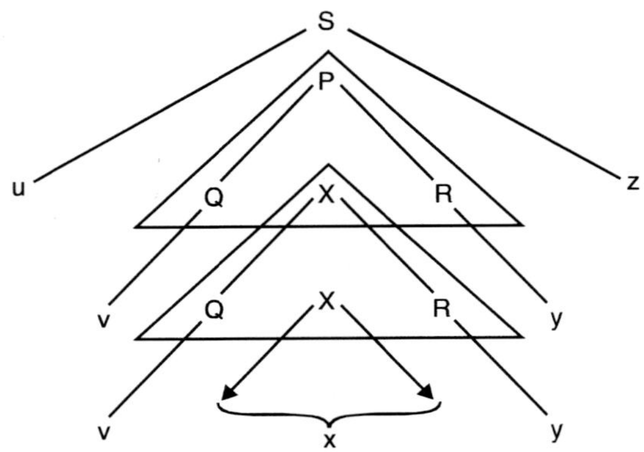

Fig. 8.7.

that is yield will be $\mu vvxyyz$.

The part μ comes from S to the left of the whole triangle. First v comes from the inside the first triangle to the left of second X. The second v comes from the stuff in the second triangle to the left of the third X. The x part comes from the third X. The first y comes from stuff in the second triangle to the right of the third X. The second y comes from the stuff in the first triangle to the right of the second X. The z as before comes from S from the stuff to the right of the first triangle.

In general if we repeat the triangle i times, we get a derivation for the word

$$\mu v^i x y^i z$$

which must be in the language generated by G.

We can also use our $\overset{*}{\Rightarrow}$ symbol to provide an algebraic proof.

$$S \overset{*}{\Rightarrow} \mu Xz \overset{*}{\Rightarrow} \mu vXyz \overset{*}{\Rightarrow} \mu vxyz = W.$$

Example 8.1. Prove that language $L = \{a^n b^n c^n / n \geq 0\}$ is not context-free language.

Solution. We will use pumping lemma for CFL's to prove the L is not CFL.

Here we are going to adopt the same approach as we follow in the pumping lemma for regular sets. That is first we assume that given language is regular then by applying pumping lemma on it we find an contradiction so automatically our assumption that given language is CFL will be wrong and in this way we will prove that language is not context-free.

So let us assume that $L = \{a^n\ b^n\ c^n / n \geq 0\}$ is context-free and m be the constant obtained from the pumping lemma. Now let us pick the string

$$W = a^m\ b^m\ c^m,\ |\ W\ | > n$$

and according to pumping lemma let us break W in substrings as

$$W = \mu vxyz,\ |\ vy\ | \geq 1,\ |\ vxy\ | \leq m$$

then lemma says that $\mu v^i x y^i z \in L$ for all $i \geq 0$.

Observations :

(*a*) But this is not possible; because if v or y contains two symbols from $\{a, b, c\}$ then μv^2xy^2z contains *a* 'b' before an 'a' or *a* 'c' before 'b' then $\mu v^2\, xy^2z$ will not be in L.

(*b*) In other case if v and y each contains only a's, b's or only c's, then $\mu v^i\, xy^iz$ can not contain equal number of a's, b's and c's. So μv^ixy^iz will not be L.

(*c*) If both v and y contains all three symbols a, b and c then μv^2xy^2z will not be in L by the same logic as in observation *a*.

So it is a contradiction to our statement (we assume that L is CFL).

So L is not context-free.

Example 8.2. Prove that language $L = \{WW/W \in \{a, b\}^*\}$ is not context-free.

Solution. Let us assume that language L is context-free and m must be constant obtained from the pumping lemma. Now

$$W = a^m\, b^m\, a^m\, b^m \in L$$

and according to pumping lemma we can write W in the following in the following form

$$W = \mu vxyz,\ |\, vy\,| \geq 1,\ |\, vxy\,| \leq m$$

then lemma says that $\mu v^ixy^iz \in L$ for all $i \geq 0$.

Observation. But this is not possible, since $|\, vxy\,| \leq m$ that is vxy contains only solid a's as follows

$$\underbrace{a_1, a_2, \ldots, a_m}_{m} \qquad \underbrace{b_1, b_2, \ldots, b_m}_{m} \qquad \underbrace{c_1, c_2, \ldots, c_m}_{m}$$

$$\longleftarrow (vxy) \longrightarrow$$

and we know that $|\, vy\,| \geq 1$ that is either v or y contains at least one a so consider $i = 2$ that is μv^2xy^2z, it will increase the number of a's but not b's and c's. So $\mu v^2xy^2z \notin L$, which is a contradiction, so L is not context-free.

Example 8.3. Show that the language $L = \{a^mb^mc^n/m \leq n \leq 2m\}$ is not context free.

Solution. Given language is :

$L = \{a^mb^mc^n/m \leq n \leq 2m\}$, we can prove that L is not content free by contradiction. Let us assume that L is context free, then it must follow the pumping Lemma for CFL's.

Let $w = a^nb^nc^{2n} \in L$ where n is a constant obtained from pumping lemma.

Then $w = uvxyz$ where $|\, vxy\,| \leq n$ and $|\, vy\,| \geq 1$ according to pumping lemma for CFL's $uv^ixy^iz = \forall\ i \geq 0$ will belongs to the L.

Observation. Here vy can not contain all the three symbols a, b and c. If vy contains only a's and b's then we can choose i such that uv^iwx^iy has more than in occurrences of a (or b) and exactly $2n$ occurrences of c. This means $uv^iwx^iy \notin L$, a contradiction by proper choice of i. Thus given language L is not content free.

Example 8.4. Prove that the language $L = \{w/w \in \{a, b, c\}^*/n_a(w) < n_b(w)$ and $n_a(w) < n_w(c)\}$ is not context free.

Solution. Let us assume

$$w = a^n b^{n+1} c^{n+1}$$
$$= uvxyz, \mid vy \mid \geq 1, \mid vxy \mid \leq n$$

then $\forall\ i \geq 0\ uv^ixy^iz \in L$

Observation. If v or y contains atleast one a, then vy contains c's. Therefore uv^2xy^2z contains atleast as many a's and c's and can not be in L. If neither v nor y contains an a, then uv^oxy^oz still contains n a's; since vy contains one of the other two symbols, uv^oxy^oz contains fewer occurences of that symbol then z does and therefore is not L. We have obtained a contradiction and may conclude that L is not context free.

Example 8.5. Prove that $L = \{0^{2^i}/i \geq 1\}$ is not CFL.

Solution. Let us assume that L is CFL. Then by pumping lemma, there are substrings u, v, x, y, z with $w = uvxyz = o^{2^r} \in L$, where $\mid vy \mid \geq 1, \mid vxy \mid \leq n$

$\Rightarrow$ for $\forall\ i \geq 0\ uv^ixy^iz \in L$.

Since $\mid vy \mid \geq 1$, it follows that $v = 0^p$ and $y = 0^q$ where $p + q \geq 1$. Since $i \geq 0$ $uv^ixy^iz \in L$, we have $\mid uv^ixy^iz \mid = 2^r + (p + q)(i - 1)$ [that is $\mid uvu^{i-1}xyy^{i-1}z \mid \Rightarrow$ $\mid uvxyz.\ v^{i-1}y^{i-1} \mid$ is a power of 2, say $2j$ for some j.

$$(p + q)(i - 1) = 2^j - 2^r$$
$$(p + q)(i - 1) = 2^j - 2^r \quad \Leftrightarrow (p + q)(i - 1) + 2^r = 2^j$$
$$\Rightarrow \quad 2^{r+1}(p + q) + 2^r = 2^j$$
$$\Rightarrow \quad 2^r(2(p + q) + 1) = 2^j$$

The right hand side of the equality is a power of 2 but the left hand side of the equality in the above expression is a multiple of an odd integer, namely $(2(p + q) + 1)$, which cannot be a power of 2. Thus it is not a CFL.

Example 8.6. Check that the language $L = \{a^m b^m c^n / m \leq n \leq 2m)$ is not CFL.

Solution. Again we can prove L is not CFL by contradiction, that is let us assume that L is CFL, then it most follow pumping lemma as follows :

$$w = a^n b^n c^{2n} \quad n \text{ is constant from pumping lemma.}$$
$$= uvxyz, \quad \mid vy \mid \geq 1, \mid vxy \mid \leq n$$

$\Rightarrow \forall\ i \geq 0,\ uv^ixy^iz \in L$

Observation :

vy cannot have all three symbols a, b, c. If we assume that vy has only a's and b's then we can choose i such that uv^ixy^iz has more than $2n$ occurrence of a or b

and exactly $2n$ occurences of c. Hence $uv^ixy^iz \notin L$, which is a contradiction so L is not CFL.

EXERCISES

1. Show that following languages on $\Sigma = \{a, b, c\}$ are not context-free.

(*i*) $L = \{a^n b^j c^k : k = j^n\}$

(*ii*) $L = \{a^i b^j : i \geq (j - 1)^3\}$

(*iii*) $L = \{a^i b^j c^k : k > i, k > j\}$

2. Show that following are not context-free languages :

(*i*) The set of all strings over $\{a, b, c\}$ in which the number of occurrences of a, b, c is the same

(*ii*) $L = \{a^i b^i c^j / i \leq j \leq 2i\}$

*****

9
CHAPTER

Turing Machine

9.1. INTRODUCTION

We have seen that neither deterministic finite automata nor pushdown automata can be regarded as truly general models for computers, since they are not capable of recognising even such simple languages as $\{a^n b^n c^n / n >= 0\}$. Turing machine is the next model of computation under machine based approach and context-sensitive or phrase-structure grammar in the corresponding model under grammatical approach.

In the 1930's (before advent of digital computers) several mathematicians began to think about what it means to be able to compute a function. Alenzo-Church and Alan turing independently arrived at equivalent conclusion. As we might phrase their common definition now. **"A function is computable if it can be computed by a Turing Machine".**

INSIDE THIS CHAPTER

TIPS

A Turing machine is a very simple machine, but logically speaking, has all the power of any digital computer. The Turing Machine (TM) is an abstract model created by Alan Turing (1912-1954), who was mathematician by trade and the inspiration behind the computers as we know them today.

The impact of the turing machine and the theories around them were revolutionary in his time. Turing machine were used to show significant results since the foundation of mathematics and are also highly relevant to computer science. Turing introduced Turing Machine to demonstrate the indecision of the halting problem.

9.2. TURING MACHINE AS PHYSICAL COMPUTING DEVICE

In order to help our understanding of the subject-matter of TMs, we can visualize a TM as a physical computing device that can be represented as a diagram as shown in Fig. 9.1 below :

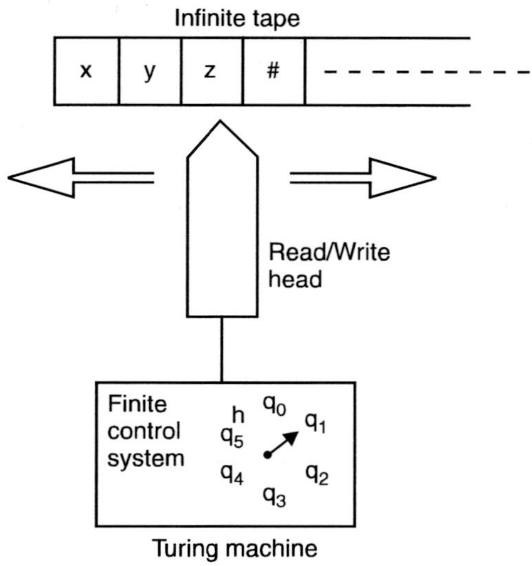

Fig. 9.1.

According to this view of TM, it consists of

(*i*) *A Tape,* with an end on the left but infinite on the right side. The tape is divided into squares or cells, with each cell capable of holding one of the tape symbols *including the blank symbol #.* At any time, there can be only finitely many cells of the tape that can contain non-blank symbols. The set of tape symbols is denoted by Γ.

As the very first step in the sequence of operations of a TM, *the input, as a finite sequence of input symbols is placed in the left-most cells of the tape.* ***The set of the input symbols*** **denoted by Σ, does not contain the blank symbol #. However, during operation of a TM, a cell may contain a *tape symbol* which is not necessarily an input symbol.**

(*ii*) *A Finite Control,* which can be in any one of the finite number of states. The states in TM can be divided in three categories that is :

(*a*) The initial state, the state of the control just at the time when TM starts its operations. The initial state is denoted by q_0.

(*b*) The *Halt state,* which is the state in which TM stops all further operations. The halt sate is *generally denoted by 'h'*. The halt state is distinct from the initial state. Thus a *TM has atleast two states.*

(*c*) Other intermediate states.

(*iii*) A tape head is always stationed at one of the tape cells and provides communication for interaction between the tape and the finite control. The head can read or scan the symbol in the cell under it. The symbol is communicated to the finite control. The control taking into consideration the symbol and its current state decides for further course of action including :

(*a*) The change of the symbol in the cell being scanned and/or

(*b*) Change of its state and/or

(*c*) Moving the head to the left or to the right. The control may decide not to move the head.

(*iv*) The course of action is called a move of the Turing Machine. In other words, the move is a function of current state of the control and the tape symbol being scanned. The change of symbol in the cell being scanned is called writing of the cell by the head. Initially, *the head scans the right-most cell of the tape.*

9.3. FORMAL DEFINITION OF TURING MACHINE

Definition A Turing machine is a six tuple of the form

$$TM = (Q, \Sigma, \Gamma, \delta, q_0, h)$$

where (*i*) Q is the finite set of states.

(*ii*) Σ is the finite set of non-blank information symbols,

(*iii*) Γ is the set of tape symbols, including the blank symbol #.

(*iv*) δ is the next-move partial function from

$$(Q \times \Gamma) \longrightarrow (Q \times \Gamma \times \{L, R, N\})$$

where $\Gamma = \Sigma \cup \#$, 'L' denotes the tape Head moves to the left adjacent cell, 'R' denotes tape Head moves to the right adjacent cell and 'N' denotes Head does not move, that is continuous scanning the same cell.

(*v*) $q_0 \in Q$, is the initial state.

(*vi*) $h \in Q$ is the 'Halt State', in which the machine stops any further activity.

Let us discuss some move of the turing machine;

(*i*) if $q \in Q$, $c \in \Sigma$ and $\delta(q, c) = (q_1, b, R)$, then Turing machine (Tm) when in state q and currently scanning symbol c, will enter in state q_1 and move towards right adjacent cell after replacing 'c' with 'b'.

(*ii*) $\delta(q_1, a) = (q_2, b, L)$ means that at present Tm is in state q_1 and scanning the symbol a, will enter in state q_2 and write 'b' in place of 'a' and move towards left adjacent cell.

(*iii*) $\delta(q_1, x) = (q_2, y, N)$ means that at present Tm is in state q_1, scanning the symbol x, will enter state q_2 and write 'y' in the place of 'x' and remain stick with current position that it does't move towards left or right.

Let us see some example for the better understanding.

Example 9.1. Consider the Turing Machine $(Q, \Sigma, \Gamma, \delta, q_0, h)$ defined below that erases all non-blank symbols on the tape, where the sequence of non-blank symbols does not contain any blank symbol # in-between :

$$Q = \{q_0, h\}$$
$$\Sigma = \{a, b\}$$
$$\Gamma = \{a, b, \#\}$$

and the next move function is defined as follows :

q : state	*σ : input symbol*	*δ(q, σ)*
q_0	a	$\{q_0, \#, L\}$
q_0	b	$\{q_0, \#, L\}$
q_0	#	$\{h, \#, N\}$
h	#	ACCEPT

Here it is assumed that tape head is scanning right most non-blank symbol first, for example #$ababba\underline{a}$#.

9.4. INSTANTANEOUS DESCRIPTION

There are following differences in Tape Head of FA/PDA and Turing machine :

(*i*) The cells of the tape of an FA or a PDA are only read/scanned but are *never* changed/*written* into, whereas the cells of the tape of TM may be written also.

(*ii*) The tape head of an FA or a PDA always moves from left to right. However, the tape head of TM can move in both the directions.

As a consequence of facts mentioned in (*i*) and (*ii*) above, we conclude that in the case of FA and PDA the *information in the tape cells already scanned do not play any role in deciding future* moves of the automation, but in the case of a TM, the information contents of all the cells, *including the ones earlier scanned* also play a role in deciding future moves. This leads to the slightly different definitions of configuration or Instantaneous Description (ID) in the case of a TM.

The ID of a Turing machine is the information in respect of :

(*i*) Contents of all the cells of the tape, starting from the right most cell upto atleast the last cell, containing a non-blank symbol and containing all cells upto the cell being scanned.

(*ii*) The cell currently being scanned by the machine and

(*iii*) The state of the machine.

For example let us assume that Turing machine TM is at state q_1 scanning the symbol g with the symbols on the tape as follows :

#	#	b	d	a	f	#	g	h	k	#	-----

↑ q_1 (under g) or

We can show the situation as follows :

$$(q_1,\ \#\#bdaf\#\underline{g}hk\#)$$

9.5. TRANSITION DIAGRAM

In some situations, graphical representation of the next-move partial function δ of Turing machine may give better idea of the behaviour of TM in comparison to the tabular representation of δ.

A Transition Diagram of the next-move functions δ of a TM is a graphical representation consisting of a finite number of nodes and directed-labelled arcs between the nodes. Each node represents a state of the TM and a label on an arc from one state (say q_1) to a state (q_2) represents the information about the required input symbol say '*a*' for the transition from q_1 to q_2 to take place and the action on the part of the control of TM. The action part consists of

(*i*) The symbol say '*b*' to be written in the current cell and

(*ii*) The movement of the tape head.

The label of an arc is generally written as *a*/(*b*, *m*), where *m* is *L*, *R* or *N*.

During the designing procedure of turning machine Tm, any string/input is assumed to be written on input taple in following form :

$$\#abbabb\underline{\#}$$

that is Head is on extreme right-blank in the starting and when machine starts processing the string at q_0 (starting state) then it is supposed to read first input symbol of the string from right to left; as follows :

$$\#abbab\underline{b}\#$$

↑

q_0 (initial state of the machine)

Example 9.2. Let TM = {Q, Σ, Γ, δ, q_0, h}

Solution.

where
$$Q = \{q_0, q_1, q_2, h\}$$
$$\Sigma = \{0, 1\}$$
$$\Gamma = \{0, 1, \#\}$$

and δ be given by the following table

	0	*1*	#
q_0	—	—	(q_2, #, *R*)
q_1	(q_2, 0, *R*)	(q_1, #, *R*)	(*h*, #, *N*)
q_2	(q_2, 0, *L*)	(q_1, 1, *R*)	(*h*, #, *N*)
h	—	—	—

The transition diagram for above TM will be as shown below, where we assume that 'q_0' is the initial state and 'h' is a final state/half state.

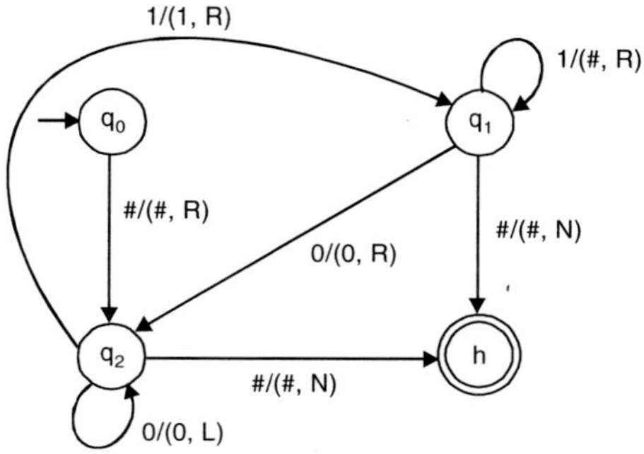

Fig. 9.2.

Example 9.3. Design a TM that recognizes the language of all strings of even length over alphabet {*a*, *b*}.

Solution. Let Turing machine is

$$\begin{aligned} \text{Tm} &= (Q, \Sigma, \Gamma, \delta, q_0, h) \\ \Sigma &= \{a, b\} \\ \Gamma &= \{a, b, \#\} \\ Q &= \{q_0, q_1, h\} \end{aligned}$$

Here h is half state when machine halts after accepting the language q_0 is initial state. Here next move partial function δ is given as follows :

	a	b	#
q_0	(q_1, a, L)	(q_1, b, L)	$(h, \#, N)$
q_1	(q_0, a, L)	(q_0, b, L)	*
h	*	*	Accept

Here '*' is for undefined move. We can construct the transition diagram as follows :

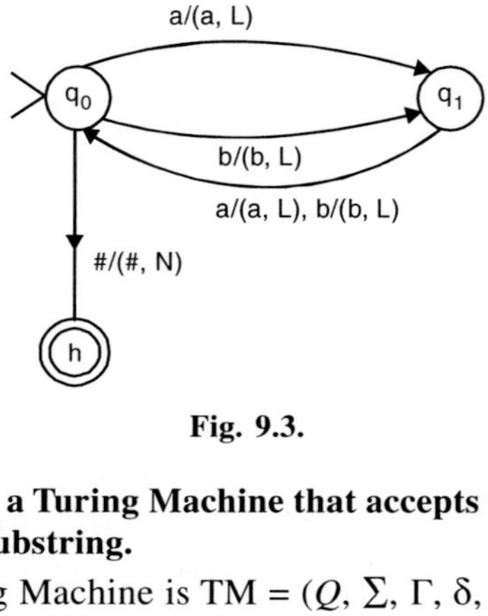

Fig. 9.3.

Example 9.4. Design a Turing Machine that accepts the language of all strings which contain *aba* as a substring.

Solution. Let Turing Machine is TM = $(Q, \Sigma, \Gamma, \delta, q_0, h)$

$$\Sigma = \{a, b\}$$

$$\Gamma = \{a, b, \#\}$$
$$Q = \{q_0, q_1, q_2, q_3, h\}$$
q_0 is initial state

and δ is defined by transition diagram as follows :

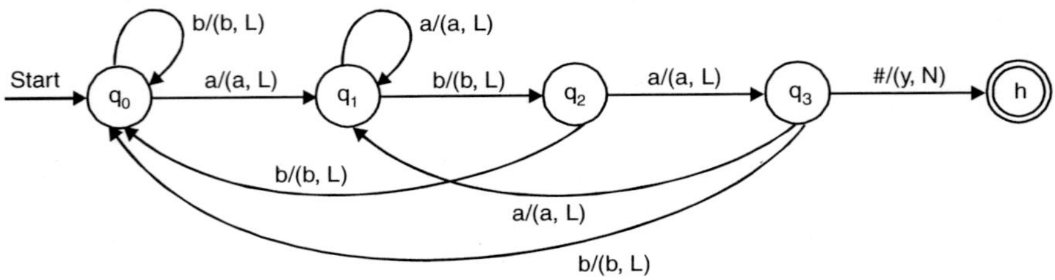

Fig. 9.4.

9.6. TURING THESIS

"The power of any computational process is captured within the class of Turing Machines."

It may be noted that Turing thesis is just a conjecture and not a theorem, hence, Turing thesis cannot be logically deduced from more elementary facts. However, the conjecture can be shown to be false, if a more powerful computational model is proposed that can recognize all the languages which are recognized by the TM model and also recognizes at least one more language that is not recognized by the TM.

The Turing Machines are designed to play atleast the following three different roles :

(*i*) *As accepting devices for languages,* similar to the role played by FAs and PDAs.

(*ii*) *As a computer of functions.* In this role, a TM represents a particular function. Initial input is treated as representing an *argument* of the function. And the (final) string on the tape when the TM enters the Halt state treated as representative of the value obtained by an application of the function to the argument represented by the initial string.

(*iii*) *As an enumerator of strings of a language*, that outputs the strings of a language, one at a time, in some systematic order that is as a list.

Example 9.5. Design a turing machine that replace every 0 and 1 with every 1 with 0 in a binary string.

Or

Prove that function $f(w) = \overline{w}$ is computable where $w \in \{0, 1\}^*$ and $\overline{w}$ the one's complement of w.

Solution. We want to design a turing machine that takes input as #101$\underline{\#}$ then its output would be #010$\underline{\#}$. This machines can be designed by the following idea; scan backwards through the input, replace every 0 with 1 and every 1 with 0, until the blank marking the left most sequence is found. In order to leave the tape in its proper format, the machine must then return its head to the blank marking the right end of the input.

Let Turing Machine is

$$\begin{aligned} Tm &= (Q, \Sigma, \Gamma, \delta, q_0, h) \\ Q &= \{q_0, q_1, h\} \\ \Sigma &= \{0, 1\} \\ \Gamma &= \{0, 1, \#\} \end{aligned}$$

q_0 is initial state.

δ can be defined by following transition diagram.

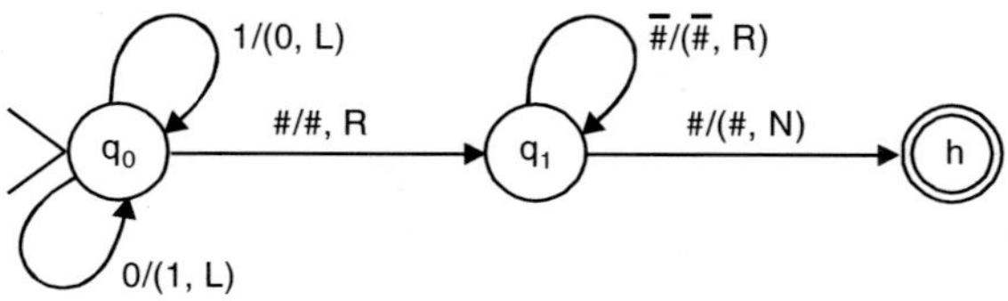

Fig. 9.5.

Here $\overline{\#}$ represent non-blank symbol (in this case either 0 or 1).

Let input string be $w = \#101100\underline{\#}$

$$\begin{aligned} \#101100\, q_0 &\vdash \#10110q_01 \\ &\vdash \#1011q_011 \\ &\vdash \#101q_011 \\ &\vdash \#10q_00011 \\ &\vdash \#1q_010011 \\ &\vdash \#q_0010011 \\ &\vdash \#q_1010011 \\ &\vdash \#0q_110011 \\ &\vdash \#01q_10011 \\ &\vdash \#010q_1011 \vdash \#0100q_111 \\ &\vdash \#01001q_11 \vdash \#010011q_1\# \vdash \#010011\#h \end{aligned}$$

9.7. TURING MACHINE AS LANGUAGE ACCEPTORS

A turing machine works as language acceptor for a language L if it is able to tell us whether a string 'w' belongs to the language L or not. In this way it gives the exact information that an automata can given. But remember, an automata is only able to accept regular languages and a turing machine can be used for some difficult structures also. We will use symbol 'y' and 'n' to show that a string is accepted by Turing machine or not. If Tm accepts a string w then it will leave symbol y on the tape and halts, that is

$$(S, \# w \underline{\#}) \vdash^* (h, \# y \underline{\#})$$

and it will leave symbol η if w is not accepted, that is

$$(S, \# w \underline{\#}) \vdash^* (h, \# n \underline{\#})$$

Now at this level we can compare, the language accepted by the regular expression, context free grammar and type-0 grammar (*I* will introduced it letter); as follows in Fig. 9.6.

Language defined by	*Corresponding Acceptor*	*Non-determinism = Determinson*	*Language closed under*	*What can be decided*	*Example of application*
Regular Expression	Finite Automata	Yes	Under union, Product, kleene star, intersection complement	Equivalence, emptiness, finiteness, membership	Text editors, sequential circuit
Context-free grammar	Pushdown automata	No	Union, product, kleene star	Emptiness finiteness membership	Programming language statements compilers
Type-0 grammar	Turing machine, post machine DPDA, *n*PDA	Yes	Union, product, Intersection, Kleene star.	Not much	Computers

Fig. 9.6.

Example 9.6. Design a turing machine that recognises the set of all strings of 0's and 1's containing at least one 1.

Solution. Designing such a turing machine is very simple. For the designing this turing machine we will move towards left until 1 is found, and then halt.

Let Turing machine be $\text{Tm} = (Q, \Sigma, \Gamma, \delta, q_0, h)$

where

$$Q = \{q_0, q_1, h\}$$
$$\Gamma = \{0, 1, \#\}$$
$$\Sigma = \{0, 1\}$$

q_0 is initial state

and transition relation δ can be represented by following transition diagram :

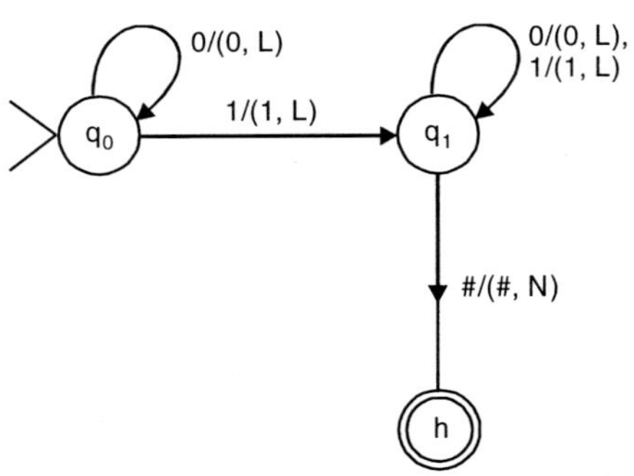

Fig. 9.7.

Here halt state *h* is reached only when at least one 1 is encountered and *h* works as accepted state.

Let us process the string : #00011$\underline{\#}$

$$\#00011q_0 \vdash \#001q_11$$
$$\vdash \#00q_111$$
$$\vdash \#0q_1011$$
$$\vdash \#0q_10011$$
$$\vdash h\#000011\#$$

9.8. COMBINING TURING MACHINE

Till now we have seen only simple examples of turing machine that do not impress us, because we have said that turing machine are ultimate computers, that is turing machines are capable of carrying out complex calculations, in spite of their limited mode of operation.

In this section we present a method to build complex turing machines by combining simple ones. In this respect a Simple Turing Machine can work as "function" for other complex turing machines. Now we see how to combine turing machine. Suppose Tm_1 is a turing machine that is not known to hang. Then Tm_1 can be made part of a large turing machine Tm as follows : Tm prepares some string as input to Tm_1, placing it near the right end of non blank portion of the tape, passes control to Tm_1 with read/write head just beyond the end of that string and finally retrieves control from Tm_1 when Tm_1 has finished its computing. It is guaranteed that Tm_1 will never interfere with the operational computation of Tm. This is like a function, when we call a function (say Tm_1) in the main function of a simple '*C*' program the control simply returns to main program after function Tm_1, performs its calculations.

9.8.1. Rules for Combining Turing Machines

Now we adopt a graphical representation to show the combining of turing machines. In fact this representation is the same as that of a finite automata. Individual machines are like the states of the automata and these machines are connected in the same way as the states of automata. The connection from one machine to another will not be pursued until the first machine halts; the other machine is then started from its initial state with the tape and head position as they were left by the first machine. This can be shown by the following figure :

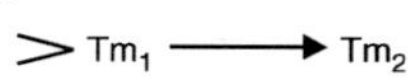

Fig. 9.8.

Here Tm_1 machine starts computing (that is initial state in finite automata) and when Tm_1 halts, control, now goes to machine Tm_2.

Now consider the following machine

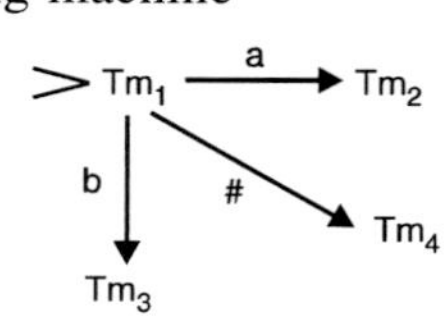

Fig. 9.9.

This machine starts computing with machine Tm_1. Now when Tm_1 halts, the symbol on the tape under the current head position can be *a*, *b* or #. Then the Fig. 9.9 shows when Tm_1 halts, the control goes to machine Tm_2, Tm_3 or Tm_4 respectively, according to the symbol under current head position which is *a* or *b* or #.

Here we are going to discuss certain symbols to represent simple machines. These symbols will be used further.

(*i*) Fig. 9.9 is the basic machine which moves towards left by one cell during scanning any input (σ) symbol. State change is also possible.

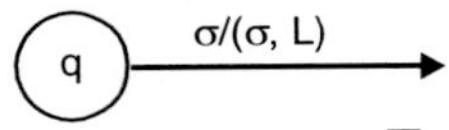

Fig. 9.10.

(*ii*) Fig. 9.10 is the basic machine moves towards right by one cell, when current input symbol is blank or non-blank (σ).

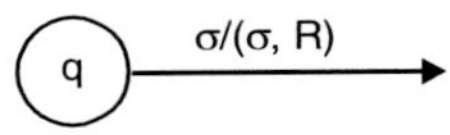

Fig. 9.11.

(*iii*) Fig. 9.11 is the basic machine, which moves towards left till the first blank symbol is encountered.

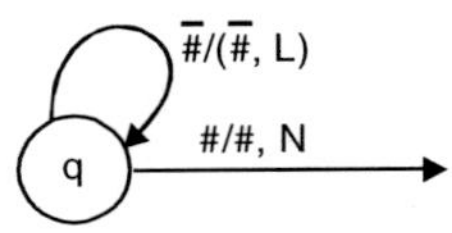

Fig. 9.12.

(*iv*) Fig. 9.13 is the basic machine, which moves towards right until the first blank symbol is encountered.

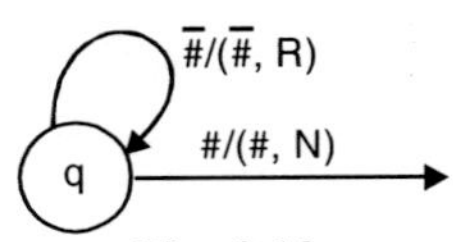

Fig. 9.13.

(*v*) Fig. 9.13 is the basic machine, which moves towards left until the first non-blank symbol is encountered.

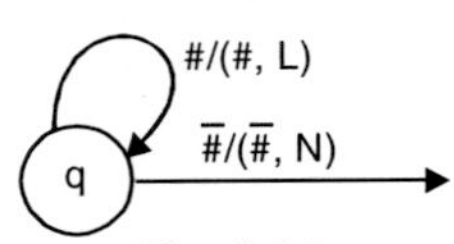

Fig. 9.14.

(*vi*) Fig. 9.14 is the basic machine, which moves towards right until the first non-blank symbol is encountered.

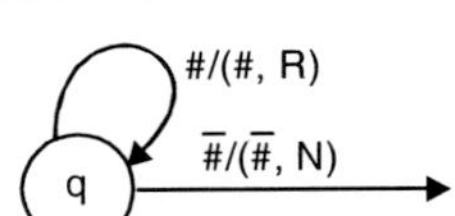

Fig. 9.15.

Example 9.7. Construct a turing machine that computes the following function.

$$f(n, m) = n + m$$

Solution. Let us decide the policy to show that input numbers : I represents 1, $\#II\#$ represents 2 and so on. We assume that input numbers (that is n and m) are written on the tape separated by #. That is if we want to add 2 and 3, the input string will be $\#II\#III\#$.

Now we want to such a turing machine which gives output as follows $f(2, 3) = 2 + 3 = 5$.

$$\#II\#III\# \vdash \#IIIII\#$$

When we analyse the output then it becomes clear that for constructing turing machine for $f(n, m) = n + m$, we have to remove #, which separates n and m.

Infact resulting turing machine works as a **left-shift machine** (say SL). It shift the each symbol of m portion of input by one cell towards left.

Let turing machine is $\quad Tm = (Q, \Sigma, \Gamma, \delta, q_0, h)$

where

$$Q = \{q_0, q_1, q_2, q_3, q_4, h\}$$
$$\Sigma = \{I\}$$
$$\Gamma = \{I, \#\}$$

q_0 is initial state

h is halt state

and δ is defined by following transition diagram.

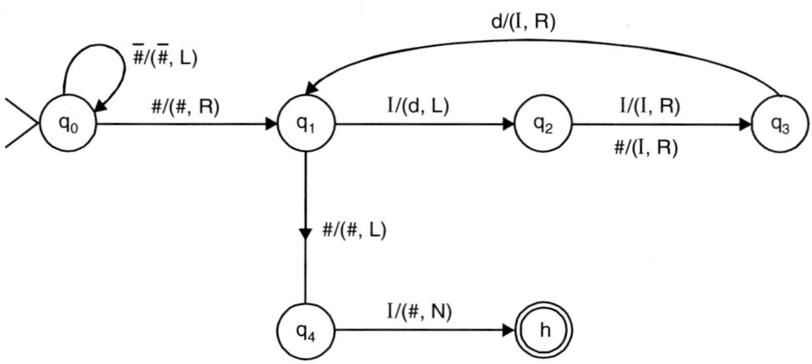

Fig. 9.16.

In the above turing machine we are not showing all the moves from every state and those move for certain input symbols are considered as, path (transition) is not available for some inputs and machine rejects the input or not processed if it is not halting at state 'h'.

Let us process a string $\#II\#II\#$

$$\#II\#IIq_0 \vdash \#II\#Iq_0I$$
$$\vdash \#II\#q_0II$$
$$\vdash \#II\#q_1II$$

$\vdash \#II\#Iq_1I$

$\vdash \#II\#q_2dI$

$\vdash \#IIIdq_3I$

$\vdash \#IIIIIq_1$

$\vdash \#IIIIq_2d$

$\vdash \#IIII\text{d}q_3$

$\vdash \#IIIII\#q_1$

$\vdash \#IIIIIq_4\#$

$\vdash \#IIII\#h$

Example 9.8. Design a Turing Machine which works as eraser.

Solution. Let us first analysed the problem, here we want to design a turing machine which works as eraser, that is if we pass input #*abb*# then output should be ##, that is null string.

Let the Turing machine be Tm = $(Q, \Sigma, \Gamma, \delta, q_0, h)$

where $Q = \{q_0, q_1, h\}$

$\Sigma = \overline{\#}$ symbols

$\Gamma = \overline{\#} \cup \#$

q_0 is initial state and

Transition relation δ is defined as follows :

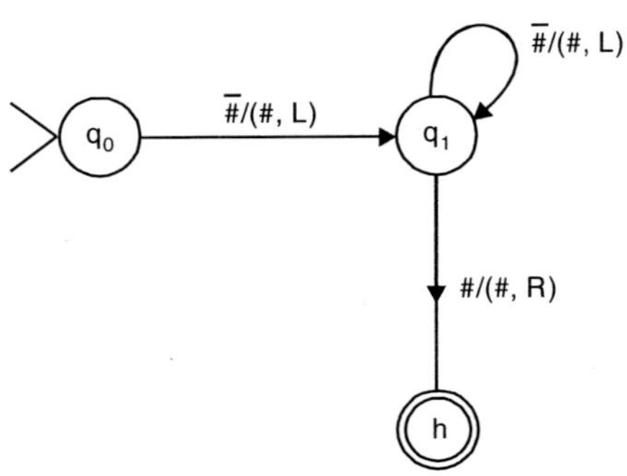

Fig. 9.17.

Here it is preassumed that there is at least one non-blank symbol in the string.

Now let us process the following string :

$\#ababaq_0 \vdash \#ababq_1\#$

$\vdash \#abaq_1\#\#$

$\vdash \#abq_1\#\#\#$

$\vdash \#aq_1\#$

$\vdash \#q_1\# \vdash \#\#h$

Example 9.9. Prove that following function is turing computable.

$$f(m) = \begin{cases} m-2, & \text{if } m > 2 \\ 1, & \text{if } m \leq 2 \end{cases}$$

Solution. We know that if any function is turing computable then there exist a turing machine for it, so we have to design a turing machine of the $f(m)$ function. For the input strings $\#I\#$ ($m = 1$), $\#\#$($m = 0$) and $\#II\#$ ($m = 2$) output will be $\#I\#$ ($m = 1$). For the input string $\#IIII\#$ ($m = 4$), output will be $\#II\#$ ($m = 4 - 2 = 2$, since $m > 2$).

Let Turing machine be $\quad \text{Tm} = (Q, \Sigma, \delta, \Gamma, q_0, h)$

where $\quad Q = \{q_0, q_1, q_2, q_3, h\}$

$\Sigma = \{I\}$

$\Gamma = \{I, \#\}$

q_0 is initial state and h is halting state.

Transition relation δ is defined as follows :

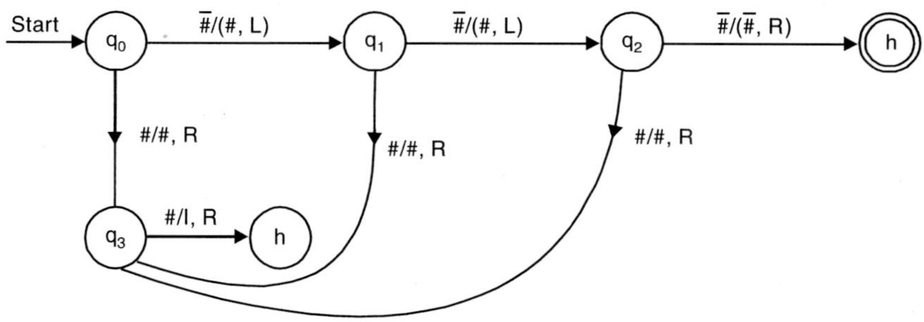

Fig. 9.18.

Let us process the following string :

$$\#IIII\#$$

$$\begin{aligned} \#IIIIq_0 &\vdash \#IIIq_1\# \\ &\vdash \#IIq_2\#\# \\ &\vdash \#II\#h \end{aligned}$$

Let us process another string $\#II\#$

$$\begin{aligned} \#IIq_0 &\vdash \#Iq_1\# \\ &\vdash \#q_2\#\# \\ &\vdash \#\#q_3\# \\ &\vdash \#I\#h \end{aligned}$$

Example 9.10. Prove that following function is turing computable.

$$f(n) = \begin{cases} n-1, & n > 0 \\ 0, & n = 0 \end{cases}$$

Solution. If function $f(n)$ is turing computable then there exist a turing machine for it, so basically we have to design a turing machine for function $f(x)$. This turing machine should give output ## for ## ($n = 0$) and #*II*# for #*III*# ($n = 3$).

Let Turing machine be

$$\text{Tm} = (Q, \Sigma, \Gamma, \delta, q_0, h)$$

where

$$Q = \{q_0, h\}$$
$$\Sigma = \{I\}$$
$$\Gamma = \{I, \#\}$$

q_0 is initial state and h is halting state

Transition relation δ is defined by following transition diagram :

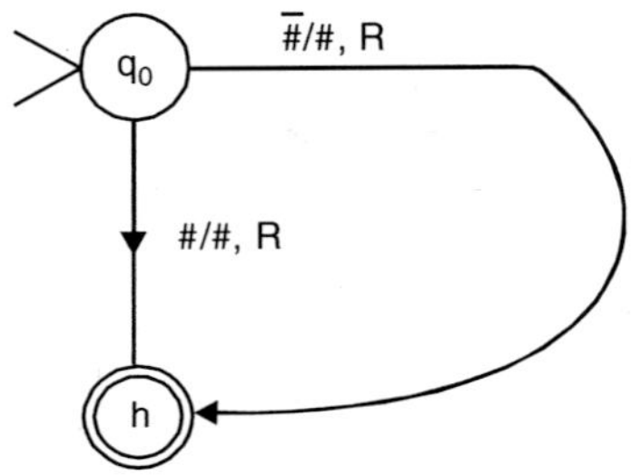

Fig. 9.19.

Let us process the following string :

$$w = \#IIII\#$$

$$\#IIIIq_0 \vdash \#III\#h$$

For string $\#q_0\# \vdash \#\#h.$

Example 9.11. Design a Turing machine for the following function :

$$f(x, y) = y.$$

Solution. Let us understand the meaning of $f(x, y) = y$, here we want to design a turing machine which takes input $x = \#II\#$ and $y = \#III\#$ as string $w = xy = \#II\#III\#$ on the input tape and output of the required turing machine is #*III*# that is y part of string.

Now let us discuss the logic for this turing machine, consider the string #*II*#*III*# move at extreme left and erase the x part and leave the tape head at extreme right.

Let turing machine be :

$$\text{Tm} = (Q, \Sigma, \Gamma, \delta, q_0, h)$$

where

$$Q = \{q_0, q_1, q_2, q_3, h\}$$
$$\Sigma = \{I\}$$
$$\Gamma = \{I, \#\}$$

q_0 is initial state and h is halting state.

δ is defined by the following transition diagram :

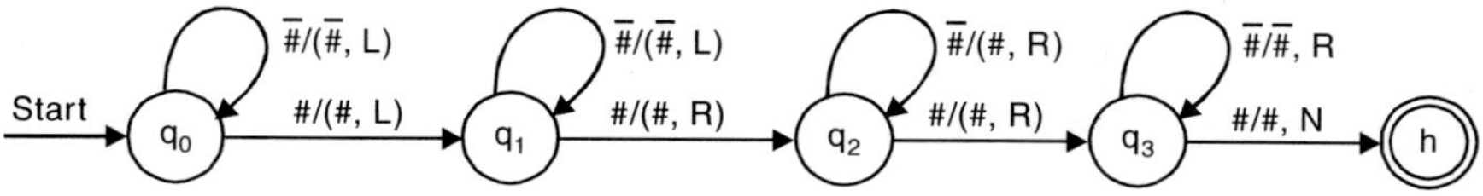

Fig. 9.20.

$$
\begin{aligned}
\#II\#Iq_0 &\vdash \#II\#q_0I \\
&\vdash \#IIq_1\#I \\
&\vdash \#Iq_1I\#I \\
&\vdash \#q_1II\#I \vdash \#Iq_2I\#I \\
&\vdash \#\#Iq_2\#I \vdash \#\#\#\#q_2I \\
&\vdash \#\#\#\#Iq_3 \vdash \#\#\#\#I\#q_3 \vdash \#\#\#\#I\#h.
\end{aligned}
$$

Example 9.12. Design a Turing machine for the following function $f(x, y) = x$.

Solution. We can solve this problem in similar fashion as we have solved example 9.13. Here we have to design a turing which takes input $w = xy$ (where $x = \#II\#$ and $y = \#III\#$) as string $\#II\#III\#$ and gives output $\#II\#$ that is only x part of w and completely erase of part of w.

Let Turing machine be $\quad Tm = (Q, \Sigma, \Gamma, \delta, q_0, h)$

$$
\begin{aligned}
Q &= \{q_0, h\} \\
\Sigma &= \{I\} \\
\Gamma &= \{I, \#\}
\end{aligned}
$$

q_0 is initial state and h halt state

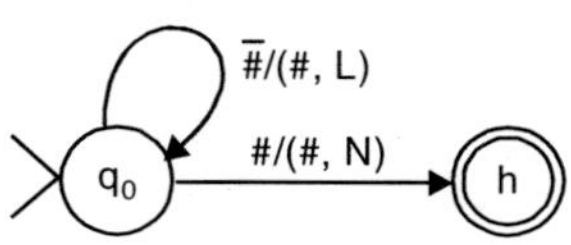

Fig. 9.21.

Let us process the string $\#II\#III\#$

$$
\begin{aligned}
\#II\#IIIq_0 &\vdash \#II\#IIq_0\# \vdash \#II\#Iq_0\#\# \\
&\vdash \#II\#q_0\#\#\# \vdash \#II\#h.
\end{aligned}
$$

Example 9.13. Design a Turing machine for the regular expression $r = aa^*$.

Solution. We have to design a turing machine which accepts strings which are in regular expression r.

$$r = aa^*$$

It is collection of strings starts with 'a' followed by any number of a's and contain only a's in the strings.

Let turing machine is $\quad Tm = (Q, \Sigma, \Gamma, \delta, q_0, h)$

$$Q = \{$$

$$\Sigma = \{a\}$$
$$\Gamma = \{a, \#\}$$

q_0 is initial state and h is halt state

δ is defined as follows :

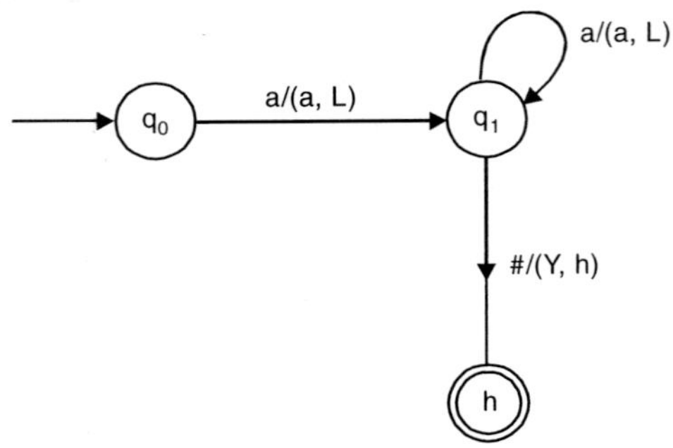

Fig. 9.22.

Y stand for yes string is accepted here halt state $\boldsymbol{Y_h}$ signifies as accepting state. Consider the string $\#aaa\underline{\#}$

$$\begin{aligned} \#aaaq_0 &\vdash \#aaq_1a \\ &\vdash \#aq_1aa \\ &\vdash \#q_1aaa \\ &\vdash Yhaaa\# \end{aligned}$$

when we get *Y* at input cell, means that string is accepted.

Example 9.14. Design a Turing machine for the language

$$\boldsymbol{L = \{(ab)^n / n \geq 0\}}$$

or for the regular expression $\boldsymbol{r = (ab)^*}$.

Solution. Let Turing machine be

$$\text{Tm} = (Q, \Sigma, \Gamma, \delta, q_0, h)$$
$$Q = \{q_0, q_1, h\}$$
$$\Sigma = \{a, b\}$$
$$\Gamma = \{a, b, \#\}$$

q_0 is initial state and h is halt state.

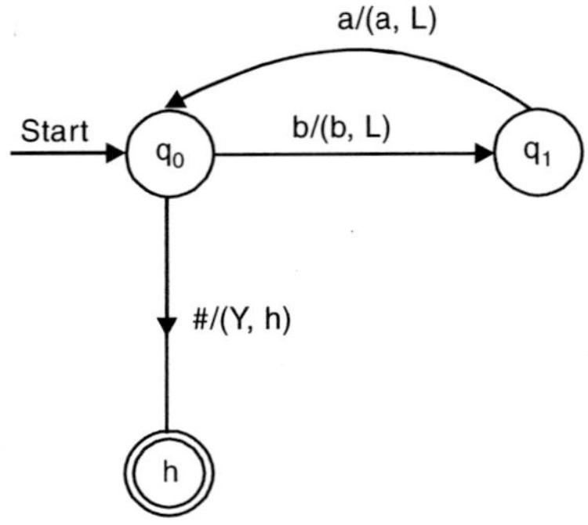

Fig. 9.23.

Note : All the move which are not shown, leads towards not accepting states.

Let us pass a string $w = \#aba\underline{b}\#$
q_0

$$
\begin{aligned}
\#ababq_0 &\vdash \#abaq_1b \\
&\vdash \#abq_0ab \\
&\vdash \#aq_1bab \\
&\vdash \#q_0abab \\
&\vdash Yhabab
\end{aligned}
$$

Y : Stands for accept, Tm accepts string and halts.

Example 9.15. Design a Turing machine which accepts the language $L = \{w \in (a, b)^*/w$ has equal number of *a*'s and *b*'s}.

Solution. The language *L* is the set of strings (included null string) in which every strings either start by *a* or *b* but number of *a*'s and *b*'s in the string are always equal. That is turing machine should accept strings like *ab*, *aaabbb*, *baba*, *abab*, and so on and it must reject strings like *a*, *b*, *abb*, *baa*, *abbb*, *abbab* and so on.

The Turing machine that accepts *L* is shown below :

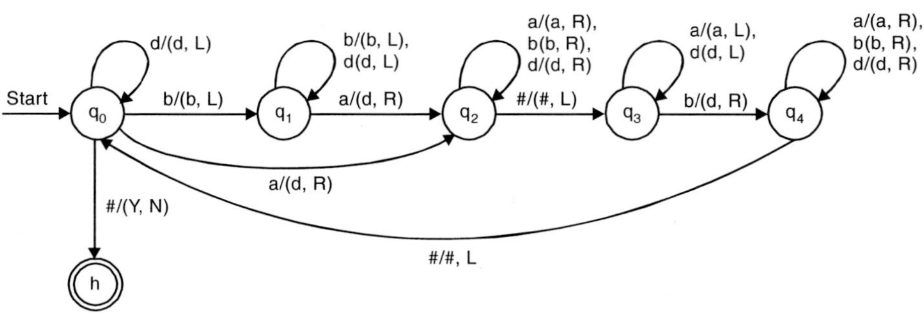

Fig. 9.24.

Y : Stands for Accept

In one main cycle, this machine changes one *a* and one *b* in two *d*'s. It does so by scanning the the non-blank part of take two times from right to left, once for an '*a*' and once for '*b*', returning each time to the right end of the tape before starting the next scan. This sequence is repeated until the supply of *a* or *b* is exhausted.

Now if machine sees all *d*'s on the non-blank part then it halts that is, it accept the string. The move which are not available in the machine leads towards not accepting state.

Following is the sequence of operation on string abba.

$$
\begin{aligned}
\#abbaq_0 &\vdash \#abbdq_2 \\
&\vdash \#abbd\#q_2
\end{aligned}
$$

$\vdash \#abbdq_3$

$\vdash \#abbq_3d$

$\vdash \#abddq_4$

$\vdash \#abdd\#q_4$

$\vdash \#abddq_0$ (one cycle is completed)

$\vdash \#abdq_0d$

$\vdash \#abq_0dd$

$\vdash \#aq_1bdd$

$\vdash \#dbq_2dd$

$\vdash \#dbdq_2d$

$\vdash \#dbddq_2$

$\vdash \#dbdd\#q_2$

$\vdash \#dbddq_3$

$\vdash \#dbdq_3d$

$\vdash \#dbq_3dd$

$\vdash \#dddq_4d$

$\vdash \#ddddq_4$

$\vdash \#dddd\#q_4$

$\vdash \#ddddq_0$

$\vdash \#dddq_0d$

$\vdash \#ddq_0dd$

$\vdash \#dq_0ddd$

$\vdash \#q_0dddd$

$\vdash Yhdddd$

Example 9.16. Design a Turing machine for the following languages :

$$L = \{a^n b^n / n \geq 1\}$$

Solution. We are going to design a turing machine which accepts a set of strings in which every string starts with '*a*', followed by any number of *a*'s, every string ends in equal number of *b*'s as *a*'s. No '*a*' is encountered after first '*b*' is read. It will not accepts the strings like $\in$, *a*, *aa*, *bb*, *abb*, *aabbb*, *aab* and so on.

Let required turing machine is :

$$\text{Tm} = \{Q, \Sigma, \Gamma, \delta, q_0, h)$$

$$Q = \{q_0, q_1, q_2, q_3, h\}$$

$$\Sigma = \{a, b\}$$

$$\Gamma = \{a, b, \#\}$$

q_0 is initial state and h is halt state.

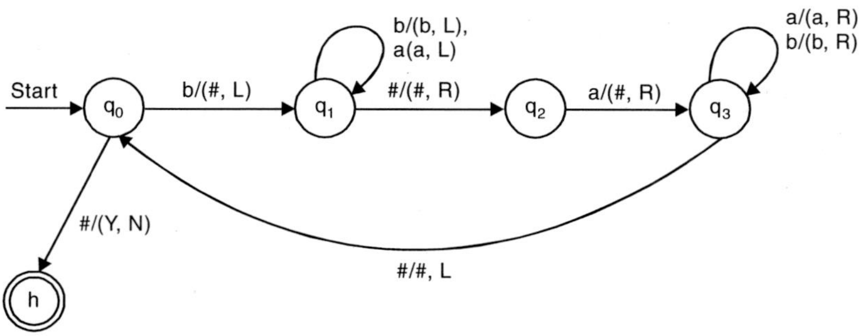

Fig. 9.25.

The moves which are unaddressed, leads towards rejecting states.

Let us pass the string $w = \#a\underline{b}\#$
$\quad\quad\quad\quad\quad\quad\quad\quad\quad\quad q_0$

$$
\begin{aligned}
\#\text{ab}q_0 &\vdash \#aq_1\# \\
&\vdash \#q_1a\# \vdash \#aq_2\# \\
&\vdash \#\#\#q_3 \vdash \#\#q_0\# \\
&\vdash \#Yh\#
\end{aligned}
$$

Example 9.17. Design a Turing machine for the following languages :
$L = \{w \in \{a, b, c\}^*/w$ has equal number of a's, b's and c's$\}$.

Solution. we can solve this problem in similar fashion as example 9.17. Here turing machine start searching from right and first search 'a' then move to extreme right and this 'a' is repalced by d. Now machine search for 'b' and if find replace it by 'd' and again move on extreme right and search corresponding 'c' and if find replace it by d, this completes one cycle. This process reepated and if all symbols are replaced by d's then string is accepted and rejected in any other case.

Turing machine is :

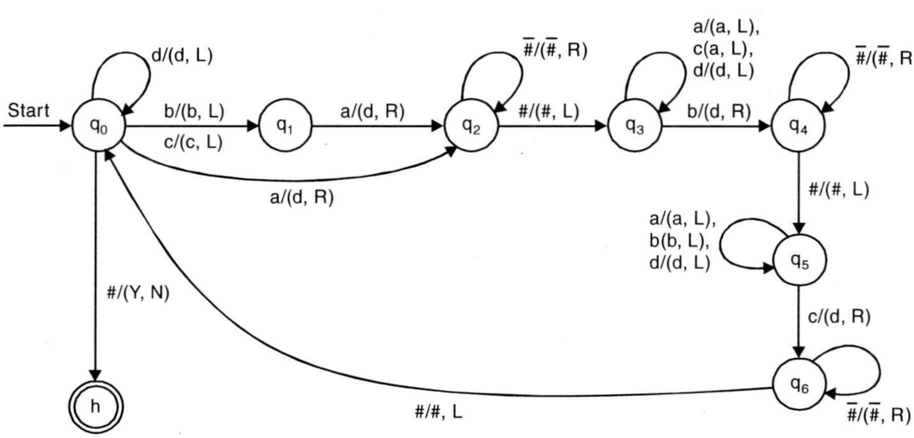

Fig. 9.26.

Let $w = \#cb\underline{a}\#$
q_0

$$
\begin{aligned}
\#cbaq_0 &\vdash \#cbd\#q_2 \\
&\vdash \#cbdq_3 \\
&\vdash \#cbq_3\text{d} \\
&\vdash \#cddq_4 \\
&\vdash \#cdd\#q_4 \\
&\vdash \#cddq_5 \\
&\vdash \#cdq_5d \\
&\vdash \#cq_5dd \\
&\vdash \#ddq_6dd \\
&\vdash \#dddq_6d \\
&\vdash \#ddddq_6 \\
&\vdash \#dddd\#q_6 \\
&\vdash \#ddddq_0 \\
&\vdash \#dddq_0\text{d} \\
&\vdash \#ddq_0dd \\
&\vdash \#dq_0ddd \\
&\vdash \#q_0dddd \\
&\vdash Yhdddd
\end{aligned}
$$

Example 9.18. Design a Turing machine for the following language :

$$L = \{ab^n / n > 0\}$$

Solution. Here we want to design a Turing machine which accepts the language *L* which consist a set of strings in which every string *w* begins with single '*a*' and followed by one or more *b*'s.

Clearly Turing machine will accept strings like *abbbb*, *ab*, and rejects strings like $\in$, *a*, *b*, aaabb and so on.

Here is required turing machine

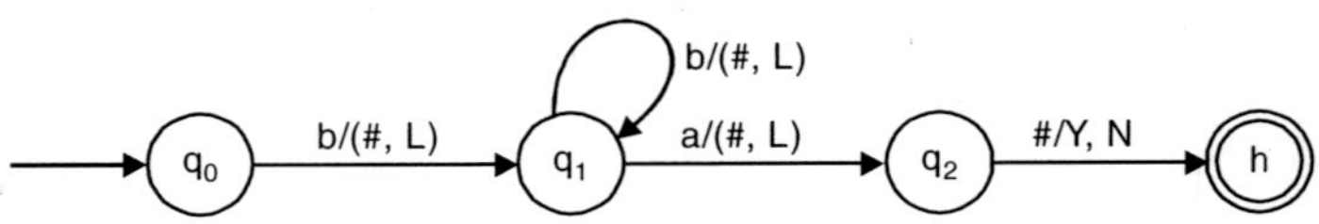

Fig. 9.27.

The moves which are not shown are considered as rejecting states.

$$
\begin{aligned}
w &= \#abb\# \\
\#abbq_0 &\vdash \#abq_1\# \\
&\vdash \#aq_1\#\# \\
&\vdash \#q_2\#\#\# \\
&\vdash Yh\#
\end{aligned}
$$

Example 9.19. Design a Turing machine which computes following function.

$f(w) = ww^R$**, where** w^R **is the reverse of string** w **.** $(w \in (a, b)^*)$**.**

Solution. Let input is #*ab*# then Tm should give output as #*abba*#.

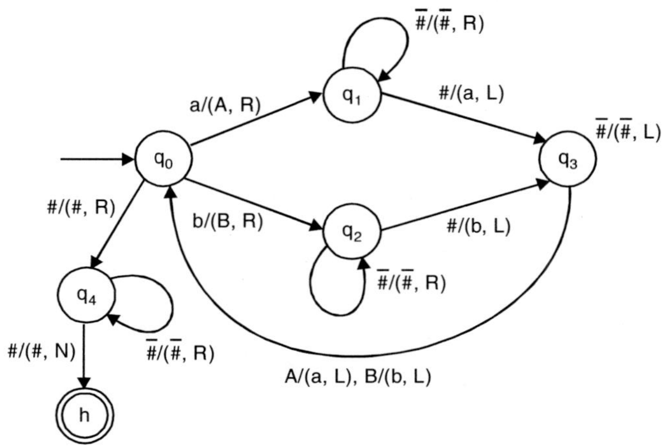

Fig. 9.28.

$$
\begin{aligned}
\#abq_0 &\vdash \#aB\#q_2 \\
&\vdash \#aBq_3b \\
&\vdash \#\mathrm{a}q_0bb \\
&\vdash \#Abq_1b \\
&\vdash \#Abbq_1 \\
&\vdash \#Abb\#q_1 \\
&\vdash \#Abbq_3a \\
&\vdash \#Abq_3ba \\
&\vdash \#Aq_3bba \\
&\vdash \#q_0abba \\
&\vdash \#aq_4bba \\
&\vdash \#abq_4ba \\
&\vdash abbq_4a \\
&\vdash \#abbaq_4 \\
&\vdash \#abba\#q_4 \\
&\vdash \#abba\#h
\end{aligned}
$$

Example 9.20. Design a Turing which works as copying machine (*c*), for $w \in (a, b)^+$**.**

Solution. We want to design a turing machine which gives output for input $w = \#ab\#$ as $\#ab\#ab\#$ that is

$$\#w\# \vdash \text{Tm } \#w\#w\#$$

The transition diagram of turing machine will be as follows :

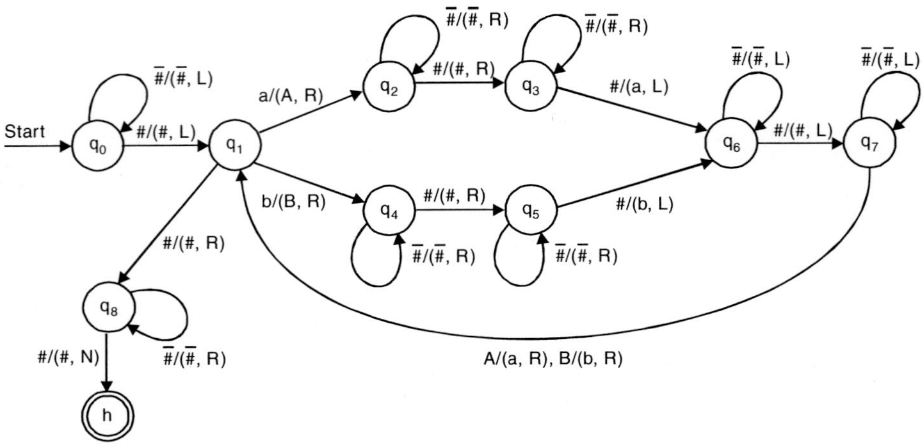

Fig. 9.29.

Let $w = \#ab\#$ is input then :

$$
\begin{aligned}
\#abq_0 &\vdash \#aq_0b \\
&\vdash \#q_0ab \\
&\vdash \#aq_1b \\
&\vdash \#Abq_2 \\
&\vdash \#Ab\#q_2 \\
&\vdash \#Ab\#\#q_3 \\
&\vdash \#Ab\#q_6a \\
&\vdash \#Abq_7\#a \vdash \#Aq_7\mathrm{b}\#a \\
&\vdash \#abq_1\#a \vdash aB\#q_4a \\
&\vdash \#aB\#aq_5 \vdash aB\#a\#q_5 \\
&\vdash \#aB\#a\#q_5 \vdash \#aB\#aq_6b \\
&\vdash \#aB\#q_6ab \vdash \#aBq_7\#ab \\
&\vdash \#ab\#q_1ab \vdash \#ab\#aq_8b \\
&\vdash \#ab\#abq_8 \vdash \#ab\#ab\#q_8 \\
&\vdash \#ab\#ab\#h.
\end{aligned}
$$

9.9. EXTENSION OF TURING MACHINES

Now it becomes clear that turing machine can perform fairly powerful computation. In order to better understand their surprising power, we shall consider the effect of extending the turing machine model in various directions. The "new, improved models" of the turing machine can in each instance be simulated by the standard model. Such results increase our confidence that the turing machine is indeed the ultimate computational device, the end of our progression to move and move powerful automata. A side benefit by these results is that we shall feel free subsequently to use the additional features when designing turing machines to solve particular problems.

The extensions of Turing machine considered are :

(*i*) There may be several tapes instead of one only, each tape having its own independent head.

(*ii*) The tape may be allowed to be infinite in both the directions.

(*iii*) There may be more than one head scanning various cells of the tape. Two or more heads simultaneously read the same cell or may attempt to write in the same cell.

(*iv*) The tape may be K-dimensional, $K \geq 2$, instead of only one-dimensional.

(*v*) For a given pair of current state and symbol under the Head, instead of atmost one possible move, there may be any finite, possible zero, number, of next moves. This is called **Non-deterministic Turing Machine.**

9.9.1. Multiple Tapes Turing Machine

One can think of turing machines that have several tapes (see Fig. 9.29). Each tape is connected to the finite control by means of a read/write head (one on each tape).

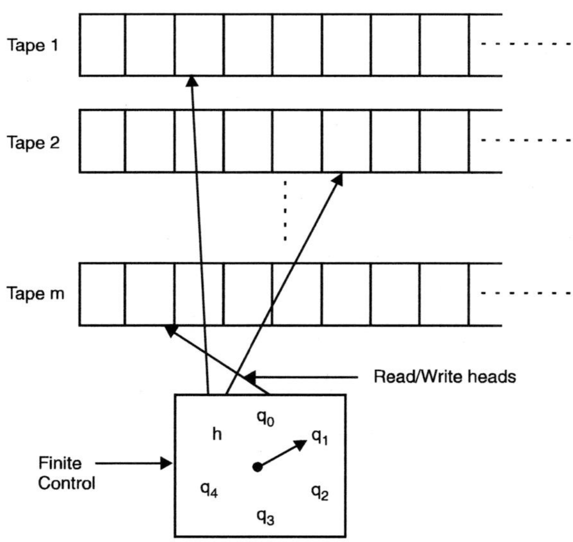

Fig. 9.30.

The machine can in one step read the symbols scanned by all its heads and then, depending on these symbols in current state, rewrite some of those scanned squares and move some of the heads to the left or right, in addition to changing the state.

For any fixed integer $m \geq 1$, an m-tape turing machine is a turing machine equipped with m tapes and corresponding heads. Thus standard turing machine studied so far in this chapter is just an m-tape turing machine, with $m = 1$.

Definition

Let $m \geq 1$ be an integer.

An m-tape turing machine is a six tuple machine as follows :

$$\text{Tm} = (Q, \Sigma, \Gamma, \delta, q_0, h)$$

where
- Q : the finite set of states of the finite control
- Σ : The finite set of input symbols
- Γ : The tape symbol where $\Gamma = \Sigma \cup \#$
- h : halt state $h \in Q$
- q_0 : The initial state, $q_0 \in Q$
- δ : The transition function is a function which maps :

$$(Q \times \Sigma^m) \rightarrow Q \times (\Sigma \cup \{L, R, N\}^m).$$

That is, for each state-q, and each m-tapes of tape symbols $(a_1, a_2, ..., a_k)$, $\delta (a_1, a_2, ..., a_k) = (P, (b_1, b_2 ...b_k))$, where P is, as before, the new state, and b_j is the action taken by Tm at tape j.

Computation takes place in all m-tapes of a m-tape Turing machine. Accordingly, a configuration of such a machine must include information about all tapes.

Example 9.21. Design an m-tape turing machine which works as copying machine.

Solution. We have already solved this problem by the standard one-tape turing machine in example 9.20. Now we are free to use m-tape in the designing of turing machine.

We will design a 2-tape turing machine, which takes input $\#w\underline{\#}$ and gives output as $\#w\#w\underline{\#}$.

Let us clear notations for the tapes first.

For the first tape machines notations used will be written as R' and L' input symbol from tape-1 is read-out as σ'. Similarly for the tape-2 notations of machines will be R^2 and L^2 and so on. Input which is read-out from tape-2 will be shown as σ^2.

This turing machine can be accomplished as follows :

(1) **Move** the heads on both tapes to the right, copying each symbol on the first tape on to the second tape, until a blank is found on the first tape. The first square of the second tape should be left blank.

(2) **Move** the head on second tape to the left until a blank is found.

(3) Again **move** the heads on both tapes to the right, this time copying symbols from the second tape on to the first tape. Halt when a blank is found on the second tape.

The required 2-tape turing machine is shown in Fig. 9.30 as follows :

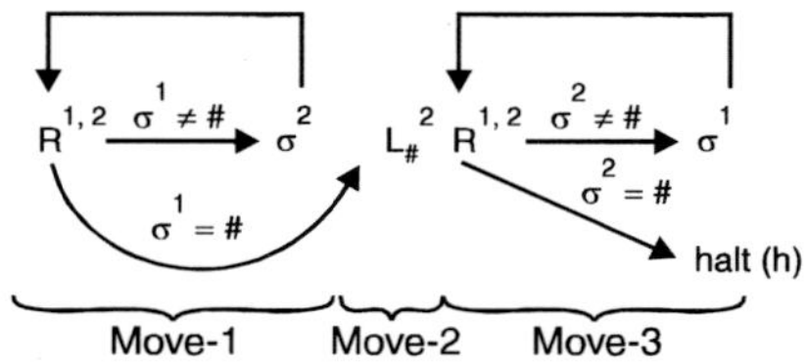

Fig. 9.31.

Here : $R \simeq$

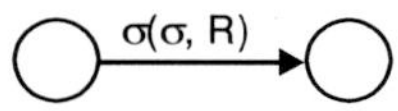

and $L_\# \simeq$

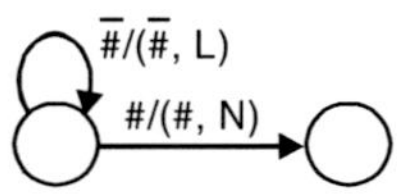

The sequence of actions can be pictured as follows :

First tape $\#w\underline{\#}$

Second tape $\#\underline{\#}$

After (move-1) : First tape $\#w\underline{\#}$

Second tape $\#w\underline{\#}$

After (move-2) : First tape $\#w\underline{\#}$

Second tape $\underline{\#}w\#$

After (move-3) : First tape $\#w\#w\underline{\#}$

Second tape $\#w\underline{\#}$

9.9.2. Two-way Infinite Tape

The extensions are distinguished from each other and from the standard Tm through different definitions of next-move relation δ and of configurations for each of the extension. So here we discuss the extension only in terms of definitions of δ and of configuration.

Like standard Tm, in this case also, the next-move is given by δ as a partial function from

$$(Q \times \Gamma) \text{ to } (Q \times \Gamma \times \{L, R, N\})$$

The following three points need to be noted in respect of configurations of Two-way Turing Machine.

***(i)* Configuration/Instantaneous Description.** In standard Tm, if there are a number of left-most positions which contain blanks, then those are included in the configuration that is if the one-way configuration tape is of the form

$$\# ab \# c \underset{\substack{\uparrow \\ q_2}}{d} ef \#\# \ldots \#$$

then the configuration in the standard Tm is written as :

$$(q_2, \#ab\#c\underline{d}ef\#\# \ldots \#)$$

However, in the two-way infinite tape Tm, both left-hand and right-hand parts of the tape are symmetrical in the sense that there is an infinite continuous sequence of blanks on each of the right hand and left hand of the sequence of non-blanks. Therefore, in case of two-way infinite tape, if the above string is on the tape then it will be in the form.

$$(q_2, \# \ldots \#\# ab \# c\underline{d}ef \#\# \ldots \#)$$

***(ii)* No Ceasing of operation without Halting.** In this case, as there is no left end of the tape, therefore, there is no possibility of jumping off the left-end of the tape.

***(iii)* The empty Tape configuration.** When at some point of time all the cells of the Tape are #'s and the state is say q, then configuration in Two-way tape may be denoted as :

$$(q, \#)$$

where only the current cell containing # is shown in configuration.

TIPS

Despite the fact that, it is possible in the new model of computer to move left as far as required. The model does not provide any additional computational capability.

9.9.3. Multiple Heads Turing Machines

In order to simplify the discussion, we assume that there are only two Heads on the tape. The tape is assumed to be one-way infinite. We explain the involved concepts with the help of an example. Let the content of the tape and the position of the two heads that is H_1 and H_2, be as given below :

$$\#\# a \underset{\underset{H_2}{\uparrow}}{b} c \# d \underset{\underset{H_1}{\uparrow}}{e} f \#\# \ldots$$

Further, let the state of the Tm be q.

The move function of the two-head one-way turing may be defined as

δ(state, symbol under Head1, symbol under Head2)

= (New state, (S_1, M_1), (S_2, M_2))

where S_i is the symbol to be written in the cell under H, (the ith Head) and M_i denotes the movement of H_i, where the movement may be L, R or N and further L denotes movement to the left, R denotes movement to the right of the current cell and N denotes 'no movement of the Head'.

Two special cases of the δ function defined above, need to be considered :

(*i*) What should be written in the current cell when both Heads are scanning the same cell at a particular time and the next moves (S_1, M_1), (S_2, M_2) for the two heads, are such that $S_1 \neq S_2$ that is symbol to be written in current cell by H_1 is not same as symbol to be written in current cell by H_2.

In such a situation, a general rule may be defined, say, as whatever is to be done by H_1 will take precedence over whatever is to be done by H_2.

(*ii*) For two-head one way tape, a configuration shall be called hanging if $\delta(q$, symbol under H_1, symbol under $H_2) = (q_1, (S_1, M_1), (S_2, M_2))$ is such that either

(*a*) Symbol under H_1 is in the left-most cell and M_1 is L, that is movement of H_1 is to be to the left,

or

(*b*) Symbol under H_2 is in the left-most cell and M_2 is L, that is movement of H_2 is to be to the left.

The above discussion can be further extended easily to the case when number of Heads is more than two. The power of Tm is not enhanced by the use of extra heads.

9.9.4. K-dimensional Turing Machine

Again to facilitate the understanding of the basic ideas involved, let us discuss initially only Two-dimensional Turing Machine. Then these ideas can be easily generated to K-dimensional case, where $K > 2$.

In the case of two-dimensional tape as shown below, we assume that the tape is bounded on the left and the bottom.

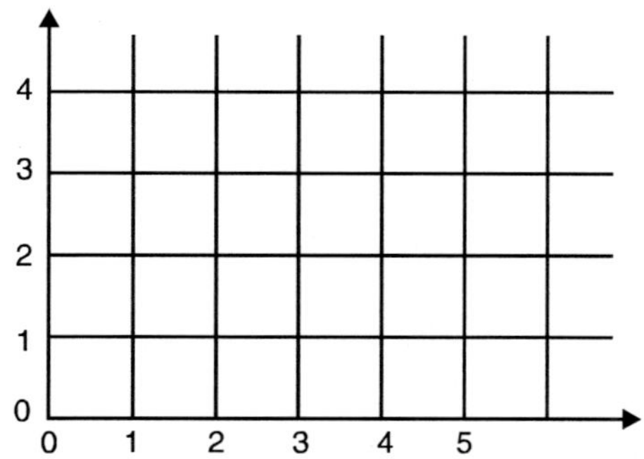

Fig. 9.32.

Each cell is given an address say (i, j) where i is the row-number of the cell and j is the column number of the cell.

A configuration of a two-dimensional Tm at a particular time may be described in terms of finitely many of the triplets of the form, (i_1, i_2, c) where for each such triplets of the form, (i_1, i_2, c) where for each such triplet (i_1, i_2) is the address of a cell and c denotes the contents of the cell. Only these cells are included in an ID, for which, the contents, are non-blank symbols.

In the configuration or ID, order of the cell which are included in an ID, **Row Major Ordering** is to be followed, that is first all the elements in the row with least index are included in the ID, followed by the elements of the row with next least index and so on. Within cells of each row, the cell with non # contents and having least column number is included first followed by the non blank cell with next least column number and so on.

Let $q \in Q$, $C_K \in \Gamma - \{\#\}$, that is C_K is a non-blank tape symbol.

The configuration at a particular instant is denoted by

$$(q, (H_1, H_2) (i_1, j_1, C_{i_1j_1}), (i_2, j_2, C_{i_2j_2}) \ldots (i_k, j_k, C_{i_kj_k}) \ldots))$$

where each of $C_{i_1j_1}$, $C_{i_2j_2}$ ... is non-blank and these are the only non-blank on the tape.

Also (H_1, H_2) denotes the location of the cell currently being scanned that is the cell under the Head.

9.9.5. Non-determinstic Turing Machine

We have already seen that when finite automata are allowed to act non-deterministically no increase in the computational power, but that non-determinism in push down automata are more powerful than deterministic ones.

We can also imagine turing machine that acts non-deterministically. In standard Tm, to each pair of current state (except the halt-state) and the symbol being scanned, there is a unique triplet comprising of the next state, unique action in terms of writing a symbol in the cell being scanned and the motion, if any, to the right or left. However, in case of NDTM (non deterministic turing machine), to each pair (q, a) with q as current state and 'a' as symbol being scanned, there may be a **finite set of triplets** $\{(q_i, a_i, m_i); i = 1, 2, \ldots\}$ of possible moves. The set of triplets may be empty that is for some particular (q, a) the Tm may not have any next move. Or alternatively the set $\{(q_i, a_i, m_i)\}$ may have more than one triplet, meaning thereby that the NDTM in the state q and scanning symbol 'a' and can choose next move from the set $\{(q_i, a_i, m_i)\}$ of next moves.

From the above discussion it can be easily understood that standard Tm is a special case of the NDTM in which for each (q, a) the set $\{(q_i, a_i, m_i)\}$ in next moves is a singleton set or empty.

In order to define the concept of NDTM and a configuration in NDTM, we assume that the tape is one-way infinite.

Proper non-determinism means that at some stage, there are atleast two next possible moves. Now, if we engage two different machines to work out further possible moves according to each of these two moves, the two can work independent of each other. This means non-determinism allows parallel computation.

Definition A non-deterministc Turing machine is a six tuple $(Q, \Sigma, \Gamma, \delta, q_0, h)$ where

Q	:	set of states
Σ	:	set of input symbols
Γ	:	set of tape symbols
δ	:	$(Q \times \Gamma) \to$ power set of $(Q \times \Gamma \times \{L, R, N\})$
q_0	:	initial state and
h	:	hut state, $h \in Q$

The concept of a configuration is same as in case of standard Tm. But the concept of 'yield in one step' denoted by Tm has different meaning. Here one configuration may yield more than one configuration.

Example 9.22. Construct an NDTM which accepts the language $\{a^n b^m : n \geq 1, m \geq 1\}$, that is language of all strings over $\{a, b\}$, in which there is at least one 'a' and one 'b' and all a's precede all b's.

Solution. The diagrammatic representation of the required NDTM is as given below : (initially tape head is at leftmost symbol, that is 'a')

In the proposed NDTM, as the motion of the head is always to the right except in the Halt state, therefore, R is not mentioned in the labels in the diagram below :

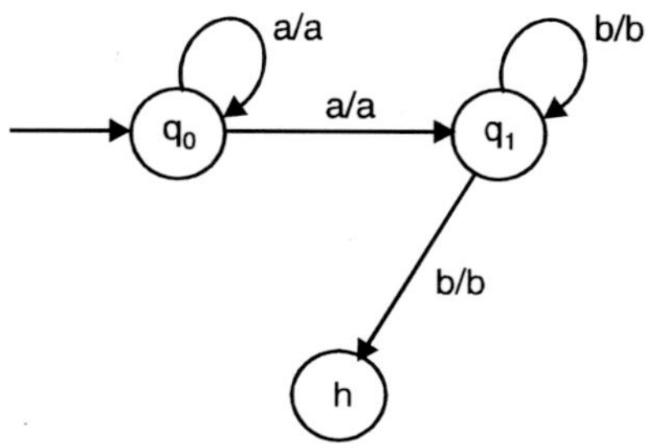

Fig. 9.33.

where the label x/y on an arc denotes that if symbol in the current cell is 'x' then contents of the cell are to be replaced by 'y'. Formally the proposed NDTM may be defined as

$$M = \{\{q_0, q_1\}, \{a, b\}, \{a, b, \#\}, \delta, q_0, h\}$$

where δ is defined as follows :

$$\delta(q_0, a) = \{(q_0, a, R), (q_1, a, R)\}$$
$$\delta(q_0, b) = \text{empty}$$
$$\delta(q_1, a) = \text{empty}$$
$$\delta(q_1, b) = \{(q_1, b, R), (h, b, N)\}$$

If the machine has no move, then it halts without accepting the string.

When turing machines are allowed to act non-deterministically, is there any increase in computational power ? We must first define what it means for a non-deterministic turing machine to compute some thing. Since a non-deterministic machine could produce two different outputs or final states from the same input, we have to be careful about what is considered to be the end result of a computation by such a machine.

For a non-deterministic machine to decide a language and compute a function, we require that all of its computations halt; we achieve this by assuming that there is no computation containing after 'm' steps, where m is an "upperbound" depending on the machine and the input. For Tm to decide a language, we only require that at least one of its possible computations end up accepting the input.

For a non-deterministic turing machine to compute a function, we require that all possible computations agree on the output. If not, we would not be able to decide which one is the right value of the function.

9.9.6. Recursively Enumerable and Recursive Languages

When a Turing machine executes an input, there are four possible outcomes of execution. Then Tm

(*i*) Halts and accept the input

(*ii*) Halts and rejects the input

(*iii*) Never halts (fall into loop), or

(*iv*) Crash.

We also know that the language accepted by Tm is known as recursively enumerable and the set of recursive languages is subset of recursive enumerable set.

The set of recursively enumerable languages are precisely those languages that can be listed (enumerated) by a turing machine. Basically, if a Tm can generate all the strings in a language, another Tm can accept all the strings that are not in the given language. (Strings that are not in the language may be rejected or may cause the turing machine to go into an infinite loop).

TIPS

A language is recursive if there exists a Turing machine that accepts every string of the language and rejects every string that is not in the language. So, we are sure about the rejection of the strings with are not in the given recursive language.

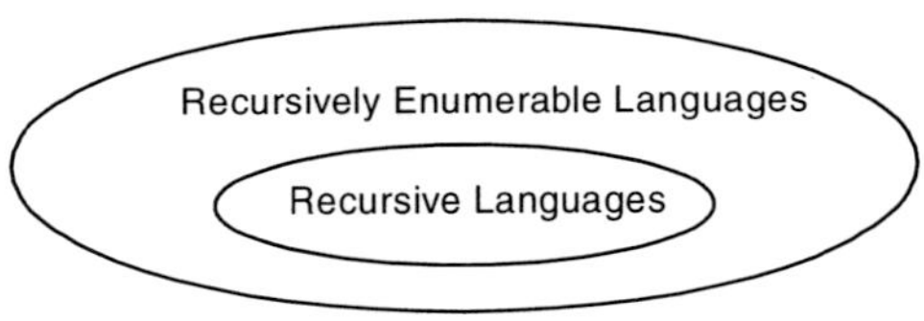

Fig. 9.34.

Theorem 9.2. If a language L is recursive, then it is recursively enumerable languages.

Proof. A language is recursively enumerable if there exists a Turing machine that accepts every string of the language and a language is recursive if there is a Tm that accepts every string of the language and does not accept strings that are not in the language.

So, every recursive language is also recursively enumerable. Hence, statement of the theorem is true.

9.9.7. Turing Machine as Enumerators

We have viewed turing machines as recognizers of languages and as computers of functions on the non-negative integers. There is a third useful view of turing machines, as generating devices. Consider a multi-tape turing machine Tm that uses one tape as an output tape, on which a symbol, once written, can never be changed, and whose tape head never moves left. Suppose also that on the output tape, Tm writes strings over some alphabet Σ, separated by a marker symbol #. We can define L(Tm), the language generated by Tm, to be set of w in Σ^* such that w is eventually printed between a pair of #'s on the output tape.

Note that unless Tm runs forever, L(Tm) is finite. Also, we do not require that words be generated in any particular order, or that any particular word be generated only once. **If L is L(Tm) for some turing machine Tm, then L is an regular set and conversely.** The recursive sets also have a characterization in terms of generators.

So finally we can say that :

TIPS

"To enumerate a set means to list the elements one at a time. To stay that a set is enumerable should mean that there is an algorithm for enumerating it, so far, we have seen Turing machines as recognizers of languages and as computers of functions on the positive integers. There is another purpose a Tm can serve. It can be a generating device.

A Turing machine Tm is said to be generating or enumerating if we are able to produce each word of any languae L separated by blank symbol (initially tape is empty). The order of the strings is not important and any string may be repeated indefiitely.

Theorem 9.2. If any language L is generated by the Turing machine, then L is accepted by some other turing machine.

or

If any language is generated by the Tm then L is recursively enumerable.

Proof. Let Tm_1 be a turing machine which generates L. Now let us construct another turing machine Tm_2 which accepts L and Tm_2 is having two tapes. The Tm_2 saves any input on the extra tape before beginning its generation process. The Tm_2 simulates Tm_1, except that every time # is written on output tape the simulation on Tm_1 pauses while Tm_2 compare its input string with the string listed just before #. If they are same, Tm_2 accepts otherwise Tm_2 continues to simulate Tm_1.

Clearly Tm_2 accepts an input string if and only if it is generated by Tm_1.

So $$L(Tm_2) = \underbrace{L(Tm_1)}_{\text{generating language by } Tm_1}$$

Theorem 9.3. If the language L is accepted by some Turing machine (*i.e.* L is recursive enumerable) then L is generated by some Turing machine.

Proof. Let Tm_1 is a Turing machine which accepts the language L. We may define another Turing machine Tm_2 which generate the words over Σ. After each new word is generated, we run a simulation of it on the machine Tm_1. If Tm_1 accepts, we print out the word on the tape inside #'s. If Tm_1 does not accepts, we skip it and go on to the next possibility from the word generator Tm_2. Again, we apply the same method for the next word and we get list of words separated by # on the tape of Tm_2.

Unfortunately, this method works if Tm_1 is guaranteed to halt on all inputs (either in final or non final states). However, it may possible that Tm_1 will loop forever for some words, in such case we shall stuck waiting forever, and we cannot go on to test the next possible input string.

To overcome this situation, we adopt the canonical order on Σ^*. The canonical order (or lexicographical order or alphabetical order) over $\Sigma = \{a, b\}$ is $\{\in, a, b, aa, ab, ba, bb, aaa, aab, ab, ...\}$ and over $\Sigma = \{0, 1\}$, we define canonical order as

$$\{\in, 0, 1, 00, 01, 10, 11, 000, 001, 010, ...\}$$

Let us number the possible input strings as w_1, w_2, ... in the usual canonical order. Let us assume that Tm_2 has four tracks. In second track it generates, in order, all the integers. Let at some point in the operation of Tm_2, track 2 has the number n on it. Now on track 3 we generate, one by one, all possible strings from w_1 to w_n. Each time we generate another input string, we copy the string from track 3 to track 4 and simulate the running of Tm_1 on it. But we run simulation for exactly n steps unless Tm_1 halts before. If n steps have not been enough to halt Tm_1 we do not waste more effort on this input string at this iteration. We erase track 4 and we go back to track 3 and generate the next input string to be tested. However, if the input string is accepted within n steps, then we print the string on track between #'s. we still erase the track 4 and go back to track 3 for the next input string to be tested.

When we go back to track 3 to get the next string, we have to be sure that we have not already tried all the input strings up to w_n. In order to be sure of this we must keep a counter on track 2 telling us how many strings we have indeed produced. If we have not gone up to n yet, then we do produce next string and repeat the process. However, if we find that we have already gone up to our limit w_n then what we must do is erase this track and increment track 2. Track 2 now has the contents $n + 1$ on it. We begin again to generate strings on track 3 from w_1 to w_{n+1} and one by one simulate on track 4 the running of them on Tm_1 for we generate all the strings on track 3 exactly $n + 1$ steps this lie. Again, if they are neither accepted nor rejected, they are abandoned temporarily. If they are accepted they are printed on track if they have been printed on track 1 already. The simulation of Tm_1 on a particular input string begins at the start State of Tm_1 even though we have once before already simulated the first n steps of the processing. May be n steps were not enough, but the $n + 1$ steps will do the trick. If Tm_1 neither accepted nor rejected in $n + 1$ steps, then we erase back 4 and get the next input text case from tracks 3, unless we have already generated up to w_{n+1}, in which case we erase tracks 3 and increment tracks 2 to $n + 2$.

Clearly, the only strings that appear on track 1 are the words that have been discovered to already be in L by having been accepted Tm_1.

Theorem 9.4. L is recursive if and only if there is a Turning machine that enumerates L in canonical order.

Proof. First we prove if L is recursive then there is a Turning machine that enumerates L is canonical order. We prove this by constructive algorithm. Let L is the recursive language accepted by the Turning machine Tm_1, clearly Tm_1 halts on every input.

Now, we construct Tm_2 such a way that Tm_2 generates the words in canonical order, one at a time, on one of its track. Then Tm_2 simulates Tm_1 on that word. If Tm_1 accepts that word, Tm_2 print that word on the other track, separated by #'s. It should be clear that Tm_1 will finish processing each word after a finite time because Tm_1 never enters into infinite loop (L is recursive and Tm_1 halts for every words). It is clear that Tm_2 enumerate the words in L is canonical order.

Now, we prove the converse *i.e.* if L is a language enumerated by some turing machine in canonical order then L is recursive. Let Tm_1 be Turing machine that enumerates the words of L in canonical order. We make a Turing machine Tm_2 that accepts L and rejects L'. Into Tm_2 we input the strings w to be tested. We then simulate the running of Tm_1 until its output (words of L) are longer than w. This will takes a finite amount of time. Now, we simply check to see whether w is among the words generated thus far by Tm_2. If it is, we accept it; if not, we reject it. It is clear that all strings are either accepted or rejected *i.e.* there is a complete decision procedure and hence L is recessive.

9.9.7.1. Non recursively Enumerable Languages

As we have seen that there are pumping lemma's to show that certain language is not regular or context free language. Unfortunately, there is no such pumping lemma for recursively enumerable languages because according to church's thesis any conceivable algorithm can be implemented by Turing machines. But there are languages to which no turing machine can accept. We show that such language exist by the certain theorems only and those languages are called non-recursively enumerable languages.

9.10. CHURCH'S THESIS

What function can not be computed by turing machines ? This is very surprising. It is believed that there are no functions that can be defined by humans, whose calculation can be described by any well-defined mathematical algorithm that people can be taught to perform, that can not be computed by Tm's. The Tm is believed to be the ultimate calculating mechanism.

In other words "*No computational procedure will be considered an algorithm unless it can be represented as a turing machine*". This statement is called Church's thesis because Alonzo Church (1936) gave many sophisticated reasons for believing it. Church's original statement was slightly different because his thesis was presented slightly before the turing invented his machines.

Church actually said that any machine that can do certain list of operations will be able to perform all conceivable algorithms. He tied together what logicians had called recursive functions and computable functions. Tm's can do all that Church asked, so they are one possible model of the universal algorithm machine Church described.

TIPS

Unfortunately, Church's thesis can not be a theorem in mathematics because ideas such as "*can ever be defined by humans*" and "*algorithm that people can taught to perform*" are not part of any known mathematics. *There are no axioms that deal with "people".*

9.11. UNIVERSAL TURING MACHINES

We can consider turing machine in both ways : The turing machine is an "unprogrammable" piece of hardware, specialized at solving one particular problem, with instructions that are "*hard-wired at the factory*".

We shall now take the opposite point of view. We shall argue that turing machines are also software. That is, we shall show that there is a certain "generic" turing machine that can be programmed, about the same way that a general purpose computer can, to solve any problem that can be solved by turing machine. The "program" that makes this generic machine behave like a specific machine Tm will have to be a description of Tm. In other words, we shall be thinking of the formalism of turing machines as a programming language, in which we can write programs. Programs written in this language can then be **interpreted by universal machine** that is to say another program in same language.

Definition To begin, we must present a general way of specifying turing machines, so that their descriptions can be used as input to other turing machines.

Universal turing machines 'U' takes two arguments, a description of a machine Tm, "Tm", and a description of an input string *w*, "*w*". We want U to have the following property : U halts on input "Tm" "*w*" if and only if Tm halts on input *w*.

$$U\ (\text{"}m\text{" "}w\text{"}) = \text{"}m\ (w)\text{"}.$$

It is the functional notation of universal turing machine.

To justify this we have to define a language whose strings are all legal representation of turing machines. We adopt the following convention. A string representing a turing machine state is of the form $\{q\}\{0, 1\}^*$; that is, the letter q followed by binary string. Similarly, a tape symbol is always represented as a string in $\{q\}\ \{0, 1\}^*$.

Let Tm = (Q, Σ, δ, S) be a turing machine, and k and l be smallest integers such that $2^k \geq |Q|$, and $2^l \geq |\Sigma| + 2$. Then each state in Q will be represented by q followed by a binary string of length k; each symbol in Σ will be likewise represented as letter a followed by a binary string of l bits. The head directions that left (L) and right (R) are included in input tape symbols to justify "+2" term in the definition of l.

Example 9.23. Consider the turing machine $m = (Q, \Sigma, \delta, S)$ where

$$Q = \{S, q, h\}$$

$\Sigma = \{\#, b, a\}$, and δ is given in following table.

States	*Symbol*	δ
S	*a*	(*q*, #)
S	#	(*h*, #)
S	*b*	(*S*, *R*)
q	*a*	(*S*, *a*)
q	#	(*S*, *R*)
q	*b*	(*q*, *R*)

Since there are three states, three symbols in $\sum$, we have $k = 2$ and $l = 3$. These are the smallest integers such that $2^k \geq 3$ and $2^l \geq 3 + 2$. The states and symbols are represented as follows :

State/Symbol	*Representation*
S	*q*00
q	*q*01
h	*q*11
#	*a*000
b	*a*001
L	*a*010
R	*a*011
a	*a*100

Thus representation of *baa#a* is

"*baa#a*" = *a*001*a*100*a*100*a*000*a*100.

The representing "Tm" of the turing machine Tm is the following strings:

"Tm" = (*q*00, *a*100, *q*01, *a*000), (*q*00, *a*000, *q*11, *a*000), (*q*00, a001, *q*00, *a*011), (*q*01, *a*100, *q*00, a011), (*q*01, a000, *q*00, a011), (*q*01, *a*001, *q*01, 011)

In the last let us see pictorial representation of universal turing machine

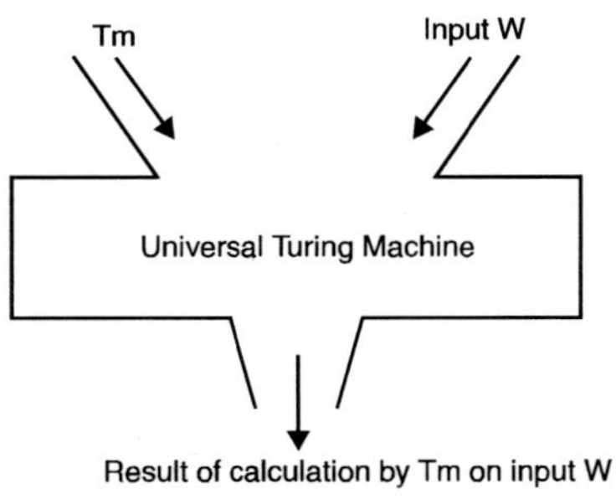

Fig. 9.35. Result of calculation by Tm on input *w*.

After discussing of above example we can have following observation :

Observation 1. A Turing machine *M* designed to solve a particular problem *P* consists, apart from the description of the set of possible states and the set of possible input etc. of mainly the description of the process in some coded form of a sequence of steps required to solve the problem in the form of the move function δ. In case of universal turing machine, the **process part** involving δ of the turing machine M, and the inputs are expressed in the code (that is language) of the Universal turing machine. This code of the process alongwith the code of the input, is stored in the memory of the UTM. And just on the lines of the control unit of a general purpose computer the control unit of UTM, reads the codes for steps, one step at a time, decodes and execute the code for each step, until the code for the final result is stored on the tape of UTM.

Observation 2. A Turing machine Tm designed to solve a particular problem P can easily be specified by

(*i*) The initial state say $q_{0\text{Tm}}$ of the Turing machine M.

(*ii*) The next-move function δ_m of Tm, which can be described by the rules of the form : if the current state of Tm is q_i and contents of cell being scanned are a_j then the next state of Tm is q_k, the symbol to be written in the current cell is a_1 and move m_f of the Tape Head may be : To-left, To-right or None.

Thus, each of these rules for a particular Tm can be specified by **quintupoles** of the form (q_i, a_i, a_1, m_f). And hence the next-move function δ_{Tm} for machine Tm is completely specified by the set.

$$\{(q_i, a_j, q_k, a_1, m_f) : q_i, q_j \in Q_{\text{Tm}};\ a_j, a_1, \in \Gamma_{\text{Tm}};\ m_f \in \{\text{To-left, To-right, None}\}\}$$

Observation 3. Next question that arises in the context of the construction of universal Turing machine, is about the number of distinct states in UTM and number of distinct inputs/tape symbols required in the UTM, so that it can solve any solvable problem.

As UTM should be able to simulate each Turing Machine, therefore, it may appear that number of distinct states and number of distinct tape symbols in the UTM, should be at least as much as is possible in any Tm, because UTM may be required to accomplish the task of any Tm. However, by proper coding techniques we may use only two symbols to represent set of symbols.

9.12. THE HALTING PROBLEM

Let assume that we have given the description of turing machine Tm and an input w, when started in the initial configuration q_0w, perform a computation that eventually halts? Using an abbreviated way of talking about the problem, we ask wheather Tm applied to w, or simply (Tm, w) halts or does not halt. The domain of this problem is to be taken as the set of all turing machines and all w; that is, we are looking for a single turing machine that, given the description of an arbitrary Tm and w, will predict whether or not the computation of Tm applied to w will halt.

We can not find the answer by simulating the action of Tm on w, say by performing it on universal turing machine, because there is no limit on the length of the computation. If Tm enters an infinite loop, then no matter how long we wait, we can never be sure that Tm is in fact in a loop. It may be simple case of very long computation. What we need is an algorithm that can determine the correct answer for any Tm and w by performing some analysis on the machine's description and the input. But it is clear that no such algorithm exists.

Definition In short halting problem is :
To determine for an arbitrary given Turing machine Tm and input w,
Whether Tm will eventually halt on input w.

9.13. UNDECIDABILITY/DECIDABILITY

We know that recursive languages are those languages which are accepted by atleast one turing machine and these sets of recursive languages are subclass of regular sets, called the recursive sets.

Definition "*A problem whose language is recursive is said to be decidable*". Otherwise problem is undecidable. That is, a problem is undecidable if there exist no algorithm that takes as input an instance of the problem and determine whether the answer to that instance is "yes" or "no".

9.13.1. Facts about Turing-Decidable and Turing Acceptable Languages

1. If L is turing-decidable then L is turing-acceptable

Proof. If m decides L, then the turing machine

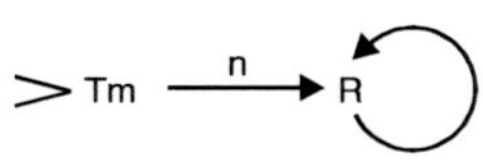

Fig. 9.36.

accepts/semidecides L.

That if Tm decides L, we enter into either state 'Y' or 'n'.

➤ If Tm halts in 'Y', this machine halts (because if $w \in L$, Tm goes to state Y and halts).

So clearly if L is decided by some turing machine Tm then L is also accepted by some turing machine.

2. If L is turing-decidable then so is $\overline{L}$.

Proof. Let Tm be a turing machine such as :

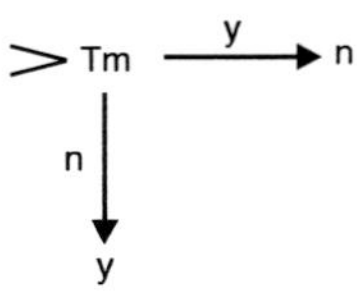

Fig. 9.37.

This machine goes to a no (n) state when Tm ended up in a yes (Y) state and vice-versa.

Therefore, there is also a turing machine which decides $\overline{L}$ (complement of the language L).

9.13.2. Turing Acceptable

We know that a turing machine Tm accepts $w \in \Sigma^*$ if Tm halts on input w, language L is turing-acceptable if there is some turing machine that accepts it.

"Let assume, given a countably infinite number of turing machines Tm_1, Tm_2, Tm_3, ..., Tm_i ... and language L is turing acceptable if at least one of the Tm_i accepts (halts on) any $w \in L$".

But the question is how do we know or discover which Tm_i (if any) accepts L. If we start testing Tm_1, Tm_2. etc. If Tm_1 doesn't accept L it will never halt. Yet, how do we know that if we run Tm_1 long enough it would not eventually halt on w ? So the issue comes down to : we would like to be able to accurately predict whether a machine will eventually halt on input w. If it were possible to predict for any turing machine and any input string whether or not that turing machine would eventually halt on that input then every turing acceptable language would also be turing-decidable. Some authors call **Turing Acceptable Language** as **Recursively Enumerable Language also.**

9.13.3. Turing-acceptable Vs Turing-decidable

Consider the language L_0 = {"Tm" "w" : Turing machine Tm accepts w}

Then let us see some points :

(1) L_0 is turing-acceptable (by definition) :

There is a machine, Tm_0 (a variant of U)

Which accepts input = "Tm" "w" such that Tm accepts input w.

(2) So if Tm accepts $w \in L$ then Tm_0 also accepts "Tm" "w" $\in L_0$.

(3) L_0 is like infinite dictionary :

$$L_0 = \{\text{“}Tm_1\text{“ “}w_{11}\text{“, “}Tm_1\text{“ “}w_{12}\text{“, “}Tm_1\text{“, “}w_{13}\text{“, ... “}Tm_1\text{“ “}w_n\text{“,}\ \ldots \text{“}Tm_2\text{” “}w_{21}\text{“, “}Tm_2\text{“ “}w_{22}\text{“, ...}\}$$

a dictionary which contains answer to questions :

Does arbitrary turing machine Tm accepts input w ?

Because to answer the question we

(a) look in the dictionary to see if Tm is paired with input w.

(b) If Tm does appear with w then Tm accepts w, else Tm doesn't accept w.

(4) So L_0 is formulized version of the halting problem because for any arbitrary Tm and w, we could just examine L_0 to see if "Tm" "w" $\in L_0$. But since L_0 is infinite, if we don't find "Tm" "w", we will never know if we have searched long enough.

(5) However, if we could answer the question in general is "Tm" "w" $\in L_0$.

(6) But L_0 is not turing-decidable $\Rightarrow$ the halting problem is not solvable.

(7) For specific Tm, w we could predict whether Tm will halt on input w that is :

(a) could see if Tm has any halt state

(b) for real simple Tm, w.

But there is no completely general method that correctly decides all cases.

Now on the basis of above discussion we can understand that in the case of a Language L which is Turing Acceptable Language but which is not Turing Decidable, there may be a Tm which halts on a large number of strings W, where $W \notin L$, but there must be at least one string $W \notin L$ on which Tm does not halt.

Similarly, in the case of a language L which is not Turing Acceptable (and hence cannot be turing decidable), it may happen that there is a Turing machine Tm which may halt for a large number of inputs W which belong to L. But there must be atleast one string $W \in L$ for which M does not halt.

Now we can make following observations :

(*i*) Each Turing Decidable Language L is necessarily Turing Acceptable.

(*ii*) However, there may be languages which are Turing Acceptable but not Turing Decidable.

(*iii*) Further, there may be languages $L \subseteq \Sigma^*$ which may neither be Turing Acceptable and hence not Turing Decidable. For a language L which is not Turing Acceptable, there cannot be any Turing Machine Tm which halts for every string W of L.

9.13.4. Reducibility

A reduction (reducibility) is a way of converging one problem into another problem in such a way that a solution to the second problem can be used to solve first problem. Such "reducibilities" come up often in every day life, even if we don't usually refer to them this way.

- The problem of understanding in French, the answer to the question : "where is the nearest hotel" reduces to the general problem of understanding French at all.
- Suppose you want to find your way around a new city. You know this would be easy if you have a map. Thus you can reduce the problem of finding your way around the city to the problem of obtaining a map of the city.

Reduction doesn't mean to make smaller, it means to transform or convert a problem X to another problem Y that is at least as hard as X. Usually Y is at least as hard as X, and so we express a reduction from X to Y as $X \leq Y$.

9.13.4.1. Use of Reducibility to Show that a Problem is Unsolvable

Reducibility is the primary method for proving that a problem is computationally unsolvable. Here is how it works :

(*a*) Reduce an unsolvable problem X to a more difficult or general problem, Y.

(*b*) Then show that if you could solve the **more general** problem, B, you could use the solution to Y to solve the **more specific** problem, X, because Y is at least as hard as X ($X \leq Y$).

(*c*) Conclude that since X is known to be unsolvable, and the solution to Y would also solve problem X then Y must be unsolvable too.

(that is a solution to the more general problem Y, would also have to cover the more specific problem X, which of course, can't be solved)

Note. *Reducibility says nothing about solving X or Y alone; only about the solvability of X in the presence or context of Y.*

- For example problem of measuring the area of a rectangle to the problem of increasing its height and width.
- The problem of solving of linear equation reduces to the problem of inverting matrices.

9.13.4.2. Reducibility and Decidability/Solvability

Reducibility also plays an important role in classifying problem as decidable or undecidable, and later on in complexity there too.

- When X is reducible to Y, solving X can't be harder than solving Y, because the solution to Y can also be used to solve X.
- In terms of the decidability/computability question, then
  - (*a*) If X is reducible to Y and Y is decidable, then X is also decidable.
  - (*b*) Equivalently, if X is undecidable/unsolvable and X reduces to Y, then Y is also undecidable. This is the key to proving that problem are undecidable or unsolvable.

Note. In this case, where X is undecidable, to reduce Y to X would prove nothing : It would simply show that Y could be decidable if we could decide X which we know we can not.

9.13.5. Properties of Recursive and Enumerable Languages

We know that a Language L is said to be recursive enumerable if there exist a Turing machine which accepts all words in L and rejects for every words in the complement of L, that is $\bar{L}$. We have already studied that number of problems can be solved by reducing one problem to another. This reduction involves combining several turing machines to form a composite machine. The state of the composite turing machine has a component for each individual component machine. Similarly the composite machine has separate tapes for each individual machine.

Given an algorithm, we can allow the composite turing machine to perform one action if the algorithm accepts and another if it does not accept. We could not do this if we were given an arbitrary turing machine than an algorithm, since if the turing machine did not accept, it might run forever, and the composite machine would never initiate the next task.

Theorem 9.5. The complement of a recursive language is recursive.

Proof. Let L be a recursive language and Tm be turing machine that halts on all inputs and accepts L. Let us construct a turing machine Tm′ from Tm so that if Tm enters a final state on input w, then Tm′ halts without accepting. If Tm halts without accepting, Tm′ enters a final state.

Since one of these two events occurs, Tm′ is an algorithm. So clearly T(Tm′) is the component of L and thus the complement of L is recursive language. Following figure shows construction of Tm′ from Tm.

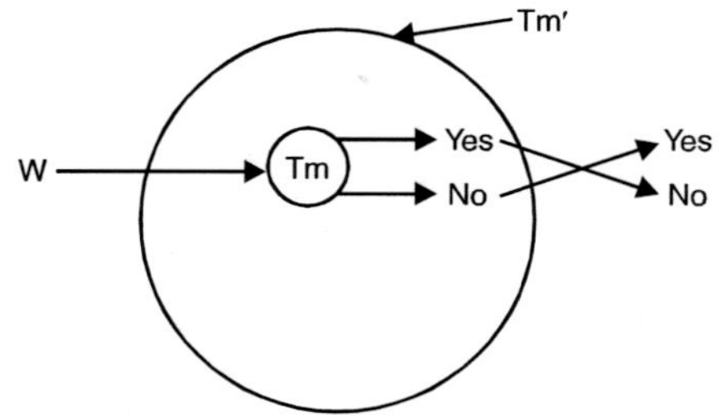

Fig. 9.38.

Theorem 9.6. The union of two recursive languages is recursive.

Proof. Let L_1 and L_2 be two recursive languages accepted by turing machines Tm_1 and Tm_2. We construct a turing machine Tm, which first simulate Tm_1. If Tm_1 accepts, then Tm accepts. If Tm_1 rejects then Tm simulates Tm_2 and accepts if and only if Tm_2 accepts. Since both Tm_1 and Tm_2 are algorithm so Tm is guaranteed to halt. Clearly Tm accepts $L_1 \cup L_2$.

So $L_1 \cup L_2$ is also recursive since there exist a turing machine Tm for it.

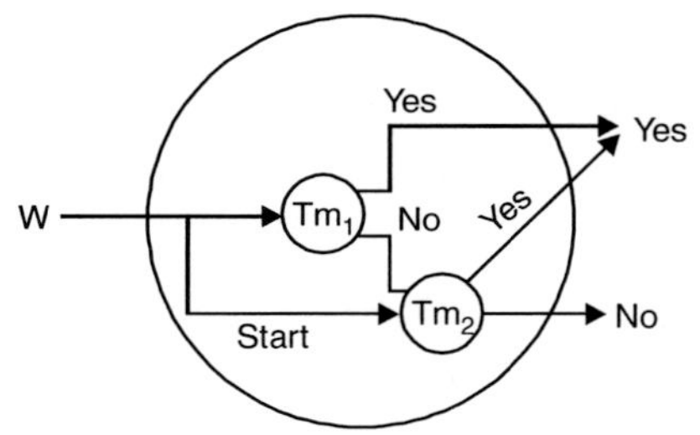

Fig. 9.39. Construction for union.

Theorem 9.7. The union of two recursively enumerable languages is recursively enumerable.

Proof. We have studied earlier that enumerating turing machine have one input tape and one output tape.

Let L_1 and L_2 are recursively enumerable languages and their enumerative turing machines are Tm_1 and Tm_2 respectively. Let us design a turing machine Tm which can simulate Tm_1 and Tm_2 simultaneously on separate tape. If either accepts, then Tm accepts as follows :

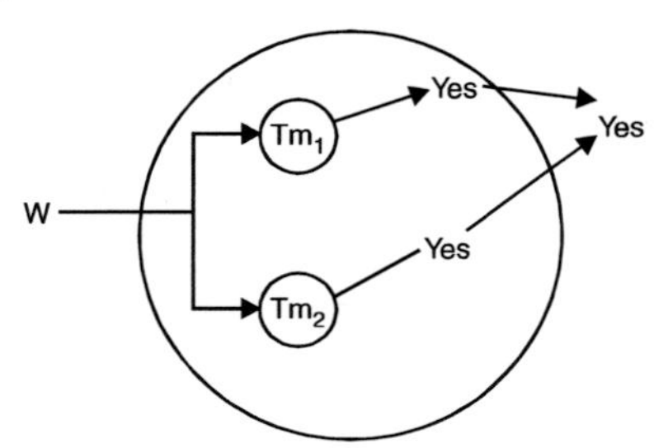

Fig. 9.40. Construction of Union.

Theorem 9.8. If a language L and its complement $\bar{L}$ are both recursively enumerable, then L (and hence $\bar{L}$) is recursive.

Proof. Let Tm_1 and Tm_2 accept L and $\bar{L}$ respectively. Let us construct a turing machine Tm which simulate Tm_1 and Tm_2 simultaneously. Tm accepts w if Tm_1 accepts and rejects w if Tm_2 will accept. Thus Tm will always say either "Yes" or "No", but never say both. Note that there is not a priority limit on how long it may take before Tm_1 or Tm_2 accepts, but it is certain that one or the other will do so. Since Tm is algorithm that accepts L, if follows that L is recursive.

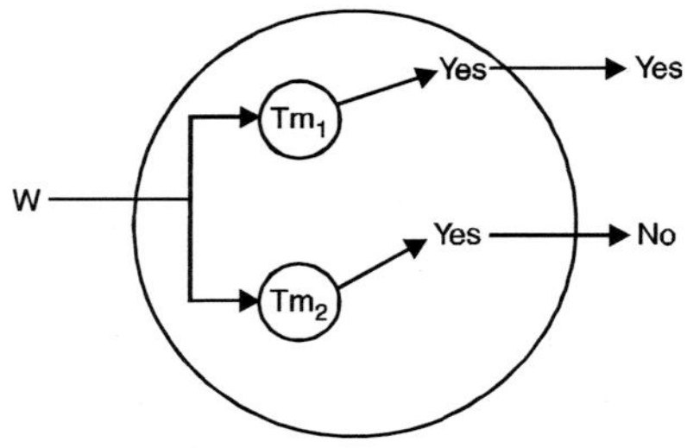

Fig. 9.41.

Theorem 9.9. If L is a recursive language than $\Sigma^* - L$ is recursive.

Proof. The required Turing machine Tm-complement can be represented by following diagram.

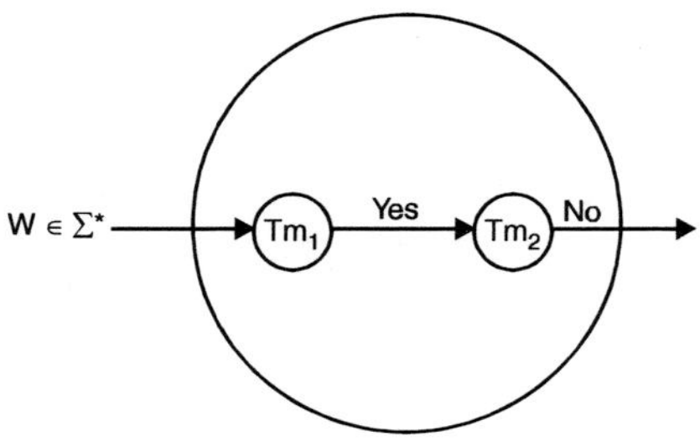

Fig. 9.42.

The machine Tm-complement functions as follows : when a string $W \in \Sigma^*$ is given an input to Tm-complement, its control passes the string to Tm_1 as input to Tm_1. As Tm_1 decides the language L, therefore, for $W \in L$ after a finite number of moves, Tm_1 outputs "Yes", which is the given as input to Tm_2, which in turn returns "No".

Similarly for $W \notin L$, Tm_2 returns "Yes". Hence there exist a turing machine Tm-complement for $\Sigma^* - L$. So it is turing-decidable, that is recursive.

9.13.6. The Post Correspondence Problem

The post correspondence problem is another undecidable problem that turns out to be a very helpful tool for proving problems in logic or in formal language theory to be undecidable.

Definition Let Σ be an alphabet with at least two letters. An instance of the post corresponding problem (for short PCP) is given by two sequences $U = (u_1, u_2, ..., u_m)$ and $V = (v_1, v_2, ..., v_m)$ of strings $u_i, v_i \in \Sigma^*$. The problem is to find whether there is a (finite) sequence

$(i_1, i_2, ..., i_p)$, with $i_j \in \{1, 2, ..., m\}$ for

$$i_j = 1, 2, ... p \text{ so that, } P \geq 1$$

$$u_{i_1}, u_{i_2} u_{i_3} ... u_{i_p} = v_{i_1} v_{i_2} ... v_{i_p}.$$

Equivalently, on instance of the PCP is a sequence of pairs

$$\begin{pmatrix} \mu_1 \\ \vdots \\ v_1 \end{pmatrix}, ..., \begin{pmatrix} \mu_m \\ \vdots \\ v_m \end{pmatrix}$$

The sequence $i_1, i_2 ... i_p$ is said to be solution to this instance of PCP.

Example 9.24. Let $\Sigma = \{0, 1\}$. Let X and Y be lists of three strings each, defined as follows :

	List X	*List Y*
i	w_i	x_i
1	1	111
2	10111	10
3	10	0

In this case PCP has a solution, let $P = 4$

$$i_1 = 2, i_2 = 1, i_3 = 1 \text{ and } i_4 = 3 \text{ then}$$

$$w_2\, w_1\, w_1\, w_3 = x_2\, x_1\, x_1\, x_3$$

$$= 101111110$$

which is the solution of instance of PCP.

Example 9.25. Prove that following instance of PCP has no solution over $\Sigma = \{0, 1\}$, X and Y be lists of three strings as follows :

	List X	*List Y*
i	w_i	x_i
1	10	101
2	011	11
3	101	011

Solution. Let us assume that this instance of PCP has solution $i_1, i_2, ..., i_p$. Clearly $i_1 = 1$ since no string beginning with $w_2 = 011$ can equal a string beginning with $x_2 = 11$; no string beginning with $w_3 = 101$ can equal a string beginning with $x_3 = 011$.

We write the string from list X the corresponding string from Y. So for we have

10

101

The next selection from X must begin with a 1. Thus $i_2 = 1$ or $i_2 = 3$. But $i_2 = 1$ will not do, since no string beginning with $w_1\, w_1 = 1010$ can equal a string beginning with $x_1\, x_1 = 101101$. with $i_2 = 3$ we have

10101

101011

Since the string from list Y again exceeds the string from list X by the single symbol 1, a similar argument shows that $i_3 = i_4 = ... = 3$. Thus there is only one sequence of choices that generates compatible strings, and for this sequence string Y is always one character longer. Thus this instance of PCP has no solution.

Example 9.26. Find the solution of following instance of PCP

$$\begin{pmatrix} abab \\ ababaaa \end{pmatrix}, \begin{pmatrix} aaabbb \\ bb \end{pmatrix}, \begin{pmatrix} aab \\ baab \end{pmatrix}, \begin{pmatrix} ba \\ baa \end{pmatrix}, \begin{pmatrix} ab \\ ba \end{pmatrix}, \begin{pmatrix} aa \\ a \end{pmatrix}.$$

Solution. The solution for this instance is

$i_1 = 1, i_2 = 2, i_3 = 3, i_4 = 4, i_5 = 5, i_6 = 5, i_7 = 6$

$$ababaaabbbaabbaababaa = ababaaabbbaabbaababaa.$$

9.13.6.1. The "Modified" PCP

It is easy to reduce undecidable language L to PCP if we first introduce an intermediate version of PCP, which we call the modified post's correspondence problem, or MPCP.

Definition In the modified PCP, there is the additional requirement on a solution that the first pair on the list X and list Y must be the first pair in the solution. More formally, an instance of MPCP is two lists

$$X = w_1, w_2, ..., w_k$$

and

$$Y = x_1, x_2, x_3, ..., x_k,$$

and a solution is a list of 0 or more integers $i_1, i_2, i_3, ...\, i_p$ such that

$$w_1, w_{i_1}, w_{i_2}, ..., w_{i_m} = x_1, x_{i_1}, x_{i_2}, ..., x_{i_p}$$

Notice that the pair (w_1, x_1) is forced to be at the beginning of the two strings, even though the index 1 is not mentioned at the front of the list that is solution. Also unlike PCP, where the solution has to have at least one integer on the solution list, in MPCP, the empty, the empty list could be a solution if $w_1 = x_1$.

Let us consider the following instance of the MPCP.

	List X	*List Y*
i	w_i	x_i
1	1	111
2	10111	10
3	10	0

Then as an instance of MPCP it has no solution. In proof, observe that any partial solution has to begin with index 1, so the two strings of a solution would begin :

$$X : 1 \ldots$$
$$Y : 111 \ldots$$

The next integer could be 2 or 3, since both w_2 and w_3 begin with 10 and thus would produce a mismatch at the third position. Thus the next index would have to be 1, yielding :

$$X : 11$$
$$Y : 111111$$

We can argue this way indefinitely. Only another 1 in the solution can avoid a mismatch, but if we can only pick index 1, the string remains three times as long as the X string, and the two strings can never become equal.

EXERCISES

1. Let Tm be the Turing machine (Q, Σ, δ, S), where

$$Q = \{q_0, q_1, q_2\}$$
$$\Sigma = \{a, \#\} \text{ and}$$

$S = q_0$ and δ is as follows :

q	σ	$\delta(q, \sigma)$
q_0	a	(q_1, L)
q_0	#	$(q_0, \#)$
q_1	a	$(q_2, \#)$
q_1	#	$(h, \#)$
q_2	a	(q_2, a)
q_2	#	(q_0, L)

Describe the operation of Tm when started as $(q_0, \#\ aaaa\underline{a})$

2. Describe what the following Turing machine does.

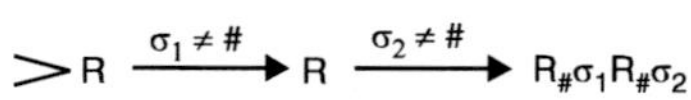

Fig. 9.43.

3. (*i*) Build a Turing machine that accepts the language

$$L = \{a^n b^{n+1}\}$$

(*ii*) Build a Turing machine that accepts the language

$$L = \{a^n b^{2n}\}$$

4. Build a Turing machine that accepts the language of all words that contain the substring *bbb*.

5. Build a turing machine that accepts the language of all words that do not contain the substring *bbb*.

6. Build a Turing machine that accepts the language ODD PALINDROME.

7. Build a Turing Machine that accepts all strings with more *a*'s than *b*'s, the language MORE.

8. Construct the Turing machine for the following languages :

(*i*) $L = L(aba^*b)$ (*ii*) $L = \{w : | w | \text{ is even}\}$

(*iii*) $L = \{w : | w | \text{ is a multiple of } 3\}$

(*iv*) $L = \{a^n b^m a^{n+m} : n \geq 0, m \geq 1\}$

(*v*) $L = \{a^n b^n a^n b^n : n \geq 0\}$

(*vi*) $L = \{a^n b^{2n} : n \geq 1\}$.

9. Prove that following functions are computable functions :

(*i*) $f(x) = 3x$ (*ii*) $f(a, b) = \begin{cases} a-b & a > b \\ 0 & a \leq b \end{cases}$

(*iii*) $f(a, b) = 2a + 3b$ (*iv*) $f(a) = a \bmod 5$

10. Construct the Turing machine for the language :

$$L = \{0^n 1^n / n \geq 1\}.$$

11. Prove that PCP with {(01, 011), (1, 10), (1, 11)} has no solution.

12. Does the PCP with $x = (b^3, ab^2)$ and $y = (b^3, bab^3)$ have a solution.

13. Find atleast three solutions to PCP defined by the dominoes :

$$\begin{pmatrix} 1 \\ 111 \end{pmatrix}, \begin{pmatrix} 10 \\ 0 \end{pmatrix}, \begin{pmatrix} 10111 \\ 10 \end{pmatrix}$$

14. Design a TSPDA for

$$L = \{ww : w \in \{a, b\}^*\}.$$

15. Implement a binary incrementer on turing machine. Use general terms to describe action taken. Would your solution be deterministic or non-deterministic ? Why ?

*****

SECTION-D

- ➤ Chomsky Hierarchies
- ➤ Computability

SYLLABUS

Chomsky Hierarchies : Chomsky Hierarchies of Grammars, Unrestricted Grammars, Context Sensitive Languages, Relation Between Languages of Classes.

Computability : Basic Concepts, Primitive Recursive Functions.

10

CHAPTER

Chomsky Hierarchies

INSIDE THIS CHAPTER

10.1. PHRASE-STRUCTURE GRAMMARS

A phrase-structure grammar is a collection of three things :

1. A finite alphabet Σ of letters called terminals.
2. A finite set of symbols called non-terminals that includes the start symbol S.
3. A finite set of production of the form

$$\text{String 1} \rightarrow \text{String 2}$$

where string 1 can be any string of terminals and non-terminals that contains at least one non-terminals and where string 2 is any string of terminals and non-terminals what so ever.

A derivation in a phrase-structure grammar is a series of working string beginning with the start symbol S, which by making substitution according to the productions, arrives at a string of all terminals, at which point generation must stop.

TIPS

The language generated by a phrase structure grammar is the set of all strings of terminals that can be derived starting at S.

Example 10.1. The following is a phrase structure grammar over $\Sigma = \{a, b\}$ with nonterminals A an B :

$$A \rightarrow BA/\in \qquad (P_1, \text{say})$$

$$B \rightarrow aB/a \qquad (P_2\text{, say})$$

$$aaaB \rightarrow ba \qquad (P_3\text{, say})$$

Solution. The first production P_1 says that we can start with A and derive number of symbols of the type B, for example

$$\begin{aligned} A &\Rightarrow BA \\ &\Rightarrow BBA \\ &\Rightarrow BBBA \\ &\Rightarrow BBBBA \\ &\Rightarrow BBBB \end{aligned}$$

The second production shows us that each B can be any string of a's (with at least one a) :

$$\begin{aligned} B &\Rightarrow aB \\ &\Rightarrow aaB \\ &\Rightarrow aaaB \\ &\Rightarrow aaaaB \\ &\Rightarrow aaaaa \end{aligned}$$

The third production (P_3) says that any time we find three a's and B, we can replace these four symbols with the two-terminal string ba.

The following is a summary of one possible derivation in this grammar :

$$\begin{aligned} A &\overset{*}{\Rightarrow} BBBBBB \\ &\overset{*}{\Rightarrow} aaaaBBBBB \\ &\Rightarrow aabaBBBB \\ &\Rightarrow aabaaaBB \\ &\Rightarrow aabbaBB \\ &\Rightarrow aabbaaaB \\ &\Rightarrow aabbba \end{aligned}$$

All CFG's are phrase-structure grammars in which we restrict ourselves as to what we put on the left side of productions. So, all CFLs can be generated by phrase-structure grammars.

10.2. CHOMSKY HIERARCHY

We can exhibit the relationship between grammars by the Chomsky Hierarchy. Noum Chomsky, a founder of formal language theory, provided an initial classification in to four languages types :

Type – 0 (*Unrestricted grammar*)

Type – 1 (*Context sensitive grammar*)

Type – 2 (*Context free grammar*)

Type – 3 (*Regular grammar*)

Type 0 languages are those generated by unrestricted grammars, that is, the recursively enumerable languages. Type 1 consist of the context-sentive languages, Type 2 consists of the context-free languages and Type 3 consists of the regular languages. Each language family of type k is a proper subset of the family of type $k - 1$. Following diagram shows the original Chomsky Hierarchy.

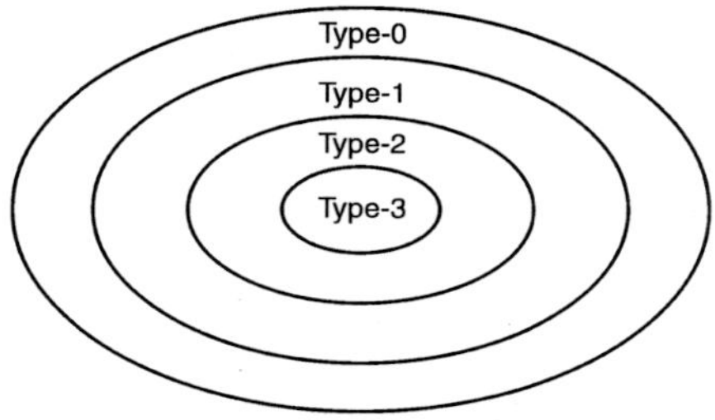

Fig. 10.1. The original Chomsky Hierarchy.

We have also met several other language families that can be fitted in to this picture. Including the families of deterministic context-free languages (L_{DCF}), and recursive languages (L_{REC}). The modified Chomsky Hierarchy can be seen in Fig. 10.2.

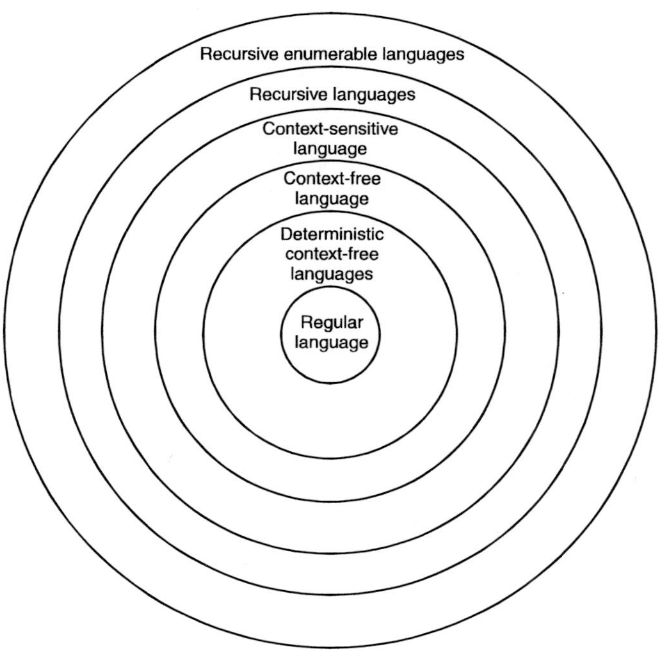

Fig. 10.2.

We know that $L = \{w : n_a(w) = n_b(w)\}$ is deterministic, but not linear. On the other hand, the language

$$L = \{a^n b^n\} \cup \{a^n b^{2n}\}$$

is linear, but not determinstic. The relationship between, linear, determinsitic context-free and non-determinstic context-free language is shown in Fig. 10.3.

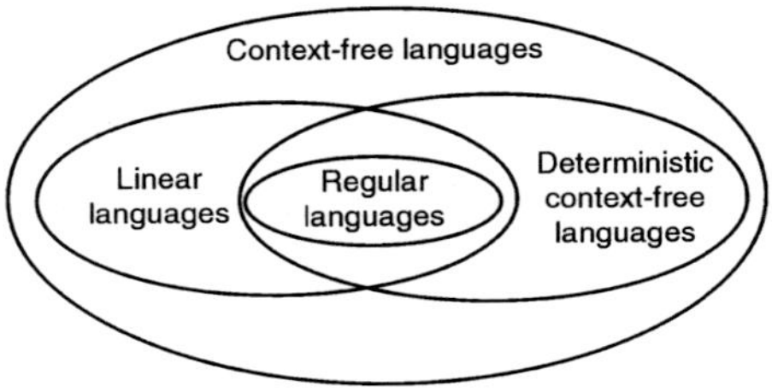

Fig. 10.3.

In the last we can summarise our discussion as follows :

Type	*Name of languages generated*	*Production Restrictions* $A \to B$	*Acceptor*
0	Unrestricted (reversively-enumerable)	A = any string with non-terminal B = any string	Turing machine
1	Context-sentive	A = any string with non-terminals B = any string as long as or longer than A	Linear bound automata
2	Context-free	A = one non-terminal B = any string	Pushdown automata
3	Regular	A = one non-terminal $B = aX$ or $B = a$, where a is a terminal and X is a non-terminal.	Finite automata

10.3. UNRESTRICTED (TYPE-0) GRAMMARS

Definition The unrestricted grammar is defined as

$$G = (V_n, V_t, P, S)$$

where V_n = a finite set of non-terminals

V_t = a finite set of terminals

S = starting non-terminal, $S \in V_n$

and P is set of productions of the following form

$$\alpha \to \beta$$

where α and β are arbitrary string of grammar symbols with $\alpha \neq \epsilon$. These grammars are known as type-0, phase-structure or unrestricted grammars.

Example 10.2. Write a grammar generating $\{a^i / i$ is a positive power of 2$\}$.

Solution. Let grammar be $G = (V_n, V_t, P, S)$

where
$$V_n = \{S, A, C, B, D, E\}$$
$$V_t = \{a, \in\}$$
$$S = \{S\}$$

and P is defined as follows :

$$S \rightarrow ACaB \quad (P_1)$$
$$Ca \rightarrow aaC \quad (P_2)$$
$$CB \rightarrow DB \quad (P_3)$$
$$CB \rightarrow E \quad (P_4)$$
$$aD \rightarrow Da \quad (P_5)$$
$$AD \rightarrow AC \quad (P_6)$$
$$aE \rightarrow Ea \quad (P_7)$$
$$AE \rightarrow \in \quad (P_8)$$

A and B serve as left and right end marks for sentential forms; C is a marker that moves through the string of a's between A and B, doubling their number by production P_2. When C hits the right end marker B, it becomes a D or E by production P_3 or P_4. If a D is choosen, that D migrates left by production P_5 until the left end marker A is reached. At the point the D becomes a C again by production P_6, and the process starts over. If an E is chosen, the right endmarker is consumed. The E migrates left by production P_7 and consumer the left endmarker, leaving a string of 2^i a's for some $i > 0$.

Example 10.3. Construct a grammar for language $L = \{a^n b^n c^n / n \geq 0\}$.

Solution. We know that it is difficult to write grammar for $L = \{a^n b^n c^n / n \geq 0\}$ recursively. So we try to write this grammar in two parts.

(*i*) First we construct grammar for $a^n x^n$

(*ii*) Then convert x^n in to $b^n c^n$.

Let grammar be
$$G = (V_n, V_t, P, S)$$
where
$$V_n = \{S, B, C\}$$
$$V_t = \{a, b, c\}$$
$$S = \{S\}$$

and P is defined as follows :

$$S \rightarrow aSBC / aBC$$
$$CB \rightarrow BC$$
$$aB \rightarrow ab$$
$$bB \rightarrow bb$$
$$bC \rightarrow bc$$
$$cC \rightarrow cc$$

Now let us derive $a^n b^n c^n$ from above discussed grammar

$$S \overset{*}{\Rightarrow} a^{n-1} S(BC)^{n-1} \quad \text{(Use } S \to aSBC\ (n-1) \text{ times)}$$

$$\Rightarrow a^{n-1}\, a\, BC\, (BC)^{n-1} \quad \text{(Use } S \to aBC\text{)}$$

$$\overset{*}{\Rightarrow} a^n B^n C^n$$

(By applying $CB \to BC$ several times since $CB \to BC$ interchange B and C)

$$\Rightarrow a^{n-1} abB^{n-1} c^n \quad \text{(By applying } aB \to ab\text{)}$$

$$\overset{*}{\Rightarrow} a^n b^n C^n \quad \text{(By applying } bB \to bb \text{ several times)}$$

$$\Rightarrow a^n b^{n-1} bcc^{n-1} \quad \text{(By applying } bC \to bc \text{ by once)}$$

$$\Rightarrow a^n b^n c^n \quad \text{(By applying } cC \to Cc \text{ several times)}$$

Example 10.4. Let $G = (V_n, V_t, P, S)$

where

$$V_n = \{S, X, Y\}$$
$$V_t = \{a, b\}$$
$$S = \{S\}$$

and P is defined as follows :

$$S \to aXYa$$
$$X \to baXYb$$
$$Y \to Xab$$
$$aX \to baa$$
$$bYb \to abab$$

Test whether $w = baabbabaaabbaba$ is in $L(G)$.

Solution. Here I am underline the substring to be replaced by the use of a production :

$$S \Rightarrow \underline{aX}Ya$$
$$\Rightarrow baa\underline{Y}a$$
$$\Rightarrow baa\underline{X}aba$$
$$\Rightarrow baa\underline{b}aX\underline{Y}baba$$
$$\Rightarrow baabbaa\underline{Y}baba$$
$$\Rightarrow baabba\underline{aX}abbaba$$
$$\Rightarrow baabbabaaabbaba$$
$$\Rightarrow w$$

So $w \in L(G)$.

10.4. CONTEXT SENSITIVE GRAMMAR

Let $G = (V_n, V_t, P, S)$ be context sentive grammar

where

V_n = finite set of non-terminals

V_t = finite set of terminals

S = starting non-terminals $S \in V_n$

and P is the set of rules called productions defined as

$$\alpha \rightarrow \beta$$

where β is at least as long as α that is clearly

$$|\alpha| \leq |\beta|$$

The term "Context-sentive" comes from a normal form for these grammars, where each production is of the form $\alpha_1 A \alpha_2 \rightarrow \alpha_1 \beta \alpha_2$, with $\beta \neq \in$. Replacement of variable A by string β is permitted in the "Context" of α_1 and α_2.

For example, language $L = \{a^n b^n c^n : n \geq 0\}$ is context sensitive. We can show this exhibiting a context sensitive grammar for the language L_1 grammar as

$$S \rightarrow abc/aAbc$$
$$Ab \rightarrow bA$$
$$Ac \rightarrow Bbcc$$
$$bB \rightarrow Bb$$
$$aB \rightarrow aa/aaA$$

"A type 0 grammer is called length-increasing if for all rules $\alpha \rightarrow \beta$ in P we have $|\alpha| \leq |\beta|$" (same as context sensitive grammar).

So we can say that $\in$ free type-0 grammar is said to be length increasing grammar.

10.5. LINEAR BOUNDED AUTOMATA

A Linear Bounded Automata (LBA) is a non-deterministic turing machine satisfying the following two conditions.

(*a*) Its input alphabet includes two special symbols $\$_1$ and $\$_2$, the left and right end marker, respectively.

(*b*) The LBA has no more left from $\$_1$ and no right from $\$_2$, nor may it print another symbol over $\$_1$ or $\$_2$.

The linear bound automata is simply a turing machine which instead of having potentially infinite tape on which to compute, is restricted to the portion of the tape containing the input plus the two tape squares holding the end markers.

LBA will be denoted as

$T_m = (Q, \Sigma, \Gamma, \delta, q_0, h, \$_1, \$_2)$ where Q, Σ, δ, S are as for non-deterministic turing machine; $\$_1$ and $\$_2$ are symbols in Γ, the left and right end markers.

The $L(T_m)$ the languages accepted by turing machine, is

$\{W/W \text{ is in } \Sigma^* \text{ and } (q_0, \$_1\, W\, \$_2) \overset{*}{\underset{T_m}{\vdash}} (h, \$_1\, y\, \$_2)\}$

Here y is for yes and h shows halting of LBA after accepting the string W.

10.6. RELATION BETWEEN CLASSES OF LANGUAGES

The four classes of languages—recursive sets, context sensitive languages, context-free languages and regular sets are often referred to as language of types 0, 1, 2 and 3 respectively.

We can find the relation between all these sets of languages through following theorems :

Theorem 10.1. Every context sensitive language is recursive.

Proof. We know that for every recursive language there exists a turing machine for it. But we have not been very specific about how one inputs a grammar into turing machine. Let us imagine that some string of delimiters separating the productions, possibly allowing the production arrow to be tape character as well.

Now let us feed this turing machine and W (input string) into universal turing machine. Because W either halts or crashes on the coded turing machine, this procedure will indeed, lead to a decision about W's membership in the language generated by G. So we prove that there exists a turing machine for every context-sensitive language, which can decide that any arbitrary string W can be derived from the given context-sensitive grammar or not, so we can say that every context-sensitive language is recursive.

Theorem 10.2. There is at least one language L that is recursive but not context-sensitive.

Proof. In the previous theorem, we indicate that there was encoding an entire context-sensitive grammar into a single string of symbols.

Let us consider the set of all context-sensitive grammars on $V_t = \{a, b\}$.

Let us assume that each grammar in the set has variable in following form

$$V_n = \{V_0, V_1, V_2, \ldots\}$$

Every context-sensitive grammar is completely specified by its productions.

We can think of them as written as a single string

$$x_1 \to y_1 \,;\, x_2 \to y_2 \,;\, \ldots,\, x_m \to y_m$$

We can represent these string in homomorphism as follows

$$h(a) = 0\,1\,0$$
$$h(b) = 0\,1^2\,0$$
$$h(R) = 0\,1^3\,0$$
$$h(i) = 0\,1^4\,0$$
$$h(V_i) = 0\,1^{i+5}\,0$$

Thus, any context-sensitive grammar can be represented uniquely by a string of the form $L\,((011^*0)^*)$. Further more, the representation is invertible in the sense that, given any such string, there is at most one context-sensitive grammar corresponding to it.

Let us introduce a proper ordering on $\{0, 1\}^+$, so we can write strings in the order w_1, w_2 etc. A given string w_j may not define a context-sensitive grammar; if it does, call the grammar G_j.

Now let us define a language.

$$L = \{W_i : W_i \text{ defines a CSG } G_i \text{ and } W_i \notin L(G_i)\}$$

Observation 1. L is recursive. To see this we construct membership algorithms. Given W_i, we check it to see if it defines a sensitive-grammar G_i. If not, then $W_i \notin L$.

If string does not define a grammar then $L(G_i)$ is recursive and we can use membership algorithm to find out if $W_i \in L(G_i)$. If it is not then $W_i \in L$. So L is recursive.

Observation 2. L is not a context sensitive. If it were there would exist some W_j such that $L = L(G_j)$. We can ask if W_j is in $L(G_j)$. If we assume that $W_j \in L(G_j)$, then by definition W_j is not in L. But $L = L(G_j)$, so we have a contradiction. Conversely if we assume that $W_j \notin L(G_j)$, then by definition $W_j \in L$ and we have another contradiction.

Taking two observations in consideration we conclude that L is not context-sensitive.

EXERCISES

1. Consider the following grammar :

$$\begin{aligned} S &\rightarrow ABS/\in \\ AB &\rightarrow BA \\ BA &\rightarrow AB \\ A &\rightarrow a \\ B &\rightarrow b \end{aligned}$$

(*i*) Derive the *abba* and *babaabbbaa* from above grammar.

(*ii*) Prove that all words with an equal number of *a*'s and *b*'s can be generated by this grammar.

2. Consider the grammar

$$\begin{aligned} S &\rightarrow ABCS/\in \\ AB &\rightarrow BA \\ BC &\rightarrow CB \\ AC &\rightarrow CA \\ BA &\rightarrow AB \\ CB &\rightarrow BC \\ CA &\rightarrow AC \\ A &\rightarrow a \\ B &\rightarrow b \\ C &\rightarrow c \end{aligned}$$

Derive the *ababcc* and *cbaabccba* from this grammar.

3. What language is generated by the grammar

$$\begin{aligned} S &\rightarrow aXYba \\ XY &\rightarrow XYbZ/\in \\ Zb &\rightarrow bZ \\ Za &\rightarrow aa \end{aligned}$$

4. Consider the following grammar

$$\begin{aligned} S &\rightarrow UVX \\ UV &\rightarrow aUX \\ UV &\rightarrow bUZ \\ YX &\rightarrow VaX \\ ZX &\rightarrow VbX \end{aligned}$$

$$Ya \to aY$$
$$Yb \to bY$$
$$Za \to aZ$$
$$Zb \to bZ$$
$$UV \to \in$$
$$X \to \in$$
$$aV \to Va$$
$$bV \to Vb$$

Derive the following words from this grammar :

(*i*) $\in$ (*ii*) aa

(*iii*) bb (*iv*) $abab$

5. Construct a grammar G generating

$$\{WW/W \in \{a, b\}^*\}$$

6. Show that $\{b^{m^2}/m \geq 1\}$ is generated by the grammar

$$S \to b$$
$$S \to X_3X_4$$
$$X_3 \to X_1X_3X_2$$
$$X_3 \to X_1X_2$$
$$X_1X_2 \to bX_2X_1$$
$$X_1b \to bX_1$$
$$X_2b \to bX_2$$
$$X_1X_4 \to X_4b$$
$$X_2X_4 \to X_5b$$
$$X_2X_5 \to X_5b$$
$$X_5 \to b$$

7. Let $G = (\{A, B, S\}, \{0, 1\}, P, S)$ where P consists of

$$S \to 0AB$$
$$A0 \to S0B$$
$$A1 \to SB1$$
$$B \to SA$$
$$B \to 01$$

Show that $L(G) = \phi$.

*****

CHAPTER 11

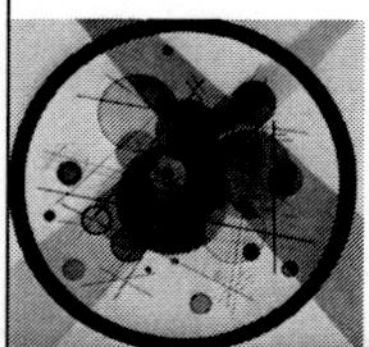

Computability and Recursive Function Theory

11.1. INTRODUCTION

INSIDE THIS CHAPTER

In the previous units, we have discussed the automata or machine models of the computational phenomenon. The automata approach to computation is **Operational** in nature, *i.e.* automata approach is concerned with the computational aspect of '**how** the computation is to be performed'

Each computational theory (rather each theory about any other phenomenon also) starts with some *assumptions*, for example, *about basic (undefined) concepts, operational capabilities and a set of statements, called axioms and postulates, which are assumed to be fundamentally true* (*i.e.* assumed to be true without any argument). In Automata Theory, the concepts like 'state' 'initial state', 'final state' and 'input' etc. are assumed to be understood, without any elaboration. Further, *the capabilities of an automata* to accept an input from the environment; to change its state on some, or even on no input; to give signal about acceptability/ unacceptability of a string; are assumed.

TIPS

In this unit, we will be concerned with Recursive Function Theory, which is a functional or declarative approach to computation. Under this approach, computation is described in terms of 'what is to be accomplished' instead of 'how to accomplish'.

In Recursive Function Theory, *to begin with*, it is **assumed** that **three types of functions** and **three structuring rules** (that is combination, composition and primitive recursion) for constructing more complex functions out of the already constructed or assumed to be constructible functions **are** so simple that ***our ability*** to construct machines to realize these functions and the structuring rules is taken as *acceptable without any argument*. **The functions, obtained by applying a finite sequence of the structuring rules to the initial functions, are called Primitive Recursive Functions.** However, with these simple functions and elementary structuring rules, though it is possible to construct very complex functions yet, even some simple functions like **division** are not constructible by the above mechanism.

11.2. INITIAL FUNCTIONS FOR NATURAL NUMBERS

It is clear that all elementary functions are all functions of natural numbers. hence, they may take zero as input, but not a negative number, and not any rational or irrational fraction.

Let $N = \{0, 1, 2, ...\}$ be a set of natural numbers. We have three initial functions over N defined below :

1. Zero Function is denoted by Z and defined as follows

$$Z(n) = 0 \text{ for } \forall\, n \in N$$

Clearly the zero function returns zero regardless of its argument.

2. Successor Function(s) is defined as follows :

$$S(n) = n + 1 \quad \text{for } \forall\, n \in N$$

Clearly successor function takes one argument and returns the succeeding number.

For example :

$$S(2) = 2 + 1 = 3 \text{ where } 2, 3 \in N$$

$$\left.\begin{aligned} Z(0) &= 0 \\ Z(2) &= 0 \end{aligned}\right\} \quad \text{where } 0, 2 \in N$$

3. Projection Function (P_i^n) is defined as follows :

$$P_i^n(a_1, a_2, ..., a_n) = a_i, \text{ where } a_i \in N \text{ for } i = 1, 2, ..., n \text{ and } i \le n.$$

Here projection function takes n arguments and returns their i^{th} argument. When it takes one argument it returns its argument as its value.

11.3. INITIAL FUNCTIONS FOR SYMBOLS

We know that set of alphabet is represented by Σ. Two initial functions are defined over Σ :

1. Null Function is defined as follows :

$$\text{NULL}(n) = \in \text{ for } \forall\, n \in \Sigma^*$$

For example : If $\Sigma = \{a, b\}$

then $\text{NULL}(a) = \in$

and $\quad \text{NULL}(b) = \in$

2. Concatenation Function is defined as follows :

$$\text{CONCATE } W(W_1) = WW_1 \text{ for } W, W_1 \in \Sigma^* \text{ and } |W_1| \geq 0.$$

For example :

$$\text{CONCATE K(IET)} = \text{KIET}$$

CONCATE a(CONCATE b (NULL (abc))

$$= \text{CONCATE } a(\text{CONCATE } b(\in))$$

$$= \text{CONCATE } a(b)$$

$$= ab$$

11.4. COMPOSITION OF FUNCTIONS

We can define a new function by the combination of two or more functions. Such defined functions are called composition function.

For example :

$$S(Z(a)) = S(t)$$

$$= 1$$

and $\quad S(S(Z(n))) = 2$ and so on.

11.5. RECURSION

Definition We know that recursion is widely used term in computer science. "A function which calls itself directly or indirectly and terminates after finite number of steps is known as recursive function."

In recursive functions, terminating point is also known as base point. A recursive function will be good if it has terminating point and it call to itself should be nearer to its terminating point. The terminating point is also known as base point.

Let g be a function of $n + 1$ arguments, g' be a function of n arguments, and h be a function of m arguments where $m \leq n + 2$. We define the function g recursively using the g' and h function as follows :

$$\text{If } (x_1, x_2, ..., x_n, 0) = g'(y_1, y_2, ..., y_n) \text{ then } g(x_1, x_2, ..., x_n, x_{n+1} + 1)$$

$$= h(x_1, x_2, ..., x_n, x_{n+1}, g(x_1, x_2, ..., x_n, x_{n+1})),$$

where g' and h are defined functions.

Example 11.1. Define m ! recursively.

Solution. Let $g(m) = m! = m * (m - 1) * (m - 2) ... * 1$

We know that $\quad 0! = 1! = 1,$

So $g(0) = 1$(it is a terminating point)

$$h(x, y) = x * y$$

$$g(n + 1) = h(n + 1, g(n)) \qquad \text{(Indirect call)}$$

Example 11.2. Define addition of two positive integers m and n recursively.

Solution. Let f is the addition function of two positive integers, so

$$f(m, n) = m + n$$

We define f to increment the argument m, n times one at a time.

So,
$$f(m, 0) = m \quad \text{(it is terminating point)}$$
$$f(m, n + 1) = f(m, n) + 1 \quad \text{(direct call)}$$
$$\begin{aligned} f(5, 2) &= f(5, 1 + 1) \\ &= f(5, 0 + 1) + 1 \\ &= 5 + 1 + 1 \\ &= 7 \end{aligned}$$

11.6. PRIMITIVE RECURSIVE FUNCTION

Definition

A function is primitive if it follows the condition :

(*i*) It is an initial function, or

(*ii*) It is obtained from recursion or composition of initial functions.

Most popular total functions are primitive recursive functions; however some total functions are not. For example, the ackerman function is total function but not primitive recursive.

Example 11.3. Show that addition function of two positive integers is primitive recursive.

Solution. In Example 11.2, we have defined the addition function f for m and n as $f(m, 0) = m, f(m, n + 1) = f(m, n) + 1$. We define f using the initial functions as follows :

$$f(m, 0) = m = P_1'(m),$$

and
$$f(m, n + 1) = f(m, n) + 1 = S(f(m, n))$$

For example, let us consider $f(7, 2)$

$$\begin{aligned} f(7, 2) &= S(f(7, 1)) \\ &= f(7, 1) + 1 \\ &= S(f(7, 0)) + 1 \\ &= f(7, 0) + 1 + 1 \\ &= 7 + 1 + 1 \\ &= 9 \end{aligned}$$

Hence, f is a primitive recursive function.

Theorem 11.1. Every primitive recursive function is turing computable.

Proof. We know that primitive functions are defined using initial functions with sometimes applying composition and recursion. So if it is possible to design turing machine (Tm) for every initial function, their composition and recursion then we can say that every primitive recursive function is turing computable. Let us see following cases :

Case 1. Turing machine for zero function $Z(x) = 0$, for $x \in N$. The Tm for zero function should work as follows :

$$\# IIII \# \xrightarrow{\text{Tm}} \#\#$$

Here IIII means that x is 4. So, here we are supposed to design an eraser turing machine :

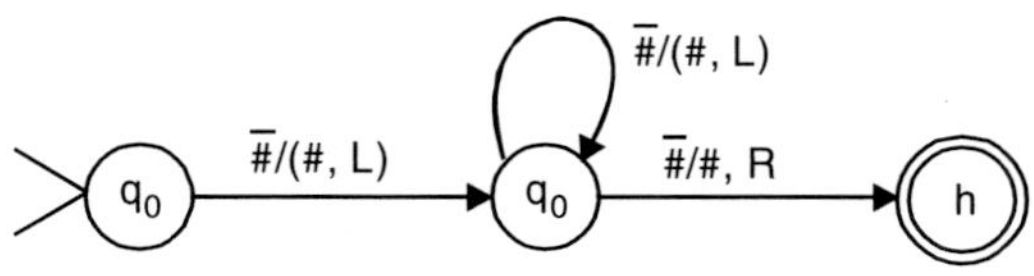

Fig. 11.1. Eraser Turing machine for zero function.

Case 2. Turing Machine for **successor function** $S(n) = n + 1$, for $n \in N$. The Tm for successor function will work as follows :

$$\# III \# \xrightarrow{\text{Tm}} \# IIII \#$$

input $\Rightarrow n = 3$ output $\Rightarrow n + 1 = 4$

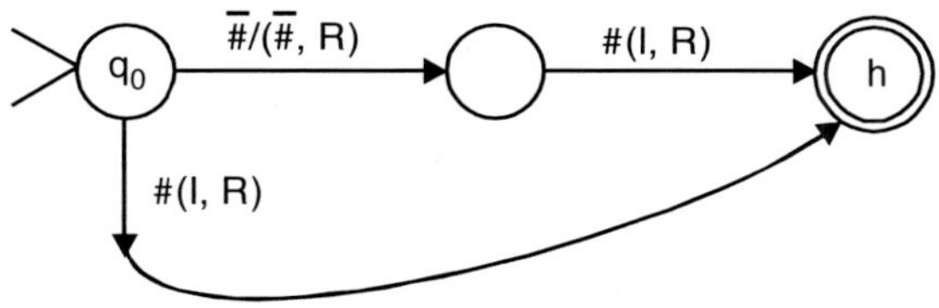

Fig. 11.2. Tm for S(n) = n + 1.

Case 3. Turing machine for **Projection function**

$$P_i^n(x_1, x_2, \ldots, x_n) = x_i \text{ for } x_1, x_2, \ldots, x_n \in N.$$

So we can easily design a Turing machine Tm for projection function :

$$\# x_1 \# x_2 \# \ldots \# x_i \# \ldots \# x_n \# \xrightarrow{\text{Tm}} \# x_i \#$$

Case 4. We can also design Turing machine for **composition function** by using the concept of combination of Turing machine.

Case 5. For recursion, we can design a Tm using multiple calls to a simple Tm designated for initial function with some fixed termination point.

Case 6. We can also combine our both cases 4 and 5 into one to design a Turing machine Tm for compsition and recursion together.

So, our theorem statement is true.

11.7. MINIMALIZATION

If g is the function of $(n \geq 1)$ arguments, then minimalization of G is defined as follows :

$$G_y(x, y) = \begin{cases} Y, & \text{the least } Y \text{ such that } G(x, y) = 0 \\ 0, & \text{otherwise} \end{cases}$$

For example, consider $g(x, y) = x + y - 2$,

So putting $g(x, y) = 0$, we get

$y = 2 - x$ if $x \leq 2$ is the result of minimalization.

So, minimalization,

$$\begin{aligned} (g(x, y)) &= 2 - x \text{ for } x \leq 2 \\ &= 0 \text{ for } n > 2 \end{aligned}$$

11.8. TOTAL RECURSIVE FUNCTION

Definition A total function is a subclass of partial function. A function is called total function if it is defined for all its arguments. Let $f(a_1, a_2, ..., a_n)$ be a function and defined on function $g(b_1, b_2, ..., b_m)$, then f is total function if every element of f is assigned to some unique element of function g.

Example 11.4. Addition of two positive integer is a total function.

Solution. Let $f(m, n) = m + n$ is the addition function from $N \times N$ to N of two positive integers m and n. This function can be defined recursively as

$$f(m, 0) = 1, \; f(m, n+1) = f(m, n) + 1.$$

Since, f is defined for all values of $m, n \in N$. So f is a total function.

Example 11.5. Multiplication of two positive integers is a total function.

Solution. Let $g(m, n)$ is multiplication function from $N \times N$ to N of two positive integers m and n, defined recursively as : $g(m, 0) = 0$, $g(m, n + 1) = m + g(m, n)$. Since, g is defined for all $m, n \in N$. So g is a total recursive function.

11.9. PARTIAL RECURSIVE FUNCTION

A function is called partial recursive if it is defined for some of its arguments. Let $f(a_1, a_2, ..., a_n)$ is a function and defined on function $g(b_1, b_2, ..., b_m)$, then f is partial function if some elements of f is assigned to almost one element of function g.

A partial function is recursive if

(*i*) It is an initial function over N, or

(*ii*) It is obtained by applying recursion or composition or minimalization on initial function over N.

11.10. μ-RECURSIVE FUNCTIONS

A function is μ-recursive if

(*a*) It is primitive recursive, or

(*b*) can be obtained from basic functions by any number of successive applications of composition, recursive definition, and minimalization on minimalizable functions

In another words the μ-recursive functions (or partial μ-recursive functions) are partial functions that take finite tuples of natural numbers and return a single natural number. They are the smallest class of partial functions that includes the initial functions and is closed under composition, primitive recursion, and the μ operator.

In computability theory, the μ operator, minimization operator, or unbounded search operator searches for the least natural number with a given property.

11.11. DEFINITION OF COMPUTABILITY

Computability refers to the ability to solve a problem in an effective manner. It is a key topic of the field of computability theory within mathematical logic and computer science. The computability of a problem is closely linked to the existence of an algorithm to solve the problem.

A computable - *In computability theory, a Turing reduction from a problem A to a problem B, named after Alan Turing, is a reduction which solves A, assuming B is already known.*

Computable : Expressible as a yes/no question that can be answered in any case by a computer in finite time.

11.11.1. Computable Function

Computable functions are the basic objects of study in computability theory. Computable functions are the formalized analogue of the intuitive notion of algorithm. They are used to discuss computability without referring to any concrete model of computation such as Turing machines or register machines. Any definition, however, must make reference to some specific model of computation but all valid definitions yield the same class of functions. Particular models of computability that give rise to the set of computable functions are the Turing-computable functions and the μ-recursive functions.

The class of computable functions can be defined in many equivalent models, including

- Turing machines
- μ-recursive functions
- Lambda calculus
- Post machines (Post-Turing machines and tag machines).
- Register machines

Although those models use different representations for the functions, their inputs and their outputs, translations exist between any two models. In the remainder of this article, functions from natural numbers to natural numbers are used (as is the case for, e.g., the μ-recursive functions).

Each computable function f takes a fixed, finite number of natural numbers as arguments. Because the functions are partial in general, they may not be defined for every possible choice of input. If a computable function is defined for a certain input, then it returns a single natural number as output (this output can be interpreted as a list of numbers using a pairing function). These functions are also called **partial recursive functions**. In computability theory, the **domain** of a function is taken to be the set of all inputs for which the function is defined.

A function which is defined for all possible arguments is called **total**. If a computable function is total, it is called a **total computable function** or **total recursive function**.

The notation $f(x_1, ..., x_k)$ indicates that the partial function f is defined on arguments $x_1, ..., x_k$, and the notation $f(x_1, ..., x_k) = y$ indicates that f is defined on the arguments $x_1, ..., x_k$ and the value returned is y. The case that a function f is undefined for arguments $x_1, ..., x_k$ is denoted by $f(x_1, ..., x_k)$.

11.11.2. Turing Computable Functions

Turing machine can be used to compute functions. How is it possible ? First we see how to input a string to a turing machine. We adopt the following policy to input any string to a turing machine.

The string W is presented in to the form of # W #, that is string, W is surrounded by blanks symbols from both sides and is placed on the left most square of tape, the head of turing machine is positioned at the right most blank symbol which immediately follows the string W.

This is shown by an underscore that is we use an underscore to show the current position of machine head in the tape. A turing machine is said to halt on input 'W' if we can reach to a halting state 'h', after performing some operations, that is if,

$T_m = (Q, \Sigma, \Gamma, \delta, q_0, h)$ is a turing machine then Tm is said to halt to on input W if and only if $(q, \#\ W\ \underline{\#})$ yields to $(h, \#\ \mu\ \underline{\#})$.

Definition

A function $f(x) = y$ is said to be computable by a turing machine Tm $(Q, \Sigma, \Gamma, \delta, q_0, s)$, if $(s, \#x\underline{\#}) \vdash^* (h, \#y\underline{\#})$ where x may be in some alphabet Σ_1^* and y may be some alphabet Σ_2^* and $\Sigma_1, \Sigma_2 \subseteq \Sigma$.

Symbol $\vdash^*$ has its usual meaning as in automata, means string x yields to y after a finite number of steps.

It means that if we give input x to the turing machine Tm, turing machine gives output as a string if it computes the function $f(x) = y$.

Example 11.6. Design a turing machine which compute the function $f(m) = m + 1$ for each m that belongs to the set of natural numbers.

Solution. Given function is $f(m) = m + 1$. Here we represents input m on the tape by a number of I's on the tape.

For example $m = 1$; input will be $\#I\#$,

For $n = 2;$

input will be #*II*# and so on.

Similarly output can be seen by the number of *I*'s on the tape.

Let Turing machine be Tm $= (Q, \Sigma, \Gamma, \delta, q_0, h)$

where
$$Q = \{q_0, h\}$$
$$\Gamma = \{I, \#\}$$

q_0 is initial state.

	σ = *I*	σ = #
q_0	(q_1, I, R)	$(q_1, \#, R)$
q_1	*	(h, I, R)
h	*	*

Here σ is input at any time.

Let input be $n = 3$ that is tape situation is #*III*#

Now let us process the given string.

(**Note :** We are processing string from right to left).

$$\#III\ q_0 \vdash \#III\#q_1$$
$$\vdash \#IIII\#h$$

Example 11.7. Prove that following function is computable

$$f(n) = n + 2$$

Solution. We know that if any function is computable then there exist a turing machine for it. "so it will be sufficient to construct a turing machine to prove any function computable".

Let Turing machine for given function be

$$\text{Tm} = (Q, \Sigma, \Gamma, \delta, q_0, h)$$
$$Q = \{q_0, q_1, q_2, h\}$$
$$\Gamma = \{I, \#\}$$

q_0 is initial state and halting state is represented by '*h*'.

Transition relations are defined as follows :

δ	*I*	#
q_0	(q_1, I, R)	$(q_1, \#, R)$
q_1	*	(q_2, I, R)
q_2	*	(h, I, R)
h	*	*

Initially turing machine is in q_0 state, when it reads *I* or # then it changes it's state to q_1 and moves towards right (right symbol is only blank). At q_1 it encountered with # and replace it by I, change its state to q_2 and moves towards right. At q_2, # is replaced by I and machine halts after moving right.

Let $n = 3$ that is #III#

then Tm, behaves as follows L

$$\begin{aligned} \#IIIq_0 &\vdash \#III\#\, q_1 \\ &\vdash \#IIII\#q_2 \\ &\vdash \#IIIII\#\, h \end{aligned}$$

Example 11.8. Design a turing machine that replace every 0 and 1 with every 1 with 0 in a binary string.

Or

Prove that function $f(w) = \overline{w}$ is computable where $w \in \{0, 1\}^*$ and $\overline{w}$ the one's complement of w.

Solution. We want to design a turing machine that takes input as #101$\underline{\#}$ then its output would be #010$\underline{\#}$. This machines can be designed by the following idea; scan backwards through the input, replace every 0 with 1 and every 1 with 0, until the blank marking the left most sequence is found. In order to leave the tape in its proper format, the machine must then return its head to the blank marking the right end of the input.

Let Turing Machine is

$$\begin{aligned} \text{Tm} &= (Q, \Sigma, \Gamma, \delta, q_0, h) \\ Q &= \{q_0, q_1, h\} \\ \Sigma &= \{0, 1\} \\ \Gamma &= \{0, 1, \#\} \end{aligned}$$

q_0 is initial state.

δ can be defined by following transition diagram.

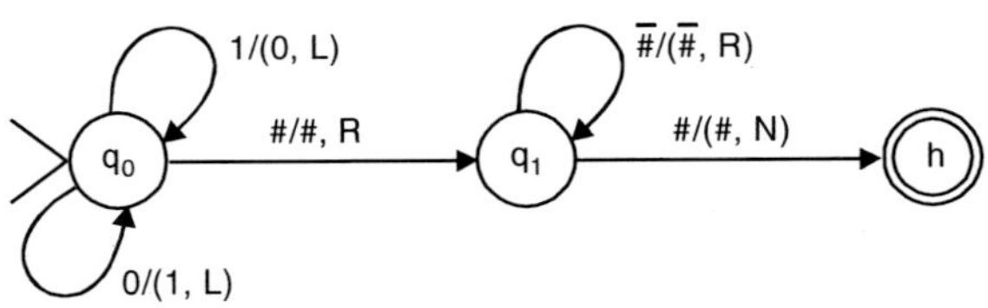

Fig. 11.3.

Here $\overline{\#}$ represent non-blank symbol (in this case either 0 or 1).

Let input string be w = #101100$\underline{\#}$

$$\begin{aligned} \#101100\, q_0 &\vdash \#10110q_01 \\ &\vdash \#1011q_011 \\ &\vdash \#101q_011 \\ &\vdash \#10q_00011 \\ &\vdash \#1q_010011 \end{aligned}$$

$$\vdash \#q_0010011$$
$$\vdash \#q_1010011$$
$$\vdash \#0q_110011$$
$$\vdash \#01q_10011$$
$$\vdash \#010q_1011 \vdash \#0100q_111$$
$$\vdash \#01001q_11 \vdash \#010011q_1\# \vdash \#010011\#h$$

11.11.3. Characteristics of Computable Functions

The basic characteristic of a computable function is that there must be a finite procedure (an algorithm) telling how to compute the function. The models of computation listed above give different interpretations of what a procedure is and how it is used, but these interpretations share many properties. The fact that these models give equivalent classes of computable functions stems from the fact that each model is capable of reading and mimicking a procedure for any of the other models, much as a compiler is able to read instructions in one computer language and emit instructions in another language.

Following are the characteristics of a procedure for computing a computable function:

(*a*) "There must be exact instructions (*i.e.* a program), finite in length, for the procedure."

(*b*) "If the procedure is given a k-tuple $\boldsymbol{x}$ in the domain of f, then after a finite number of discrete steps the procedure must terminate and produce $f(\boldsymbol{x})$."

(*c*) "If the procedure is given a k-tuple $\boldsymbol{x}$ which is not in the domain of f, then the procedure might go on forever, never halting. Or it might get stuck at some point, but it must not pretend to produce a value for f at $\boldsymbol{x}$."

EXERCISES

1. Prove that the proper subtraction of two integers is a primitive recursive function. Proper subtraction is defined as

$$\text{sub}(x, y) = x - y \text{ if } x > y$$
$$= 0, \quad \text{otherwise}$$

2. Prove that absolute function over positive integer is primitive recursive function, where $N = \{0, 1, 2, ...\}$.

3. Prove that the function, which finds the minimum of two positive integers is primitive recursive function.

4. Prove that $f(n)$ is primitive recursive function where $f(n)$ is defined as :

$$f(n) = \begin{cases} x/2, & \text{if } x \text{ is even} \\ (x-2)/2, & \text{if } x \text{ is odd} \end{cases}$$

5. Show that exponentiation function exp $(x, y) = x^y$ is primitive recursive function.

6. Find the Godel numbering for a, ab, abc, ccc, $abcd$ where $\Sigma = \{a, b, c, d\}$.

*****

SHORT QUESTIONS WITH ANSWERS

SHORT QUESTIONS WITH ANSWERS

CHAPTER-1

Question 1:

What is alphabet ?

Answer :

"Alphabets are defined as a finite set of symbols"

An example of an alphabet is a set of decimal numbers.

$$\Sigma = \{0, 1, 2, 3, 4, 5, 6, 7, 8, 9\}$$

Here Σ is representation for the alphabet.

Question 2:

What is String ?

Answer :

A string (or sometimes word) is a finite sequence of symbols selected from some alphabet". For example, if Σ $\{a, b\}$ is an alphabet then *abab* is string over alphabet Σ. A string is generally denoted by w.

The "length" of the string is denoted by $|w|$ and it is, the number of positions for the symbol in the string. For instance string

$$w = 01101 \text{ has length}$$

$$|w| = 5.$$

Question 3:

What is Language ?

Answer :

A set of strings all of which are chosen from some Σ^*, where Σ is a particular alphabet, is called a language". If Σ is an alphabet, and $L \subseteq \Sigma^*$, then L is said to be language over alphabet Σ.

Question 4:

Write the definition of DFA ?

Answer :

A deterministic finite automata is a quintuple

$$M = (Q, \Sigma, \delta, q_0, F)$$

where

Q : Is a non-empty finite set of states presents in the finite control. $(q_0, q_1, q_2, ...)$

Σ : Is a non-empty finite set of input symbols which can be passed to the finite state machine. $(a, b, c, d, e, ...)$

q_0 : Is a starting state, one of the state in Q.

F : Is a non-empty set of final states or accepting states, set of final states belongs to Q.

δ : Is a function called transition function that takes two arguments a state and a input symbol, it returns a *single* state.

Let 'q' is the state and 'a' be input symbol passed to the transition function as :

$$\delta(q, a) = q'$$

Question 5:

What are the various limitations of finite automata ?

Answer :

(*a*) The defining characteristic of FA is that they have only a finite number of states. Hence, finite automata can only "count" a finite number of input scenarios.

(*b*) Finite automata don't have any auxiliary memory so it works in "process and forget" mode. It can store anything at particular state. There is no finite automaton that recognizes these strings:

(*i*) The set of binary strings consisting of an equal number of 1's and 0's

(*ii*) The set of strings over '(' and ')' that have "balanced" parentheses

(*c*) Finite automata can recognize and process regular languages.

CHAPTER-2

Question 1:

What is regular expression ?

Answer :

The set of regular expression is defined by the following rules :

1. Every letter of Σ can be made into a regular expression, null string, $\in$ itself is a regular expression.

2. If r_1 and r_2 are regular expression, then

   (*i*) (r_1) (*ii*) r_1r_2

   (*iii*) $r_1 + r_2$ (*iv*) r_1*

   (*v*) r_1^+ are also regular expression.

3. Nothing else is regular expression.

Question 2:

State and prove Arden's theorem.

Answer :

Let P and Q be two regular expression over alphabet Σ. If P does not contain null string $\in$, then

$$R = Q + RP$$

has a unique solution that is $R = QP*$

It can be understand as : $R = Q + RP$

Put the value of R in R.H.S.

$$R = Q + (Q + RP)P = Q + QP + RP^2$$

When we put the value of R again and again we got the following equation.

$$R = Q + QP + QP^2 + QP^3 \ldots$$

$$R = Q(1 + P + P^2 + P^3 \ldots$$

$$R = Q(\in + P + P^2 + P^3 + \ldots$$

$$R = QP*$$

(By the definition of closure operation for regular expression)

Question 3:

What is the necessary condition for two regular expression being equivalent ?

Answer :

Two expressions with variables are equivalent if whatever languages we substitute for the variables, the result of the two expressions are the same language.

Question 4:

Regular expression and DFA have equivalent expressive power, justify ?

Answer :

Regular expressions and finite automata have equivalent expressive power:

(*i*) For every regular expression r, there is a corresponding FA that accepts the set of strings generated by r.

(*ii*) For every FA, M there is a corresponding regular expression that generates the set of strings accepted by M.

CHAPTER-3

Question 1:

Describe Moore Machine.

Answer :

Mathematically moore machine is a six-tuple machine and defined as

$$M_0 = (Q, \Sigma, \Delta, \delta, \lambda', q_0)$$

M_0 is the six-tuple moore machine in which

Q : A nonempty finite set of state in M_0.

Σ : A nonempty finite set of input symbols.

Δ : A nonempty finite set of outputs.

δ : It is transition function which takes two arguments as in finite automata, one is input state and another input symbol. The output of this function is a single state, so clearly δ is the function which is responsible for the transitions in M_0.

q_0 : is the initial state of M_0 and $q_0 \in Q$.

λ' : is a mapping function which maps Q to Δ, giving the output associated with each state.

Let M_0 be a moor machine and Σ is $a_1; a_2, a_3, ..., a_n, n \geq 0$ then output of M_0 for Σ is $\lambda'(q_0), \lambda'(q_1), \lambda'(q_2).... \lambda'(q_n)$, where $q_0, q_1, q_2, ... q_n$ is the sequence of states such that

$$\delta(q_{i-1}, a_i) = q_i \text{ for } 1 \leq i \leq n$$

Question 2:

Describe Mealy Machine.

Answer :

Mathematically Mealy machine is defined as follows :

$$M_e = (Q, \Sigma, \Delta, \delta, \lambda', q_0)$$

Q : A nonempty finite set of state in M_e.

Σ : A nonempty finite set of input symbols.

Δ : A nonempty finite set of outputs.

q_0 : initial state of M_e.

δ : It is a transition function which takes two arguments as in moore machine, one is input state and another is input symbol. The output of this function is a single state, so clearly δ is the function which is responsible for the transition in M_e.

λ' : is a mapping function which maps $Q \times \Sigma$ to Δ, giving the output associated with each transition.

Question 3:

Differentiate Mealy and Moore Machines.

Answer :

S. No.	Moore Machine	Melay Machine
1.	Its output depends only on present state.	Its output depends on the transition (input) and present state.
2.	Its transition function is excess into Q (finite set of states).	Its transition function is excess into Δ (output alphabet).
3.	If input string is of length n the output string is of length $n + 1$.	It input string is of length n then the output string is of length n.
4.	At the different input on the same state, its output is same.	At the different input on the same state, its output is also different.

Question 4:

Mealy and Moore Machines machine are finite machines, Justify.

Answer :

Both mealy and Moore machines are finite machines since they follow all the properties of finite machines. In case of Moore machine we consider output for every corresponding state, in case of mealy we consider output for every transition but all finite machines gives either yes or no as answer.

CHAPTER-4

Question 1:

State the pumping lemma for regular languages.

Answer :

Statement : Let L be a regular language. Then there exist a constant n (which depends on L) such that for every string w in L, such that $| w | \geq n$, we can break w into three substrings, $w = xyz$, such that :

(1) $y \neq \in$

(2) $| xy | \leq n$

(3) For all $i \geq 0$, the string $xy^i z$ is also in L.

Question 2:

What is main application of pumping lemma for regular expression?

Answer :

The pumping lemma is extremely useful in proving that certain sets are non-regular.

Question 3:

What is homomorphism ?

Answer :

A string homomorphism is a function on strings that works by substituting a particular string for each symbol.

Suppose Σ and Σ' are alphabets. Then function

$$h : \Sigma \longrightarrow \Sigma'$$

is called a "homomorphism".

Question 4:

What is inverse homomorphism ?

Answer :

Homomorphism may also be applied "backwards," and in this made they also preserve regular language. That is, suppose h is a homomorphism from some alphabet Σ to strings in another (possible the same) alphabet Σ'. Let L be a language over alphabet Σ'. Then $h^{-1}(L)$, read "h inverse of L," is the set of strings w in Σ^* such that $h(w)$ is in L.

Question 5:

Describe the distinguishable states.

Answer :

Distinguishable states. Two states q_1 and q_2 of DFA are called distinguishable if

$\hat{\delta}(q_1, w) \in F$ and $\hat{\delta}(q_2, w) \notin F$ or vice versa; for all $w \in \Sigma^*$.

Question 6:

Describe the indistinguishable states.

Answer :

Indistinguishable states. Two states q and r of DFA are called indistinguishable if

$$\hat{\delta}(q, w) \in F \Rightarrow \hat{\delta}(r, w) \in F$$

and $\quad \hat{\delta}(q, w) \notin F \Rightarrow \hat{\delta}(r, w) \notin F$

for all $\quad w \in \Sigma^*$.

Question 7:

What is Myhill-Nerode theorem ?

Answer :

Statement of the Myhill-Nerode Theorem.

The Myhill-Nerode theorem says the following three statements are equivalent :

1. The set L, a subset of Σ^*, is accepted by a DFA. (We know this means L is regular language).
2. L is the union of some of the equivalence classes of a right invariant (with respect to concatenation) equivalence relation of finite index.
3. Let equivalence relation R_L be defined by : xR_Ly if and only if for all z in Σ^*, xz is in L exactly when yz is in L. Then R_L is of finite index.

CHAPTER-5

Question 1:

What is CFG ?

Answer :

"A grammar $G = (V_n, V_t, P, S)$ is said to be context-free"

where V_n : A finite set of non-terminals, generally represented by capital letters, $A, B, C, D, \ldots$

V_t : A finite set of terminals, generally represented by small letters, like, $a, b, c, d, e, f, \ldots$

S : Starting non-terminal, called start symbol of the grammar. S belongs to V_n.

P : Set of rules or productions in CFG.

G is context-free and all productions in P have the form

$$\alpha \to \beta$$

where $\alpha \in V_n$ and $\beta \in (V_t \cup V_n)^*$

Question 2:

What is CFL ?

Answer :

If $G = (V_n, V_t, P, S)$ is a CFG, the language of G, denoted by $L(G)$, is the set of terminal strings that have derivations from the start symbol. That is

$$L(G) = \left\{ w \text{ in } V_t \,/\, S \overset{*}{\underset{G}{\Rightarrow}} w \right\}$$

If a language L is the language of some context-free grammar, then L is said to be context-free language, or CFL.

Question 3:

What is Bacos Nour Form BNF?

Answer :

While linguists were studying CFG's computer scientists began to describe programming languages by notation called Bacus-Naur form or Bacus Normal Form, which is the CFG notation with minor changes in format and some shorthand. For example, the above grammar can be rewritten in BNF as

$$S \to bA/aB$$

$$A \to bAA/aS/a$$

$$B \to aBB/bS/a$$

Hence BNF is a shorthand notation for context free grammar.

Question 4:

Describe left most and right most derivations?

Answer :

In order to restrict the number of choices of replacement of variables, if at each step we replace the "left most" variable by one of its production bodies such a derivation is called a "left most derivation", and we indicate that a derivation is left most by using relation $\underset{lm}{\Rightarrow}$ and $\overset{*}{\underset{lm}{\Rightarrow}}$ for one or many steps, respectively.

Similarly, it is possible that at each step the "right most" variable is replaced by one of its bodies. If so, we call the derivation "right most" and use symbols $\underset{rm}{\Rightarrow}$ and $\overset{*}{\underset{rm}{\Rightarrow}}$ to indicate one or many right most derivation steps, respectively.

Question 5:

What is Parse Tree ?

Answer :

Let $G = (V_n, V_t, P, S)$ be a context-free grammar. An ordered tree for this CFG, G is a derivation tree if and only if it has the following properties.

(*a*) The root is labeled by the starting non-terminal of the CFG that is S.

(*b*) Every leaf of the orderd ordered tree has a label from $V_t \cup \{\in\}$.

(*c*) Every Interior node of ordered tree has a label from V_n.

(*d*) Let us assume that a vertex has label $X \in V_n$, and its children are labeled (from left to right) $y_1, y_1, y_3, ..., y_n$, then production must contain a production of the form

$$X \rightarrow y_1, y_2, ..., y_n.$$

(*e*) A leaf labeled $\in$ has no siblings, that is, vertex with a child labeled $\in$ can have no other children. Clearly if the leaf is labeled $\in$, then it must be the only child of its parent.

Question 6:

What is ambiguity ?

Answer :

A CFG is called ambigious if for at least one word in the lanauge that it generates, there are two possible derivations of the word that correspond to different syntax trees. If a CFG is not ambiguous, it is called unambiguous.

CHAPTER-6

Question 1:

How useful symbols can be identified ?

Answer :

A symbols Y in a context-free grammar is useful it and only if :

(*a*) $Y \overset{*}{\Rightarrow} w$, where $w \in L(G)$ and w in V_t*, that is Y leads to a string of terminals. Here Y is said to be "generating".

(*b*) If there is a derivation $S \overset{*}{\Rightarrow} \alpha Y \beta \overset{*}{\Rightarrow} w$, $w \in L(G)$, for same α and β, then Y is said to be reachable.

Question 2:

What is Unit production ?

Answer :

A production of the form

Non-terminal $\rightarrow$ One non-terminal

that is a production of the form $A \rightarrow B$ (where A and B, both are non-terminals) is called unit production. Unit production increase the cost of derivation in a grammar. Following algorthim can be used to eliminate the unit production.

Question 3:

What is CNF ?

Answer :

If a CFG has only production of the form

Non-terminals $\rightarrow$ string of exactly two non-terminals or of the form

Non-terminals $\rightarrow$ one terminal

is said to be chomsky normal form or CNF.

Question 4:

What is GNF ?

Answer :

For every context free language L without $\in$, there exist a grammar in which every production is of the form $A \rightarrow aV$, where 'A' is a variable, 'a' is exactly one terminal and 'V' is the string of none or more variables, clearly $V \in V_n^*$.

"In other words if every production of the context free grammar is of the form $A \rightarrow aV/a$, then it is in Greibach Normal Form."

Question 5:

What is regular grammar ?

Answer :

A regular grammar may be left linear or right linear.

"If all production of a CFG are of the form $A \rightarrow wB$ or $A \rightarrow w$, where A and B are variables and $w \in V_t^*$, then we say that grammar is right liner".

"If all the production of a CFG are of the form $A \rightarrow Bw$ or $A \rightarrow w$, we call it left linear."

A right or left linear grammar is called a regular grammar.

CHAPTER-7

Question 1:

What is PDA ?

Answer :

A pushdown automata is a system, which is mathematically defined as follows :

$$P = (Q, \Sigma, \Gamma, \delta, s, F)$$

where Q : non-empty finite set of states

Σ : non-empty finite set of input symbols

Γ : is finite set of pushdown symbols

s : is the initial state, $s \in Q$

F : is the set of final states and $F \subseteq Q$

δ : it is a transition function or transition relation which maps.

$$(Q \times \Sigma^* \times \Gamma^*) \rightarrow (Q \times \Gamma^*)$$

Question 2:

What is DPDA ?

Answer :

A pushdown automata $P = (Q, \Sigma, \Gamma, \delta, S, F)$ is said to be deterministic if it is an automation as defined in definition of PDA, subject to the restrictions that for every $q \in Q$, $a \in (\Sigma \cup \{\in\})$ and $b \in \Gamma$.

(*a*) $\delta(q, a, b)$ contains at most one element,

(*b*) if $\delta(q, \in, b)$ is not empty, then $\delta(q, c, b)$ must be empty for every $c \in \Sigma$.

Question 3:

What is the relation between CFG and PDA ?

Answer :

PDA can recognise any language for which there exists a CFG. "That is class of language accepted by pushdown automata is exactly the class of context-free languages".

Question 4:

Write the steps of constructing PDA equivalent of CFG.

Answer :

Let $G = (V_n, V_t, P, S)$ be a context-free grammar; we must construct a pushdown automata P such that $L(P) = L(G)$. The machine we construct has only two states, p and q, and remains permanently in state q after its first move. Also, p uses V_n, the set of non-terminals and V_t the set of terminals, as its stack alphabet. We let $P = (Q, \Sigma, \Gamma, \delta, S, F)$

where
$$Q = (p, q\}$$
$$\Sigma = V_t$$
$$\Gamma = V_n \cup V_t$$

that is the set of terminals and non-terminals.

$$S = p$$

and transition relation δ is defined as follows :

(1) $((p, \in, \in), (q, S))$ (as S is strating non-terminal of CFG)

(2) $((q, \in, A), (q, x))$ for each rule $A \rightarrow x$ in CFG.

(3) $((q, a, a), (q, \in))$ for each $a \in V_t$.

CHAPTER-9

Question 1:

What is Turing Machine ?

Answer :

$$\text{TM} = (Q, \Sigma, \Gamma, \delta, q_0, h)$$

where (*i*) Q is the finite set of states.

(*ii*) Σ is the finite set of non-blank information symbols,

(*iii*) Γ is the set of tape symbols, including the blank symbol #.

(*iv*) δ is the next-move partial function from

$$(Q \times \Gamma) \longrightarrow (Q \times \Gamma \times \{L, R, N\})$$

where $\Gamma = \Sigma \cup \#$, '*L*' denotes the tape Head moves to the left adjacent cell, '*R*' denotes tape Head moves to the right adjacent cell and '*N*' denotes Head does not move, that is continuous scanning the same cell.

(*v*) $q_0 \in Q$, is the initial state.

(*vi*) $h \in Q$ is the 'Halt State', in which the machine stops any further activity.

Question 2:

What are different roles of Turing machines ?

Answer :

The Turing Machines are designed to play atleast the following three different roles :

(*i*) *As accepting devices for languages,* similar to the role played by FAs and PDAs.

(*ii*) *As a computer of functions.* In this role, a TM represents a particular function. Initial input is treated as representing an *argument* of the function. And the (final) string on the tape when the TM enters the Halt state treated as representative of the value obtained by an application of the function to the argument represented by the initial string.

(*iii*) *As an enumerator of strings of a language*, that outputs the strings of a language, one at a time, in some systematic order that is as a list.

Question 3:

What is nondeterministic Turing machine ?

Answer :

A non-deterministc Turing machine is a six tuple $(Q, \Sigma, \Gamma, \delta, q_0, h)$ where

Q : set of states

Σ : set of input symbols

Γ : set of tape symbols

δ : $(Q \times \Gamma) \rightarrow$ power set of $(Q \times \Gamma \times \{L, R, N\})$

q_0 : initial state and

h : hut state, $h \in Q$

Question 4:

What is Church's Thesis ?

Answer :

In other words "*No computational procedure will be considered an algorithm unless it can be represented as a turing machine*". This statement is called Church's thesis because Alonzo Church (1936) gave many sophisticated reasons for believing it. Church's original statement was slightly different because his thesis was presented slightly before the turing invented his machines.

Church actually said that any machine that can do certain list of operations will be able to perform all conceivable algorithms. He tied together what logicians had called recursive functions and computable functions. Tm's can do all that Church asked, so they are one possible model of the universal algorithm machine Church described.

Question 5:

What is Universal Turing machine ?

Answer :

To begin, we must present a general way of specifying turing machines, so that their descriptions can be used as input to other turing machines.

Universal turing machines 'U' takes two arguments, a description of a machine Tm, "Tm", and a description of an input string w, "w". We want U to have the following property : U halts on input "Tm" "w" if and only if Tm halts on input w.

$$U (\text{“}m\text{” “}w\text{”}) = \text{“}m\ (w)\text{”}.$$

It is the functional notation of universal turing machine.

Question 6:

What is halting problem ?

Answer :

In short halting problem is :

To determine for an arbitrary given Turing machine Tm and input w,

Whether Tm will eventually halt on input w.

Question 7:

What is PCB ?

Answer :

Let Σ be an alphabet with at least two letters. An instance of the post corres-ponding problem (for short PCP) is given by two sequences $U = (u_1, u_2, ..., u_m)$ and $V = (v_1, v_2, ..., v_m)$ of strings $u_i, v_i \in \Sigma^*$. The problem is to find whether there is a (finite) sequence

$(i_1, i_2, ..., i_p)$, with $i_j \in \{1, 2, ..., m\}$ for

$$i_j = 1, 2, ... p \text{ so that, } P \geq 1$$

$$u_{i_1}, u_{i_2} u_{i_3} ... u_{i_p} = v_{i_1} v_{i_2} ... v_{i_p}.$$

Equivalently, on instance of the PCP is a sequence of pairs

$$\begin{pmatrix} \mu_1 \\ \vdots \\ v_1 \end{pmatrix}, \ldots, \begin{pmatrix} \mu_m \\ \vdots \\ v_m \end{pmatrix}$$

The sequence $i_1, i_2 \ldots i_p$ is said to be solution to this instance of PCP.

CHAPTER-10

Question 1:

What is Chomsky Hierarchy ?

Answer :

We can exhibit the relationship between grammars by the Chomsky Hierarchy. Noum Chomsky, a founder of formal language theory, provided an initial classification in to four languages types :

Type – 0 (*Unrestricted grammar*)

Type – 1 (*Context sensitive grammar*)

Type – 2 (*Context free grammar*)

Type – 3 (*Regular grammar*)

Question 2:

What is unrestricted grammar ?

Answer :

The unrestricted grammar is defined as

$$G = (V_n, V_t, P, S)$$

where V_n = a finite set of non-terminals

V_t = a finite set of terminals

S = starting non-terminal, $S \in V_n$

and P is set of productions of the following form

$$\alpha \rightarrow \beta$$

where α and β are arbitrary string of grammar symbols with $\alpha \neq \in$. These grammars are known as type-0, phase-structure or unrestricted grammars.

Question 3:

What is linear bound automata ?

Answer :

A Linear Bounded Automata (LBA) is a non-deterministic turing machine satisfying the following two conditions.

(*a*) Its input alphabet includes two special symbols $\$_1$ and $\$_2$, the left and right end marker, respectively.

(*b*) The LBA has no more left from $\$_1$ and no right from $\$_2$, nor may it print another symbol over $\$_1$ or $\$_2$.

Question 4:

What is Context Sensitive Grammar?

Answer :

Let $G = (V_n, V_t, P, S)$ be context sentive grammar

where V_n = finite set of non-terminals

V_t = finite set of terminals

S = starting non-terminals $S \in V_n$

and P is the set of rules called productions defined as

$$\alpha \rightarrow \beta$$

where β is at least as long as α that is clearly

$$|\alpha| \leq |\beta|$$

The term "Context-sentive" comes from a normal form for these grammars, where each production is of the form $\alpha_1 A \alpha_2 \rightarrow \alpha_1 \beta \alpha_2$, with $\beta \neq \in$. Replacement of variable A by string β is permitted in the "Context" of α_1 and α_2.

CHAPTER-11

Question 1:

What is partial recursive function ?

Answer :

We know that **partial recursive functions** are Turing Computable, it means that there exist a turing machine Tm for every partial recursive function.

Question 2:

What is total recursive function ?

Answer :

All those partial recursive functions for which Turing machine halts are called **total recursive functions.** From the definition of both it is clear that total recursive function is the subset of partial recursive function.

Question 3:

What is primitive recursive function ?

Answer :

Primitive Recursive function is another subclass of partial recursive function and can be defined using initial functions. Initial functions are defined over a set of natural numbers $N = \{0, 1, 2, ...\}$ and some alphabet of symbols.

Question 4:

What is recursion ?

Answer :

We know that recursion is widely used term in computer science. "A function which calls itself directly or indirectly and terminates after finite number of steps is known as recursive function."

Question 5:

What is μ-Recursive functions ?

Answer :

A function is μ-recursive if

(*a*) It is primitive recursive, or

(*b*) can be obtained from basic functions by any number of successive applications of composition, recursive definition, and minimalization on minimalizable functions

Question 6:

Define computability.

Answer :

Computability refers to the ability to solve a problem in an effective manner. It is a key topic of the field of computability theory within mathematical logic and computer science. The computability of a problem is closely linked to the existence of an algorithm to solve the problem.

A computable - *In computability theory, a Turing reduction from a problem A to a problem B, named after Alan Turing, is a reduction which solves A, assuming B is already known.*

Computable : Expressible as a yes/no question that can be answered in any case by a computer in finite time.

❑❑❑

INDEX

A

B

C

D

E

F

G

H

I

K

L

M

N

P

R

S

T

U

W

*****